TABLE OF LOGARITHMS

Natural numbers	0	1	2	3	4	5	6	7	8	9	Proportional parts								
											1	2	3	4	5	6	7	8	9
55	7404	7412	7419	7427	7435	7443	7451	7459	7466	7474	1	2	2	3	4	5	5	6	7
56	7482	7490	7497	7505	7513	7520	7528	7536	7543	7551	1	2	2	3	4	5	5	6	7
57	7559	7566	7574	7582	7589	7597	7604	7612	7619	7627	1	2	2	3	4	5	5	6	7
58	7634	7642	7649	7657	7664	7672	7679	7686	7694	7701	1	1	2	3	4	5	5	6	7
59	7709	7716	7723	7731	7738	7745	7752	7760	7767	7774	1	1	2	3	4	4	5	6	7
60	7782	7789	7796	7803	7810	7818	7825	7832	7839	7846	1	1	2	3	4	4	5	6	6
61	7853	7860	7868	7875	7882	7889	7896	7903	7910	7917	1	1	2	3	4	4	5	6	6
62	7924	7931	7938	7945	7952	7959	7966	7973	7980	7987	1	1	2	3	3	4	5	6	6
63	7993	8000	8007	8014	8021	8028	8035	8041	8048	8055	1	1	2	3	3	4	5	5	6
64	8062	8069	8075	8082	8089	8096	8102	8109	8116	8122	1	1	2	3	3	4	5	5	6
65	8129	8136	8142	8149	8156	8162	8169	8176	8182	8189	1	1	2	3	3	4	5	5	6
66	8195	8202	8209	8215	8222	8228	8235	8241	8248	8254	1	1	2	3	3	4	5	5	6
67	8261	8267	8274	8280	8287	8293	8299	8306	8312	8319	1	1	2	3	3	4	5	5	6
68	8325	8331	8338	8344	8351	8357	8363	8370	8376	8382	1	1	2	3	3	4	4	5	6
69	8388	8395	8401	8407	8414	8420	8426	8432	8439	8445	1	1	2	2	3	4	4	5	6
70	8451	8457	8463	8470	8476	8482	8488	8494	8500	8506	1	1	2	2	3	4	4	5	6
71	8513	8519	8525	8531	8537	8543	8549	8555	8561	8567	1	1	2	2	3	4	4	5	5
72	8573	8579	8585	8591	8597	8603	8609	8615	8621	8627	1	1	2	2	3	4	4	5	5
73	8633	8639	8645	8651	8657	8663	8669	8675	8681	8686	1	1	2	2	3	4	4	5	5
74	8692	8698	8704	8710	8716	8722	8727	8733	8739	8745	1	1	2	2	3	4	4	5	5
75	8751	8756	8762	8768	8774	8779	8785	8791	8797	8802	1	1	2	2	3	3	4	5	5
76	8808	8814	8820	8825	8831	8837	8842	8848	8854	8859	1	1	2	2	3	3	4	5	5
77	8865	8871	8876	8882	8887	8893	8899	8904	8910	8915	1	1	2	2	3	3	4	4	5
78	8921	8927	8932	8938	8943	8949	8954	8960	8965	8971	1	1	2	2	3	3	4	4	5
79	8976	8982	8987	8993	8998	9004	9009	9015	9020	9026	1	1	2	2	3	3	4	4	5
80	9031	9036	9042	9047	9053	9058	9063	9069	9074	9079	1	1	2	2	3	3	4	4	5
81	9085	9090	9096	9101	9106	9112	9117	9122	9128	9133	1	1	2	2	3	3	4	4	5
82	9138	9143	9149	9154	9159	9165	9170	9175	9180	9186	1	1	2	2	3	3	4	4	5
83	9191	9196	9201	9206	9212	9217	9222	9227	9232	9238	1	1	2	2	3	3	4	4	5
84	9243	9248	9253	9258	9263	9269	9274	9279	9284	9289	1	1	2	2	3	3	4	4	5
85	9294	9299	9304	9309	9315	9320	9325	9330	9335	9340	1	1	2	2	3	3	4	4	5
86	9345	9350	9355	9360	9365	9370	9375	9380	9385	9390	1	1	2	2	3	3	4	4	5
87	9395	9400	9405	9410	9415	9420	9425	9430	9435	9440	0	1	1	2	2	3	3	4	4
88	9445	9450	9455	9460	9465	9469	9474	9479	9484	9489	0	1	1	2	2	3	3	4	4
89	9494	9499	9504	9509	9513	9518	9523	9528	9533	9538	0	1	1	2	2	3	3	4	4
90	9542	9547	9552	9557	9562	9566	9571	9576	9581	9586	0	1	1	2	2	3	3	4	4
91	9590	9595	9600	9605	9609	9614	9619	9624	9628	9633	0	1	1	2	2	3	3	4	4
92	9638	9643	9647	9652	9657	9661	9666	9671	9675	9680	0	1	1	2	2	3	3	4	4
93	9685	9689	9694	9699	9703	9708	9713	9717	9722	9727	0	1	1	2	2	3	3	4	4
94	9731	9736	9741	9745	9750	9754	9759	9763	9768	9773	0	1	1	2	2	3	3	4	4
95	9777	9782	9786	9791	9795	9800	9805	9809	9814	9818	0	1	1	2	2	3	3	4	4
96	9823	9827	9832	9836	9841	9845	9850	9854	9859	9863	0	1	1	2	2	3	3	4	4
97	9868	9872	9877	9881	9886	9890	9894	9899	9903	9908	0	1	1	2	2	3	3	4	4
98	9912	9917	9921	9926	9930	9934	9939	9943	9948	9952	0	1	1	2	2	3	3	4	4
99	9956	9961	9965	9969	9974	9978	9983	9987	9991	9996	0	1	1	2	2	3	3	3	4

OPTICAL SPECTRA

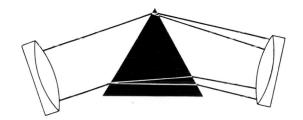

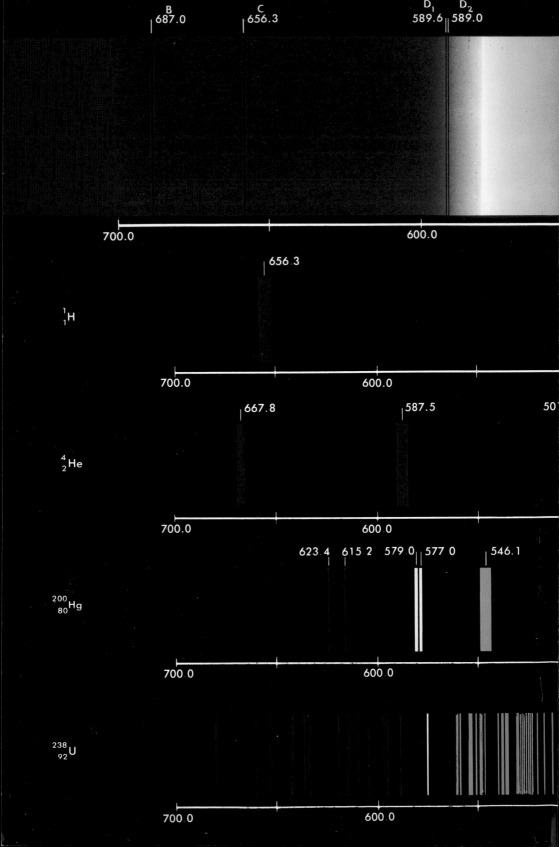

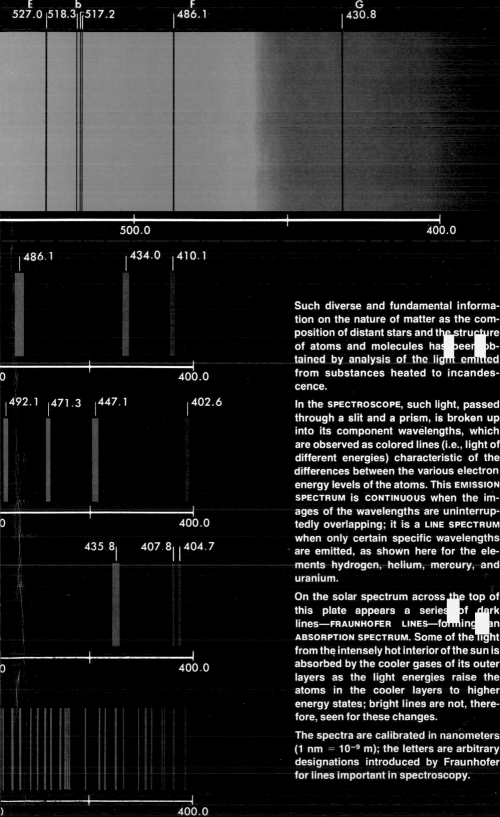

500.0 400.0

486.1 434.0 410.1

400.0

492.1 471.3 447.1 402.6

400.0

435.8 407.8 404.7

400.0

400.0

Such diverse and fundamental information on the nature of matter as the composition of distant stars and the structure of atoms and molecules has been obtained by analysis of the light emitted from substances heated to incandescence.

In the SPECTROSCOPE, such light, passed through a slit and a prism, is broken up into its component wavelengths, which are observed as colored lines (i.e., light of different energies) characteristic of the differences between the various electron energy levels of the atoms. This EMISSION SPECTRUM is CONTINUOUS when the images of the wavelengths are uninterruptedly overlapping; it is a LINE SPECTRUM when only certain specific wavelengths are emitted, as shown here for the elements hydrogen, helium, mercury, and uranium.

On the solar spectrum across the top of this plate appears a series of dark lines—FRAUNHOFER LINES—forming an ABSORPTION SPECTRUM. Some of the light from the intensely hot interior of the sun is absorbed by the cooler gases of its outer layers as the light energies raise the atoms in the cooler layers to higher energy states; bright lines are not, therefore, seen for these changes.

The spectra are calibrated in nanometers (1 nm = 10^{-9} m); the letters are arbitrary designations introduced by Fraunhofer for lines important in spectroscopy.

EXPERIMENTAL
COLLEGE PHYSICS

EXPERIMENTAL COLLEGE PHYSICS

A Laboratory Manual

MARSH W. WHITE, Ph.D.

Professor of Physics
The Pennsylvania State University

KENNETH V. MANNING, Ph.D.

Associate Professor of Physics
The Pennsylvania State University

THIRD EDITION

McGRAW-HILL BOOK COMPANY

NEW YORK   TORONTO   LONDON

1954

EXPERIMENTAL COLLEGE PHYSICS

Library of Congress Catalog Card Number: 53-8014

3 0   DODO   8 3 2 1 0

ISBN 07-069749-3

PREFACE

The title of this book calls attention to the fact that its approach to the subject is the true scientific one of experiment and logical reasoning. Although the book is primarily a laboratory manual of experiments in physics, it is not merely a set of directions for performing experiments. An introductory chapter emphasizes the objectives of laboratory work, the intangible values to be obtained from experimenting, and includes instructions concerning tabulation of data, graphical methods, and the techniques of analyzing and interpreting data. Each chapter includes a concise statement of the important theory underlying the experiments, descriptions of the apparatus used, and an unusually complete list of suggestive questions and problems. The directions for the actual experimental procedures are complete enough for the average thoughtful student, but a studious attempt has been made to design the steps in the manipulation of the apparatus so that the reasoning powers of the student are utilized and developed to the maximum extent.

An especial emphasis is placed in each experiment upon the desirability of having the student attempt to evaluate the errors involved in the measurements. Throughout the book the proper use of significant figures is strictly adhered to, including the data given in problems.

The present form of the book represents a development of twenty years' use with laboratory classes including students taking almost every kind of course. The experiments have therefore been tested adequately by actual laboratory instruction.

A feature of this edition is the division of the book into chapters, each of which presents the basic theory of a limited number of topics. A single chapter may include several experiments covering related topics, or those ranging in difficulty from an elementary treatment to the more sophisticated experiments. The numbering system for the experiments makes it easily possible to utilize additional experiments involving the basic theory included in the chapter. A feature of this edition is the statement of the *Method* of each experiment, wherein there is included a very concise statement designed to outline the general method of attack in performing the experiment and analyzing the results.

Emphasis is placed in the experiments on the preparation of a carefully planned record of the observed data. In most experiments the directions for the *Procedure* are given first. They are followed by the instructions for making sample calculations, drawing of graphs, and the analysis and interpretation of the results obtained.

In this edition a number of new experiments have been added and a few which were not extensively used have been omitted. Entirely new figures have been prepared and the lists of problems have been greatly amplified. All the questions and problems are designed specifically to illustrate the theory of the experiments and not to duplicate the textbook type of problems.

The apparatus utilized in the experiments is all of the standardized type now generally available.

It is impossible to acknowledge properly all of those who have helped, consciously or otherwise, in the preparation of the material for this book. So many of the colleagues of the authors have collaborated in the work at various times and so deeply have impressions been formed from other teachers and textbooks that it would be hopeless to attempt to mention here all the sources of the present effort. We are pleased to acknowledge the permission of the Central Scientific Co. to utilize portions of their *Selective Experiments in Physics*. Dr. Seville Chapman and his publishers have graciously given permission for us to include in our introductory chapter some of the material from his laboratory manual.

An important contribution to this edition was the expert and critical reading of the manuscript by Prof. Pearl I. Young. Her stimulating suggestions have been invaluable in the attempt to present the material in good form and style.

The instrument companies who so kindly loaned pictures of apparatus deserve particular thanks, especially the Central Scientific Company and the Leeds and Northrup Company.

The questions and problems given with each experiment have been taken largely from the authors' files. They have been accumulated and tested over a number of years. It is too much to expect that all of them are entirely original. If the authors have inadvertently appropriated some from other books, they trust that this statement will serve, however inadequately, as an explanation of their presence.

Marsh W. White
Kenneth V. Manning

CONTENTS

PART I—GENERAL

CHAPTER 1

INTRODUCTION; GRAPHICAL ANALYSIS

The laboratory part of a course in college physics may be highly valuable or it may be worthless, depending on the attitude taken toward it. If the student approaches the laboratory with the thought that it offers a personal opportunity to learn by means of actual observation some of the principles of physics, to do some independent thinking, to become familiar with modern measuring equipment, and to learn the fundamentals of preparing a technical report on the results—these hours may be very profitable.

In the elementary laboratory one is not likely to discover any new laws of physics or disprove any of the old ones; most of the material in the course has been known for decades and some for centuries. But the student has not known about it for decades; with the proper mental approach he may be able to enjoy the thrill of the original investigator as he performs some of the standard experiments.

Tangible Objectives of Laboratory Courses. The experiments in a course in beginning college physics vary in difficulty; some are short to perform but difficult to understand. Some experiments are long and tedious in the collection of data but simple and straightforward in application. They are alike in one particular: each provides the maximum of value only to the student who tackles it knowing exactly what he is trying to gain from that particular experiment. For this reason it is desirable that the appropriate principles be presented in the lecture and recitation part of the course before the experiment is performed. In any event the short theory section that precedes each experiment should be mastered before the student comes to the laboratory.

The work in the physics laboratory has been designed with the goal of providing certain definite benefits to the student: (1) to supplement the factual knowledge and technical vocabulary of the lecture and recitation work by firsthand manipulation of apparatus and personal application of the principles studied there; (2) to train students in the procedures of thoughtful experimentation, careful observation, honest and discriminatory recording of data, critical analysis, and correlation and interpretation of experimental results; (3) to furnish experience in the use of graphical analysis; (4) to provide for studies of the precision of measurements; (5) to train in the use of laboratory apparatus and techniques used in physical measurements; and (6) to become familiar with the art of drawing valid and worthwhile conclusions from observed data.

Intangible Values Obtained from Laboratory Work. *Understanding of the Attitude and Method of Science.* For an appreciation of the operation of the scientific method—observation, hypothesis, verification, modification—one must study examples of its application. Although no short single experiment is an ideal example of the scientific method, because not enough cross checking is possible, each experiment may be considered as part of the verifying data. We might ask ourselves when compiling these small increments: To what extent can they be applied in other sciences, particularly to the social sciences?

1

Understanding of the Exactness and Limits of Knowledge about Any Particular Matter. Many people go through life with amazingly vague ideas about the physical principles behind everyday phenomena; in fact, some people never realize that there are persons in the world to whom such things are clear. From the concrete examples shown in the laboratory the student should continually ask if similar applications in everyday life are understood. What kind of machine is a pair of grass-clipping shears? What makes the colored film in an oil slick on the pavement after a rain? What is the efficiency of the pulley with which the movers are taking the piano to the second floor? Why is there a row of lighted candles in the air over the lawn opposite the dining room?

Respect for the Prestige and Reliability of Authority. After performing certain of the experiments in college physics the student should check his values against those in a standard reference book. Many times the standard value will be known to one or two more significant figures than those obtained in the laboratory. It is profitable to inquire how they were obtained with such accuracy. If the Coast and Geodetic Survey, for example, says that the value of g for the locality of the laboratory is 980.124 ± 0.005 cm/sec^2 and in the experiment it was difficult to check the 980 cm/sec^2, what kind of apparatus did they use to obtain the next three figures? How do they measure the value with such accuracy? Sometimes such a chain of reasoning leads to the incidental gathering of interesting related material: Who are the people who make up the Coast and Geodetic Survey? Would it be a good place for a career in government service? What other constants do they keep up to date?

Examine, for instance, in different source books the values of the coefficient of friction for oak on oak. Record all the values. Why do they not agree with each other? Would anyone be justified in giving the values of coefficient of friction to more than two significant figures? Are the values the student obtains for oak on oak more reliable for the chosen sample than those one could get from a handbook? For what type of values should one scrutinize the date of issue? Would the values given for an aluminum alloy be more accurate if obtained from a handbook or from a manufacturer's catalogue?

Development of Character and Sense of Social Responsibility. When physics instructors are asked (as they often are) for recommendations for students, the subject of character is always stressed. Often character traits are broken down and a prospective employer asks about several: dependability, promptness, honesty, judgment, and the ability to work with other people.

One person who can give employers an accurate estimate of the character traits of students in these respects is the laboratory instructor. If you are deficient in some of these traits, cultivate them in laboratory work, in accordance with the following suggestions:

1. Be thoroughly dependable in reading data and in recording them to the correct number of significant figures.

2. Be prompt in arriving at the laboratory station, getting to work, completing the assignment, and writing up the report. If you must be absent, make an appointment to perform the experiment at some other time.

3. Be honest in stating individual ideas. Copy no data, conclusions, or computations from any source. If a final result seems outside the limits of permissi-

ble variation from standard, present the result frankly and make the best possible explanation of the discrepancy.

4. Use judgment concerning the part of the experiment to perform first, when an option is given. If one set of equipment must be used by several groups, arrange the work so that the apparatus may be used without holding up some-one else.

5. Bear a proportionate share of the responsibility of performing the experiment. Do not depend on a partner to supply all the answers. Refrain from undue noise and disturbance.

Care of Apparatus. Much equipment used in the physics laboratory is costly and fragile. Although normal wear and tear on apparatus is anticipated, the student is expected to treat it with care and to report immediately any apparent malfunctioning. The apparatus will ordinarily be found in a neat and workable condition and should be similarly left.

Tabulation of Data. The tabulation of data obtained during the course of an experiment offers one of the best opportunities for learning correct habits of treating technical data. Most students in technical curricula engage in this form of activity to some degree the rest of their professional lives. The following suggestions apply not only to the college physics laboratory but also to most similar situations.

1. Be painstakingly neat in recording data.

2. Take data with ink or sharp pencil in final form with no intention of recopying. Any copying of "original" data leads to errors and might arouse suspicion that the data have been tampered with.

3. If a mistake is made in recording data, cancel the wrong value with a fine line drawn through it, leaving the original value legible.

4. Use a standard form of tabulation with like values in columns, not in rows. Any suggested data forms in the manual or furnished with extra instruction are to serve as guides. In many cases it is desirable to make the data-sheet headings before coming to the laboratory.

5. Identify in some fashion on the data sheet apparatus used for measuring and equipment used in testing. Such precautions are valuable in later checking by the student or by the instructor.

6. Write at the top of the column the quantity being measured and add the unit immediately below. Unexplained symbols should not be used to replace the names of the quantities being recorded.

7. Note directly on the data sheet any unusual features of the equipment that might affect the accuracy of the data.

8. Keep a record of any peculiarities of equipment or arrangement that are not obvious from the manual or are departures from it.

9. The data sheet should be so clear and complete that another person could take it and perform the computations.

Recording Large and Small Numbers. Whenever very large or very small numbers are recorded it is better to eliminate the use of awkward zeros by the use of powers of ten. For example, the speed of light, about $18\overline{6},000$ mi/sec, should be written 1.86×10^5 mi/sec. Or the wavelength of yellow light, 0.00005893 cm, may be written as 5.893×10^{-5} cm. It is even more desirable to make use whenever possible of the common decimal prefixes such as the following: micro (μ) = 10^{-6}, milli (m) = 10^{-3}, centi (c) = 10^{-2}, deci (d) = 10^{-1}, kilo (k) = 10^3, mega = 10^6. Thus a capacitance of 1.59×10^{-6} farad may be written 1.59 μf; a length of 0.00468 m is better stated as 4.68 mm; a frequency of 3.50×10^8 cycles/sec is commonly given as 350 megacycles/sec.

Use of the Slide Rule. The precision of most laboratory experiments is such that a slide rule may profitably be used for computations. The student is expected to utilize this timesaver whenever possible. The physics laboratory offers an excellent opportunity for practice in computations with this versatile device. For a novice the simpler forms of slide rule are preferable and an inexpensive rule is quite satisfactory for most purposes.

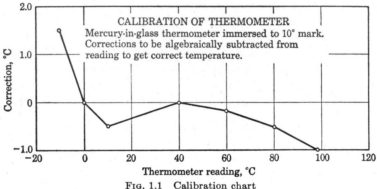

FIG. 1.1   Calibration chart

Presentation of Graphical Data. *Significance of Graphical Presentation.* Physical laws and principles express relationships between physical quantities. These relationships may be expressed (1) in words, as is commonly done in the formal statement; (2) by means of symbols in an equation; or (3) by the pictorial representation called a *graph*. The choice of the means of expression is dictated by the use to be made of the information. If calculations are to be made, the equation form is ordinarily the most useful. The graph, however, presents to the trained observer a vivid picture of the way in which one quantity varies with another.

A graph shows more clearly than a tabulation in adjacent columns the manner in which the dependent variable is related to the independent one. Graphs may also be used (1) to obtain pairs of values other than those plotted, (2) to smooth out inaccuracies in data, and (3) to indicate unsuspected trends and critical points.

Types of Graphs. Graphs may be divided into four general types. The first type, shown in Fig. 1.1, is a record of observed values at many points. Because the exact nature of the variation is unknown and may be irregular, the adjacent points are connected by straight lines. This type of graph is often used for statistical and calibration data.

The second type of graph is shown in Fig. 1.2; this is the type usually drawn to show the relation between two variables. The observed points are plotted and a smooth curve is drawn that approximately fits the observed data. This type of curve represents an empirical relationship between the quantities plotted.

The tacit assumption is made that the points not on the curve are displaced because of experimental inaccuracy.

The third type of graph, Fig. 1.3, represents theoretical relationships; the data are obtained by substituting arbitrary values in an equation expressing the theoretical relationship. Such a curve follows a mathematical form (straight line, sine curve, hyperbola) and the inclusion of specific points is meaningless.

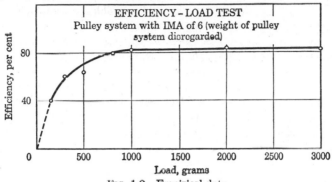

FIG. 1.2   Empirical data

In many cases a curve of this type is included with the second type to show the variation of the experimental data from the theoretical.

A fourth type of graph is a computational one prepared to be used in place of tabular data to pick off values for use in calculations. An example would be a curve showing the variation of pressure with altitude or the variation of density of a liquid or gas with temperature. Such graphs are always drawn to a very large scale on high-quality paper with many fine divisions. Because of the labor and care involved in their original preparation the original is usually filed and a print used.

Shapes of Curves. The shape of a plotted curve may be irregular or smooth. Although the usual graph is plotted on rectangular coordinate paper, curves are frequently drawn on other standard graph paper or on specially constructed grids. Polar-coordinate paper is useful in showing some variations, such as the luminous intensity of a source of light.

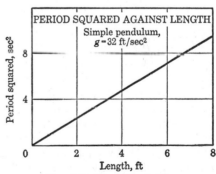

FIG. 1.3   Theoretical data

When the curve is a *straight line*, it is easily drawn, easily identified, and, in many respects, more useful than any other form of curve. The variables are therefore often rearranged in such a form that the resulting curve is a straight line. The variation of the period of a simple pendulum with the length may be shown by plotting period squared against the length. The inverse relationship of pressure and volume of a gas at constant temperature is shown by plotting the pressure against the reciprocal of the volume. The variation of the viscosity of lubricating oils with temperature is shown by plotting the viscosity and temperature along special logarithmic scales that make the data plot as straight lines. If the curve is known to be a straight line, only enough points need be plotted to locate the line. For a theoretical curve two points, or one point and the slope, are sufficient. When experimental data are plotted, the error is minimized by plotting as many points as can readily be obtained. If the type of relationship is unknown, it is common to assume a power law and to plot the data on logarithmic paper; the probable mathematical expression of the relationship may then often be deduced from the slope of the curve.

Drawing the Curve. Inasmuch as the curve has so great a significance the utmost care should be used in drawing it. Except in the rare case when adjacent points are connected by straight lines, curves are smoothed through an average of the points. The smooth continuous curve passes as closely as possible to all points but disregards those that are obviously far from the curve. As a rough approximation one may say that the best line will be the one that divides the points so that the sum of their displacements from the line will be a minimum.

The smoothing is done by laying a transparent scale or curve along the points and finding sections of the transparent guide that seem to match the direction of the points. As an aid in drawing smooth curves, it is often helpful to put one's eye nearly in the plane of the paper and to view the plotted data points at a glancing angle. It is not unusual to have to draw several trial curves before arriving at the final one. Many scientists make a practice of plotting the data points on the ruled side of fairly thin graph paper. Then they turn the paper over so that the points show through on the reverse side and do all the practice smoothing on that side lightly in pencil. When the final smooth curve has been arrived at, it is traced on the ruled side.

Specific Rules for Drawing Graphs on Rectangular-coordinate Paper. If a graph is to impart its full meaning, it must be constructed in accordance with standard rules so that it will have the same significance to every person who inspects it. Fairly rigid rules for the preparation of graphs have been adopted by representatives of the leading scientific and engineering societies. Some of the rules in present-day use are listed below for the guidance of the student in his presentation of laboratory data. The graphs included in the laboratory manual also serve as examples of correct form.

1. Plot the independent variable along the abscissa scale (X axis) and the dependent variable along the ordinate scale (Y axis). Draw heavy lines for these axes.

2. Choose a size of graph that bears some relation to the accuracy with which the plotted data are known. In general the curve should fill most of the sheet; if the spread from the lowest value to the highest value is small and the data are reliable to only two significant figures, do not spread out the graph to permit points to be read from the curve to three significant figures. Plot all data in the quadrant that is indicated by the signs of the numerical values. In many cases it is not necessary that the intersection of the two axes represent the zero values of both variables.

3. Choose scales for the main divisions on the graph paper that are easily subdivided. Values 2, 5, and 10 are best, but 4 is sometimes used; never use 3, 7,

or 9, since these make it very difficult to read values from the graph.   The divisions on the abscissa and ordinate scales need not be alike.

4. If the values are exceptionally large or small, use some multiplying device that permits using a maximum of two digits to indicate the value of the main division.   The load may be plotted as "thousands of tons," the time in "microseconds."   A multiplying factor such as $\times 10^{-2}$ or $\times 10^5$ placed at the right of the largest value on the scale may be used.   See Fig. 1.6.

5. In the white space *outside* the scale put on a clear complete label; letter everything on the graph so that it can be read from the lower right-hand corner. It is good form to write all labels in capital letters; some authorities capitalize only the first word and proper names; in either case a consistent style should be chosen.   The scale label includes the name of the quantity plotted and its unit, separated by a comma.   Abbreviate all units in standard form.

6. Locate each point in its proper place and encircle it.   (In printed graphs the dots do not appear.)   If several curves appear on the same sheet and the points might interfere, use squares and triangles to surround the dots of the second and third curve, respectively.

7. Draw the best smooth curve through the average of the points; ignore any points that are obviously erratic.   Draw the curve up to but not through the circles.   Indicate by dashes any extrapolated portions that extend materially past the terminal data points.

8. At an open space near the top of the paper state the title of the graph, that is, write a clear complete legend.   Omit any unnecessary words, such as "Curve showing."   After the main legend add any specific features needed to identify the number or the nature of the test equipment or conditions.

9. When more than one curve is plotted on the same sheet, differentiate them clearly.   If possible label them by (*a*) and (*b*) and explain in the sublegends.   If they are differentiated by different symbols, such as circles or squares, or by dotted and dashed lines, make a neat key in a clear space on the graph sheet. Most authorities prefer not to write along the curve.

10. In graphs for technical reports (not for publication) it is good practice to put the initials of the experimenter and the date in the lower right-hand corner.

11. For reports of student experiments select standard-quality conventional graph paper of $8\frac{1}{2}$ by 11 inch size, with convenient rulings, such as 20 per inch, or 10 per centimeter; do *not* use quadrille (cross-ruled) paper.

12. Substitute log, semilog, or polar-coordinate paper for the rectangular-coordinate paper when the data will show to better advantage.   Graphs illustrating many of the rules mentioned above are presented as Figs. 1.4 and 1.6.

Analysis and Interpretation of Graphs.   One of the great advantages of graphical analysis is the simplicity with which new information can be obtained directly from the graph by observing its shape, slope, and intercepts.

Shape.   The shape of a graph indicates immediately whether one variable increases or decreases as the other one increases and it also distinguishes between intervals of slow and rapid variation.   In the case of a straight-line graph one can easily determine the form of the variation.   Even in this case a careful distinction must be made between *linear variation* and *direct proportion*.   Linear variation means simply a first-degree relationship and is indicated by any straight-line graph. A direct proportion, on the other hand, means that the variables are directly proportional and therefore simultaneously zero, which occurs only for straight-line graphs that *pass through the origin of coordinates.*

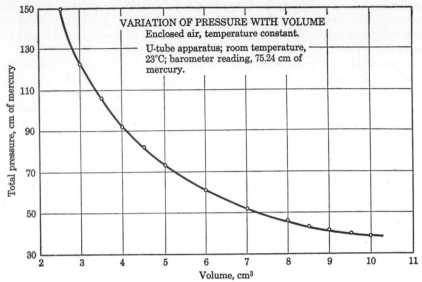

FIG. 1.4   Proper form for experimental data

Slope. In discussing the slope of a graph we must distinguish carefully between *physical* slope and *geometric* slope. In the case of a linear graph the physical

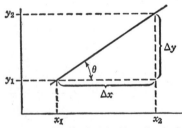

FIG. 1.5   Method of finding physical slope

slope may be found by drawing a large triangle, as shown in Fig. 1.5, and dividing Δy by Δx *using for each the scales and units that have been chosen for those axes.* The result is independent of the choice of scales and may express a significant fact about the relationship between the plotted variables. For example, the slope of a graph of velocity against time gives the acceleration. The units of the acceleration will appear from the calculation. In a typical case the physical slope m of a graph such as that in Fig. 1.6 is given by

$$m = \frac{y_2 - y_1}{x_2 - x_1} = \frac{(4.7 - 0.8) \times 10^2 \text{ cm/sec}}{(\frac{18}{40} - \frac{2}{40}) \text{ sec}} = \frac{\overline{3}90 \text{ cm/sec}}{\frac{16}{40} \text{ sec}} = \overline{9}80 \text{ cm/sec}^2$$

In contrast to the physical slope the geometric slope is defined to be $\tan \theta$, where θ is the angle between the line and the X axis. This slope depends upon the inclination of the line and hence depends upon the choice of scales. In graphical analysis, one is always concerned with the *physical* rather than the *geometric* slope.

In the case of curved-line graphs the slope will vary from point to point. Its value at any point is defined as the slope of the tangent line at that point.

Intercepts. Significant information is sometimes revealed by the intersection of a graph with a coordinate axis. Thus the intercept of a velocity-time graph with the velocity axis gives the value of the velocity when the time was zero, that is, when the experimenter chose to begin counting time. Similarly, the intercept on the time axis gives the value of time when the velocity was zero; a negative intercept on the time axis indicates the length of time that the body was in motion before the experimenter began to count time.

Use of the Appendix and Reference Books. The student is expected to familiarize himself with the material in the Appendix to this book. Some of the items

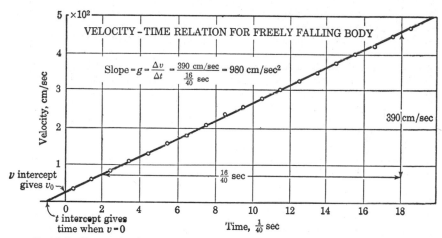

FIG. 1.6 Demonstration graph, showing interpretation by slope and intercepts

will be of concern only occasionally during the progress of the course; others are of more frequent utility. It is well at the beginning of the term to leaf through the Appendix to obtain an idea of the material to be found there.

A few reference books are mentioned here; these books contain tables of data and other valuable material that may assist in the study of the experiment and the writing of the report.

Handbook of Chemistry and Physics, Chemical Rubber Publishing Company
Smithsonian Physical Tables, The Smithsonian Institution
Dictionary of Applied Physics, by Glazebrook. The Macmillan Company
Mechanical Engineers' Handbook, by Marks. McGraw-Hill Book Company
Standard Handbook for Electrical Engineers, by Knowlton. McGraw-Hill Book Company

Questions and Problems: 1. You are required to produce two generalizations: (a) the average weight of a hen's egg and (b) the average weight of a piece of classroom chalk. You want your conclusion to be as scientific as possible with a limited time and with limited material. How would the two determinations differ in number of observations required? in exactness of hypothesis possible? in difficulty of verification? in the exactness with which you may state the average weight?

2. What information would be required to estimate and with what accuracy could you estimate: (a) the number of eggs needed in the school cafeteria during the term; (b) the number of boxes of chalk needed in the school classrooms during the term?

3. In what type of reference material might you find a check on the accuracy of your conclusions and estimates in questions 1 and 2?

4. A group of people witness a traffic accident. A housewife and an engineering student are required to testify concerning the time sequence of events, the relative speed of the vehicles, and similar observational details. In which testimony might the court place the greater confidence? Why?

5. You are required to determine by direct measurement and computation the ratio of the circumference to the diameter of a circle. How will you explain the variation in your result from the accepted value 3.14159?

6. You are furnished 10 small metal cylinders of nominal diameter $\frac{1}{2}$ in. and nominal length 2 in. You measure to three significant figures the diameters and the lengths and compute the volumes. Prepare a neat data sheet showing possible values of the data and also the average length, diameter, and volume.

7. In which of the following cases would it be advisable to identify clearly the exact piece of equipment used:

(a) A meterstick used to measure the length of a laboratory table.

(b) A meterstick used over a knife-edge to balance weights when the weight of the meterstick must be considered.

(c) A thermometer used to determine room temperature in order to check the speed of sound in a metal.

(d) A thermometer used in an experiment to check the density of water at 4°C.

(e) An ammeter used to measure relative currents in nearly equivalent circuits.

(f) An ammeter used to measure the current in silver plating.

8. Rewrite the following data in powers of 10 so that each quantity has only one digit to the left of the decimal point: Speed of light, 299,780 km/sec; wavelength of yellow light, 0.00005893 cm; distance to the sun, 93,000,000 miles; coefficient of linear expansion of copper, 0.0000094 per °F.

9. Convert the following values into the units indicated and express them in powers of 10: 250 microfarads, to farads; 47.25 megohms, to ohms; 1234 megacycles, to cycles; 0.0140 millihenry, to henrys; 1875.2 millimeters, to meters; 980 kilograms, to centigrams.

10. Perform the following operations on the slide rule and also by long hand and compare the times required: (a) $\sqrt{3714}$; (b) 125^3; (c) 468×185; (d) $3290/980$.

11. The following data were obtained by observing a falling body:

Speed, ft/sec...	2.50	5.68	9.00	12.2	15.0
Time, sec......	0	0.100	0.200	0.300	0.400

Plot a curve showing the relationship between speed and time. Use time as the independent variable. What does the shape of the graph show? Calculate the slope of the graph. When did this body start to fall (when was its speed zero)?

12. The following data were obtained by observing the temperature of a can of oil as it cooled:

Time, min.............	0	8	16	24	32	40
Temperature, °F.......	200	121	77	52	38	30

Plot a curve showing the relationship of temperature and time. What does the shape of this graph show? Calculate the slope at 8 min and at 28 min. Comment on these slopes.

13. A meterstick is placed alongside a yardstick and the lengths in centimeters of various numbers of inches are observed. What is the significance of the shape and slope of the curve of reading in centimeters plotted against length in inches? What would one expect the value of this slope to be?

14. From some recent list of statistics, such as the *World Almanac* or the *Information Please Almanac*, tabulate American aircraft production every five years from 1920 to the present time. Plot the values in the form of a graph.

15. The heating effect of an electric current in a rheostat is found to vary directly with the square of the current. What type of graph is obtained when the heat is plotted as a function of current? How could the variables be adjusted so that a linear relation would be obtained?

16. The current in a variable resistor to which a given voltage is applied is found to vary inversely with the resistance. What is the shape of the current-resistance curve? How could these variables be changed in order that a straight-line graph could be obtained?

CHAPTER 2

SIGNIFICANT FIGURES; ERRORS; UNCERTAINTY OF MEASUREMENTS

No physical measurement is ever *exact*. The accuracy is always limited by the degree of refinement of the apparatus used and by the skill of the observer. Even though the best apparatus available is used and the observer has the highest skill, there is always a point at which the reading becomes uncertain.

Most physical measurements, whether of length or of mass or of time, involve the reading of some scale. The fineness of graduation of the scale is limited and the width of the lines marking the boundaries is by no means zero. In every case, therefore, the final digit of the reading must be *estimated* and hence is, to some degree, doubtful. Nevertheless, this doubtful digit is significant in the sense that it gives meaningful information about the quantity being measured. A *significant figure* is defined as one that is known to be reasonably trustworthy. One and only one estimated or doubtful figure is retained and regarded as significant in reading a physical measurement.

In the reading of a length measurement made by means of an ordinary meterstick, a certain result might be recorded as 6.54 cm. In this reading the final digit 4 is an estimate of a part of a millimeter division on the scale. Perhaps the estimate should be as high as 6 or as low as 2; in any case it indicates something about the length and thus is useful. This reading has three significant figures.

The location of the decimal point has nothing to do with the number of significant figures. The reading just cited could be written as 65.4 mm or as 0.0654 m. Although the decimal point has been shifted, the number of significant figures is three in each case. The presence of a zero in a datum is sometimes troublesome. If it is used merely to indicate the location of the decimal point, it is not called a significant figure, as in 0.0654 m; if it is between two significant digits, as in a temperature reading of 20.5°, it is always significant. A zero digit at the end of a number tends to be ambiguous. In the absence of specific information we cannot tell whether it is there because it is the best estimate or merely to locate the decimal point. In such cases the true situation can best be expressed by writing the correct number of significant figures multiplied by the proper power of 10. Thus a student measurement of the speed of light, $18\overline{6},000$ mi/sec, may be written 1.86×10^5 mi/sec, since this value includes only three significant figures. It is just as important to include a zero as to include any other digit that is significant. If in a reading of a meterstick one can estimate a fraction of a millimeter, a reading of 20.00 cm is quite proper; it means that the last zero is the considered reading of the observer. In such a case it would be very poor practice to record the reading as 20 cm since this would imply that the reading was taken only to the nearest centimeter. The recorded figure should always express the degree of accuracy of the reading.

In computations involving measured quantities the process is greatly simplified without any loss of accuracy if figures that are not significant are dropped.   Consider a rectangle whose sides are measured as 10.77 cm and 3.55 cm; the last and doubtful digits are in italics.   When these lengths are multiplied, any operation by a doubtful digit remains doubtful.   For example

$$
\begin{array}{r}
10.77 \\
3.55 \\
\hline
5385 \\
5385 \\
3231 \\
\hline
38.\textit{0335}
\end{array}
$$

In the result only *one* doubtful digit is retained and we have 38.2 cm² as the area. A useful, but very rough, rule states that in multiplication and division the *result should have as many significant figures as the least accurate of the factors.*   In some borderline cases the answer should have one more significant figure than the least accurate of the factors.   For example, in the multiplication 9.8 × 1.28 = 12.5, the answer must have three significant figures to be as accurate as the least accurate of the factors.   This rule must be applied with judgment by the experimenter.

In addition and subtraction, carry the operation only as far as the first column that has a doubtful figure.

When insignificant figures are dropped, retain the last figure unchanged if the first figure dropped is smaller than 5.   Increase it by 1 if the first figure dropped is greater than 5.   If the dropped figure is exactly 5, leave the preceding digit unchanged if it is even but increase it by 1 if it is odd.   Thus 3.*455* becomes 3.*46*; 3.*485* becomes 3.*48*; 6.79*01* becomes 6.79*0*.

When there is doubt as to the number of significant figures, we shall overscore the last significant digit: thus 38,$\overline{4}$00 ft.   Here there are three significant digits. The zeros are present merely to locate the decimal point.

Errors.   The uncertainty of a reading is technically called *error*.   (The word "error" carries no implication of mistake or blunder; it means the uncertainty between the measured value and the standard value.)   An error that tends to make a reading too high is called a positive error and one that makes it too low a negative error.

Errors may be grouped into two classes, systematic and random.   A *systematic error* is one that always produces an error of the same sign, for example, one that would tend to make all the observations too small.   A *random error* is one in which positive and negative errors are equally probable.

Systematic errors may be subdivided into three groups: instrumental, personal, and external.   An *instrumental error* is caused by faulty or inaccurate apparatus, for example, an undetected zero error in a scale, an incorrectly adjusted watch, or a meter with undue friction.   *Personal errors* are due to some peculiarity or bias of the observer.   Probably the most common source of personal error is the tendency to assume that the first reading taken is correct and to look with suspicion on any variation from this reading.   Other personal errors may be due to fatigue, eye strain, or the position of the eye relative to a scale.   *External errors* are caused by external conditions (wind, temperature, humidity, vibration);

examples are the expansion of a scale as the temperature rises or the swelling of a meterstick as humidity increases. Corrections may be made for systematic errors when they are known to be present.

Random or erratic errors occur as variations that are due to a large number of factors each of which adds its own contribution to the total error. Inasmuch as these factors are unknown and variable it is assumed that the resulting error is a *matter of chance* and therefore positive and negative errors are equally probable. Because random errors are subject to the laws of chance their effect in the experiment may be lessened by taking a large number of observations.

Since the variations in the observations are governed by chance, one may apply laws of statistics to them and arrive at certain definite conclusions about the magnitude of the errors. These laws lead to the simple conclusion that the value having the highest probability of being correct is obtained by dividing the sum of the individual readings by the total number n of observations. This value is the *arithmetic mean*, a.m. The arithmetic mean represents the best value obtainable from a series of observations.

The difference between an observation and the arithmetic mean a.m. is called the deviation d and the average deviation a.d. is a measure of the accuracy of the observation. The *average deviation* is the sum of the deviations (without regard to sign) divided by the number n of observations, or a.d. $= \Sigma d/n$. It is known from the theory of probability that an arithmetic mean computed from n equally reliable observations is on the average more accurate than any one observation by a factor of $\sqrt{n}/1$. Consequently, the *average deviation* A.D. of the *mean* of n observations is given by

$$\text{A.D.} = \frac{\text{a.d.}}{\sqrt{n}} \tag{1}$$

For example, in the measurement of a block of wood, suppose several trials give the following data

Length cm	Deviation cm
12.32	−0.03
12.35	0.00
12.34	−0.01
12.38	+0.03
12.32	−0.03
12.36	+0.01
12.34	−0.01
12.38	+0.03
Mean: 12.35	$\Sigma d = 0.15$

$$\text{a.d} = \frac{0.15 \text{ cm}}{8} = 0.02 \text{ cm}$$

$$\text{A.D.} = \frac{\text{a.d.}}{\sqrt{n}} = \frac{0.02 \text{ cm}}{\sqrt{8}} = 0.01 \text{ cm}$$

The best average value of this set of observations should be written 12.35 ± 0.01 cm.

Percentage Error. The importance of an error in an experimental value is not in its absolute value but in its *relative* value. By *percentage error* is meant the number of parts out of each 100 parts that a number is in error

$$\text{Percentage error} = \frac{\text{error}}{\text{standard value}} \times 100\%$$

For example, if the length of a 220-yd race track is uncertain by 1 yd, the numerical error is 1 yd and the percentage error is approximately 0.5 per cent. Percentage errors are usually wanted to only one or two significant figures, so that the method of mental approximation or a rough slide-rule computation is quite sufficient for most purposes.

It frequently happens that the *percentage difference* between two quantities is desired when neither of the quantities may be taken as a "standard value." In such cases their average value may be used in place of a standard value.

Rules for Computation. The following rules are safe to apply in making computations and in estimating the errors in the results:

Rule I. In addition and subtraction, carry the result only through the first column that contains a doubtful figure.

Rule II. In multiplication and division, carry the result to the same number of significant figures that are in the factor with the least number of significant figures.

Rule III. The *numerical* error of a sum is the sum of the *numerical errors* of the individual quantities.

Rule IV. The *percentage* error of the product or quotient of several numbers is the *sum* of the *percentage errors* of the several quantities entering into the calculation.

Certain numbers that commonly appear in calculations have a peculiar relationship in that they appear *by definition* rather than by measurement. Such numbers are assumed to have an unlimited number of significant figures. The numbers 2 and π in the expression $2\pi r$ for the circumference of a circle are examples of such numbers; also many conversion factors, such as 60 sec/min.

Experiment 2.1

ERRORS AND SIGNIFICANT FIGURES

Object: To study errors and the propagation of errors when experimental data are arithmetically manipulated.

Method: The distribution of errors is examined by carefully weighing a number of rough 100-gm masses to determine whether the error is systematic or random. The same length is measured by different scales having progressively smaller graduations; fractional parts of the least scale division are estimated each time. The uncertainties of these estimates are evaluated and comparison made of the errors. A study is made of the density of a block of wood. The length, breadth, and thickness are each measured a number of times and the averages, the deviations, and the average deviations (a.d. and A.D.) are calculated. From these data the volume and its uncertainty are obtained. The block is weighed on a trip-scale balance and the uncertainty of the measurement is estimated The density and its uncertainty are calculated.

Apparatus: Triple-beam balance; 10 or more rough 100-gm masses; wooden block, about 25 by 10 by 5 cm; special meterstick, Fig. 2.1; trip-scale balance.

FIG. 2.1   Special meterstick

The special meterstick has four faces: one with no graduations, to show the length of the simple meter; the second face calibrated in tenths of meters, or decimeters; the third calibrated in hundredths of a meter, or centimeters; and the fourth graduated in thousandths of a meter, or millimeters.

Procedure: 1. Select 10 or more rough iron masses marked 100 gm, but actually only approximately that. Weigh each of them on a sensitive triple-beam balance to the nearest centigram. The balance has a hardened-steel knife-edge that may be lowered by a knurled-screw beam arrest onto an agate bearing. The arrest should always be raised when heavy masses are added to the pan. Check the zero position of the balance. Adjust the balance screw so that, with no load in the pan and the riders all set at zero, the beam swings equally on either side of the zero of the scale. In weighing it is not necessary to allow the beam to come to rest, but the masses should be adjusted to make the oscillations symmetric about the zero. Record the mass of each of the "weights" and note the positive or negative numerical error between the observed mass and the value stamped on it, assuming the balance readings to be correct. (Can this assumption be justified?) Are these errors random or systematic? If the error is systematic try to explain its origin. Calculate the percentage error of each mass and determine the average percentage error of the group.

2. Use the special meterstick to measure the block, making the same measurement successively by means of scales of progressively finer graduations. First, measure the length using the face with no graduation. This measurement is only an *estimate* of a fractional part of a meter. Try to evaluate the reliability of this estimation; that is, by what fraction of a meter you are able to rely on such a measurement. For example, can you be reasonably certain that the distance is 0.2 m, rather than 0.3 m? In this case the error would be 0.1 m. Or can you be reasonably sure that the distance is 0.25 m rather than 0.30 m? In this case the error would be 0.05 m.

Record this length and the three following sets of data in columns as follows: (*a*) observed length, including the estimated portion of the smallest scale division; (*b*) the value of the smallest scale division being read; (*c*) the fractional part of the scale division that can be estimated by the eye with reasonable certainty (for example, 0.1 of a scale division); (*d*) the numerical uncertainty of the observation, which will simply be (*c*) expressed as an actual length, for example 0.1 m; (*e*) the percentage uncertainty involved in the estimation, which will be obtained by dividing (*d*) by (*a*) and multiplying by 100 per cent.

Repeat the measurement of the length of the block, this time using the face of the stick calibrated in decimeters. Estimate the uncertainty, and record all the data, exactly as before. Repeat, using the face graduated in centimeters. Repeat, using the millimeter scale.

Discuss the relative accuracy obtained and the variation in the error caused by estimating fractional parts of scale divisions as the same length is measured with scales of progressively decreasing lengths of scale divisions.

3. Measure and record the length of the block using the millimeter scale. Take nine observations at different places on the block in order that a fair average may be obtained. In each reading estimate fractional parts of millimeter divisions. (Remember to include the proper number of zeros when the observed length seems to fall exactly on a scale division.) The best technique for making careful measurements with a rule involves placing the rule with the graduations immediately adjacent to the object to be measured and arranging the scale to coincide with one edge of the block at the 1-cm or 10-cm mark, Fig. 2.2. This arrangement eliminates some of the error due to parallax and to the worn condition of the zero end of most rules. The length of the block shown in Fig. 2.2 would be recorded as 2.38 cm.

FIG. 2.2   Use of rule to measure length

Calculate the mean of the observations of the length. Find the deviation from the mean and calculate the average of the deviations. What is this quantity called and what is its significance? Compute the A.D. of the mean length. Attach this A.D. to the mean length with a ($\pm$) sign and call this final result the observed length of the block.

4. Repeat Step 3 for the breadth and thickness of the block and determine the A.D. of each.

Calculate the volume of the block, recording the result to the proper number of significant figures. Determine the numerical and percentage errors of this volume, using for the respective errors of the length, breadth, and thickness the A.D.'s found above.

5. Determine the mass of the block by means of the trip-scale balance. Estimate the uncertainty in the determination of the mass by noting the smallest change of mass that will produce a readable change in the balance. Compute the percentage uncertainty. Compare the percentage uncertainty of the mass with the percentage uncertainty of the volume. Is it negligible? If so, was it worth while to weigh the block as accurately as you did? How can preliminary estimates of the uncertainty of one factor be used to determine the apparatus and procedure that should be used in measuring another quantity?

Review Questions: 1. Explain what is meant by the following terms: error, significant figure, systematic error, random error, relative error, and percentage error. 2. State the rules for the proper number of significant figures to be retained in addition; in subtraction; in multiplication; in division. 3. Which more clearly specifies the uncertainty of a quantity, the numerical or the percentage error? 4. Show how errors are propagated in addition; in subtraction; in multiplication; in division. 5. Explain what is meant by a.d.; by A.D. What relation exists between a.d. and A.D.? 6. How is the A.D. obtained for the simple case of a series of measurements of the length of a body?

Questions and Problems: 1. Classify the following as to whether they are systematic or erratic errors: (a) incorrect calibration of scale; (b) personal equation or prejudice; (c) expansion of scale due to temperature changes; (d) estimation of fractional parts of scale divisions; (e) displaced zero of scale; (f) pointer friction; (g) lack of exact uniformity in object repeatedly measured; (h) parallax.

2. In physical measurements where an accuracy of about 1 per cent is obtainable, should three-, four-, five-, or seven-place logarithm tables be used?   State reasons.

3. Determine to one significant figure the approximate percentage error in each of the following:

Observed Value	Standard Value
108.	105.
262.	252.
46.2	49.5
339.	336.
460.	450.
0.000011120	0.000011180
2.94×10^{10}	3.00×10^{10}

4. Express the following incorrectly stated quantities with the proper number of significant figures multiplied by 10 raised to the proper power.   Retain only one uncertain figure in both the number and the error.

Speed of light = 299,876,000 ± 90,000 m/sec

Mechanical equivalent of heat = 41,830,000 ± 40,000 ergs/cal

Electrochemical equivalent for silver = 0.00111800 ± 10^{-7} gm/coulomb

Charge on one electron = 4.8030×10^{-10} esu, good to 0.2 per cent

5. Express the following incorrectly stated quantities with the proper number of significant figures: 3.456 ± 0.2; 746,000 ± 20; 0.002654 ± 0.00008; 6523.587 ± 0.3; 716.4 ± 0.2; 12.671, good to 5 parts in 1000; 9876.52, good to 0.2%.   Assume that the errors are correctly stated.

6. The masses of three bodies, together with their respective errors, were recorded as follows: m_1 = 3147.226 ± 0.3 gm; m_2 = 8.3246 gm ± 0.10%; m_3 = 604.279 gm, error 2 parts in 5000.   Assuming that the errors are correctly given, (a) indicate any superfluous figures in the measurements; (b) compare the precision of the three quantities; (c) find their sum; (d) properly record their product.

7. Assuming that the following numbers are written with the correct number of significant figures, make the indicated computations, carrying the answers to the correct number of significant figures: (a) add 372.6587, 25.61, and 0.43798; (b) $24.01 \times 11.3 \times 3.1416$; (c) $3887.6 \times 3.1416/25.4$.

8. The following computations are arithmetically correct but the results are not properly recorded because no attempt has been made to eliminate figures that have no significance.   Assume that the last digit in each of the numbers on the left of the equality sign is doubtful and rewrite the answers so that all the figures retained have significance.

(a) $(1.732)(1.74) = 3.01368$

(b) $(10.22)(0.0832)(0.41) = 0.34862464$

(c) $(6.23)^2 = 38.8129$

(d) $628.7/7.8 = 80.602$

(e) $1624 + 478.27 + 1844.4 + 87.2 = 4033.87$

(f) $(38.4 + 14.26)(0.87) = 45.8142$

(g) $(17.34 - 17.13)(14.28) = 2.9988$

(h) $(81.4)(1.628)/0.00000440 = 30,118,800$

(i) $(43.1)(70.2)(24.5) = 74,127.690$

9. The measured dimensions of a rectangular block are 2.267 ± 0.002 in., 3.376 ± 0.002 in., and 0.207 ± 0.001 in.   Compute the volume of the block and record the result with the correct number of significant figures.

10. The diameter of a shaft is measured to be 2.14 ± 0.02 in. Compute its cross-sectional area and express the numerical uncertainty of the result.

11. The distance to a village is recorded as 3.4 mi but the last figure is doubtful. Properly express this distance in feet.

12. Two measurements, 10.20 ± 0.04 and 3.61 ± 0.03 are made. What is the error in the result when these data are added? when divided?

13. The length of one edge of a cubical block of iron is found to be (10.3 ± 0.2) cm and its mass (8065 ± 8) gm. Determine its density to the proper number of significant figures and express the error in the result in grams per cubic centimeter.

14. Suppose that a quantity enters into a computation in the fourth power and that it is subject to an error of 6%. How large a percentage error will result in the answer on account of the uncertainty of this quantity? How large will the percentage error be if the quantity enters into the computation in the cube root?

15. The masses of two bodies were recorded as (1) m_1 = 3147.278 ± 0.3 gm and (2) m_2 = 1.3246 gm ± 0.4%. Assuming that the errors are properly stated, (a) write the numbers properly, omitting any superfluous figures; (b) find the sum; (c) find the product (each to the proper number of significant figures); (d) calculate the numerical error of the sum and of the product.

16. A block has the following approximate characteristics: length, 25 cm ± 0.3%; breadth, 5 cm ± 0.5%; thickness, 4 cm ± 1%; mass, 300 gm. What *numerical* error in the measurement of the mass is allowable if the error introduced by the mass into the density determination is to be one-tenth of that introduced by the volume?

17. The following values were obtained for the length of a block: 20.56, 20.50, 20.48, and 20.54 cm. (a) To how many significant figures should their average be carried? (b) What is the a.d.? (c) What is the A.D.? (d) Properly record the average of the quantities and its numerical error.

18. For the length of a cubical block of wood the following observations, distributed over the cube, were obtained: 12.32, 12.35, 12.34, 12.38, 12.32, 12.36, and 12.38 cm, respectively. Calculate the a.d. and the A.D. of these values. What would be the percentage error of the volume as calculated from these data? If the mass were measured to be 1502 ± 3 gm, what would be the percentage error of the density? To how many significant figures should it be recorded?

CHAPTER 3

MEASUREMENT OF LENGTH; VERNIER SCALES AND MICROMETER SCREWS

For most measurements with a ruled scale, it is desirable to estimate fractions of the smallest division on the scale. Two common scale attachments that increase the accuracy of these estimates are the vernier scale and the micrometer screw. Because of the difficulty of holding a linear scale against a curved surface or against a narrow width, calipers are used. A *caliper* is an instrument with two jaws, straight or curved, used to determine the diameters of objects or the distances between two surfaces. A caliper with a vernier scale is called a *vernier caliper;* a caliper with a micrometer screw is called a *micrometer caliper.*

The Vernier Principle. The vernier is an auxiliary scale, invented by Pierre Vernier in 1631, which has graduations that are of different length from those on the main scale but that bear a simple relation to them. The vernier scale of Fig. 3.1 has 10 divisions that correspond in length to 9 divisions on the main

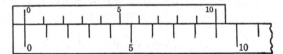

FIG. 3.1   Main and vernier scales

scale. Each vernier division is therefore shorter than a main-scale division by $\frac{1}{10}$ of a main-scale division. In Fig. 3.1 the zero mark of the vernier scale coincides with the zero mark of the main scale. The first vernier division is $\frac{1}{10}$ main-scale division short of a mark on the main scale, the second division is $\frac{2}{10}$ short of the next mark on the main scale, and so on until the tenth vernier division is $\frac{10}{10}$, or a whole division, short of a mark on the main scale. It therefore coincides with a mark on the main scale.

If the vernier scale is moved to the right until one mark, say the sixth as in Fig. 3.2, coincides with *some* mark of the main scale the number of tenths of a

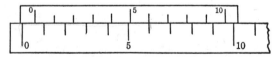

FIG. 3.2   Main scale with attached vernier, reading 0.6

main-scale division that the vernier scale is moved is the number of the vernier division that coincides with *any* main-scale division. (It does not matter with which main-scale mark it coincides.) The sixth vernier division coincides with a main-scale mark in Fig. 3.2; therefore the vernier scale has moved $\frac{6}{10}$ of a main-scale division to the right of its zero position. The vernier scale thus tells the fraction of a main-scale division that the zero of the vernier scale has moved beyond any main-scale mark. In Fig. 3.3 the zero is to the right of the second mark on the main scale and the fourth mark of the vernier scale coincides with a main-scale mark. The reading is 2.0 divisions (obtained from the main scale up

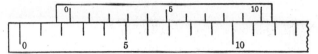

FIG. 3.3   Main scale with attached vernier, reading 2.4

to the vernier zero) and 0.4 division (obtained from the vernier coincidence), or 2.4 divisions.

The foregoing example illustrates the simplest and commonest type of vernier scale. Instruments are manufactured with many different vernier-scale to main-scale ratios. The essential principle of all vernier scales is, however, the same, and the student who masters the fundamental idea can easily learn by himself to read any special type.

In brief, the general principle of the vernier scale is that a certain number n of divisions on the vernier scale is equal in length to a *different* number (usually one less) of main-scale divisions. In symbols

$$nV = (n - 1)S \qquad (1)$$

where n is the number of divisions on the vernier scale, V is the length of one division on the vernier scale, and S is the length of the smallest main-scale division.

The term *least count* is applied to the smallest value that can be read directly from a vernier scale. It is equal to the difference between a main-scale and a vernier division. It can be expressed by rearranging Eq. (1), thus

$$\text{Least count} = S - V = \frac{1}{n}S \qquad (2)$$

When you have occasion to use a new type of vernier scale, first determine the least count of the instrument. In order to make a measurement with the instrument, read the number of divisions on the main scale before the zero of the vernier scale and note which vernier division coincides with a mark of the main scale. Multiply the number of the coinciding vernier mark by the least count to obtain the fractional part of a main-scale division to be added to the main-scale reading.

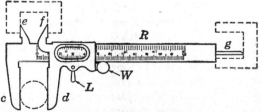

FIG. 3.4   Vernier caliper

The Vernier Caliper. A widely used type of vernier caliper is shown schematically in Fig. 3.4. The instrument has both British and metric scales and is provided with devices to measure internal depths and both inside and outside diameters. The jaws c and d are arranged to measure an outside diameter, jaws e and f to measure an inside diameter, and the blade g to measure an internal depth. The knurled wheel W is used for convenient adjustment of the movable jaw and the latch L to lock the jaw in position.

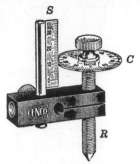

FIG. 3.5   Micrometer screw

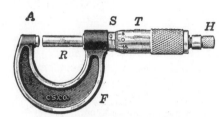

FIG. 3.6   Micrometer caliper

The Micrometer Screw.   A micrometer screw is another device for measuring very small distances. It consists essentially of a carefully machined screw R, Fig. 3.5, to which is attached a circular scale C. A linear scale S provides for observation of the forward motion of the screw. The distance the screw moves forward for one turn, the pitch of the screw, is known. The circular scale enables one to read the fractions of turns, and the linear scale enables one to record the whole number of turns. The least count of a micrometer screw is the pitch of the screw divided by the number of divisions on the circular scale.

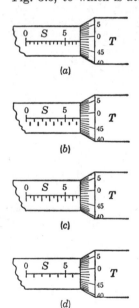

(a)

(b)

(c)

(d)

FIG. 3.7   Four methods of marking a micrometer scale

The Micrometer Caliper.   The micrometer caliper, Fig. 3.6, is used for the precise measurement of small lengths. It consists of a micrometer screw mounted in a strong frame F. The object to be measured is placed between the end of the screw and the projecting end A of the frame, called the anvil. The linear scale S is marked on the arm upon which the screw turns and the circular scale is engraved on the movable sleeve or thimble T. One type of metric micrometer caliper has the linear scale graduated in millimeters, a screw having a pitch of 0.50 mm, and 50 divisions on the circular scale. The least count of this instrument is 0.50 mm/50 = 0.010 mm. Inasmuch as it requires *two* revolutions of the screw to make it advance a distance of 1 mm, it is necessary to note whether the screw has advanced more or less than one-half of a main-scale division. Various methods are used for marking the half-millimeter distances as illustrated in Fig. 3.7a–d. In these diagrams the S reading is 6.5 mm and the T reading is 48.4 hundredths of a millimeter. The complete reading is, therefore, $6.5 + 0.484$ or 6.984 mm.

Many micrometer calipers are provided with an auxiliary milled head, H in Fig. 3.6, which is arranged to slip on the screw as soon as a certain force is exerted on the object to be measured. This arrangement is intended to ensure that the screw is always tightened on the measured object by the same amount. When no such head is provided, great care must be taken not to force the screw.

Experiment 3.1

VERNIER AND MICROMETER DEVICES

Object: To study the principles of vernier scales and micrometer screws and to use vernier and micrometer instruments in the measurement of length.

Method: Various instruments with vernier scales or micrometer screws are examined. The least count is determined and readings taken with each instrument. Vernier and micrometer calipers are used to measure certain lengths. The relative errors made in measuring a given length by various devices are estimated.

Apparatus: Various instruments having vernier scales, both linear and circular; vernier caliper with metric and British scales; metric and British micrometer calipers; meterstick; small metal cylinder and other objects for measurement.

Procedure: 1. Examine the vernier caliper. For each of its scales determine the length S of the smallest main-scale division, the length V of each vernier division, and the number n of divisions on the vernier scale. Determine the least count for each scale. Practice reading the scales by setting the zero in various places. When you are confident of your reading, have the instructor check it.

2. Close the jaws of the caliper and take a zero reading. This zero reading (with proper algebraic sign) must be subtracted from all readings made with this instrument. Measure the length and diameter of the small metal cylinder.

Record readings for both metric and British scales. For each setting make a rough freehand sketch, similar to Fig. 3.8, to show the *essential* data connected with the particular setting of the instrument. Do not sketch the instrument to scale or include all the graduations. Show merely the few lines necessary to indicate (a) the value of the main-scale division just before

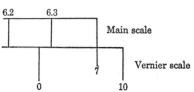

FIG. 3.8 Sketch of scales for recording data

the zero of the vernier, (b) the value of the main-scale division just after the zero of the vernier, (c) the total number of vernier divisions, and (d) the number of the vernier division that coincides with some main-scale mark.

Compare the corresponding length as measured on the two scales, using the relationship 1 in. = 2.54 cm.

3. As in Steps 1 and 2 determine the least count of the vernier scale and the reading on other instruments, such as the barometer with metric and British scales, and apparatus with circular scales, such as a spectrometer or a torsion apparatus.

4. Examine the metric and British micrometer calipers. Determine and record for each instrument (a) the value of the smallest division of the linear scale, (b) the number of revolutions of the sleeve required to advance the screw one

division on the linear scale, (c) the pitch of the screw, (d) the number of divisions on the circular scale, and (e) the least count. Observe and record the zero reading of each of the instruments and subtract it (with the proper algebraic sign) from all future readings with that instrument. Use each micrometer caliper to measure the diameter of the metal cylinder at several places along the cylinder and the diameter of a wire. Check the reading of the metric instrument against that of the British instrument.

5. Measure as accurately as the apparatus will permit the diameter of the metal cylinder, first with a meterstick, then with a vernier caliper, and finally with a micrometer caliper. Calculate the percentage uncertainty introduced by the necessity of estimating the last significant figure in each case. Discuss the significance of the relative errors involved in these data.

Experiment 3.2

VERNIER SCALES AND MICROMETER SCREWS

Object: To study the principles of vernier scales and micrometer screws and to use certain vernier and micrometer instruments in the measurement of various lengths.

Method: Various instruments with vernier scales or micrometer screws are examined, the least counts determined, and readings taken. Vernier and micrometer calipers are used to measure certain lengths. The relative errors made in measuring a given length by various measuring devices are estimated. The volume of a cylindrical cup is calculated from measurements made on it with a vernier caliper and this volume is compared with the value obtained by filling it with water whose mass is measured.

Apparatus: Various instruments having vernier scales, both linear and circular; vernier caliper with metric and British scales; metric and British micrometer calipers; micrometer screw; instruments to which micrometer screws are attached, such as expansion apparatus or micrometer microscope; small metal cylinder and other objects to be measured; measuring cup with glass cover, Fig. 3.9; trip scales; set of masses; meterstick.

Procedure: 1. Examine the vernier caliper. For each of its scales determine the length S of the smallest main-scale division, the length V of each vernier division, and the number n of divisions on the vernier scale. Determine the least count of the instrument. Practice reading the scales by setting the zero in various places. When you are confident of your reading, have your instructor check it.

FIG. 3.9 Cylindrical vessel

2. Close the vernier caliper and take a zero reading. This zero reading (with proper algebraic sign) must be subtracted from all readings with the instrument. Measure the length and diameter of a metal cylinder. Record readings for both metric and British scales. For each setting make a rough, freehand sketch, similar to Fig. 3.8, to show the essential data connected with the particular setting of the instrument. Do not attempt to sketch the instrument to scale nor include all the graduations. Show merely a few lines to indicate (a) the value of the main-scale division just before the zero of the vernier, (b) the value of the main-scale division just after the zero of the vernier, (c) the total number of vernier

divisions, and (*d*) the number of the vernier division that coincides with some main-scale mark.

3. Determine the least count of and learn to read the scale of various instruments with vernier scales, including an instrument with a circular scale. Make sketches of the settings as in Step 2.

4. Examine the metric and British micrometer calipers. Determine and record for each instrument (*a*) the value of the smallest division of the linear scale on the arm, (*b*) the number of revolutions of the sleeve required to advance the screw one division on the linear scale, (*c*) the pitch of the screw, (*d*) the number of divisions on the circular scale, and (*e*) the least count. Record the zero reading of each of the instruments and subtract it (with the proper algebraic sign) from all future readings with that instrument. Use each instrument to measure the diameter of a wire. Place the anvil against the wire and turn the screw *slowly* by means of the knurled knob *H* until the end of the screw makes contact with the wire and the ratchet clicks *once*. Further turning will slightly increase the force. Check the two measurements, using the relation 1 in. = 2.54 cm.

5. Determine the least count of a micrometer screw and use it to measure the thickness of a thin coin. For the initial reading place the base of the micrometer screw on a glass plate and adjust the screw so that its tip just touches the plate. Place the coin on the plate and reset the screw so that its tip just touches the coin. The difference in the two readings is the thickness of the coin.

6. Measure as accurately as the apparatus will permit the diameter of the metal cylinder, first with a meterstick, then with a vernier caliper, and finally with a micrometer caliper. Calculate the percentage uncertainty introduced by the necessity of estimating the last significant figure in each case. Discuss the significance of the relative errors involved in these data.

7. Use the vernier caliper to measure the internal diameter and depth of the measuring cup, Fig. 3.9. Take several measurements of each dimension. Calculate the volume of the cup, $V = \pi r^2 h$. Weigh the cup, first empty and then level full of water. Use the glass plate to level the water.

Each gram of water has a volume very nearly 1.00 cm³. From this relationship determine the volume of the water and compare this value of the volume with that computed from the dimensions of the cylinder.

Review Questions: 1. What is a vernier scale? a vernier caliper? 2. Explain the essential vernier principle. 3. Define least count and show how to obtain the least count of any vernier device. 4. What is a micrometer screw? a micrometer caliper? Explain the essential micrometer principle. 5. Show how to obtain the least count of a micrometer screw.

Questions and Problems: 1. Discuss the relative merits of the metric and British systems for use with vernier and micrometer scales.

2. Can the accuracy with which a scale may be read by the unaided eye be increased by dividing the scale more and more finely, for example, by ruling a millimeter scale into tenths of a millimeter?

3. How may the error caused by backlash in a micrometer screw be almost entirely eliminated?

4. State some of the sources of error that one might make in measuring the length of a cylinder with a vernier caliper.

5. For each of the vernier scales shown in Figs. 3.10 to 3.15 determine the least count and the reading as the scale is set.

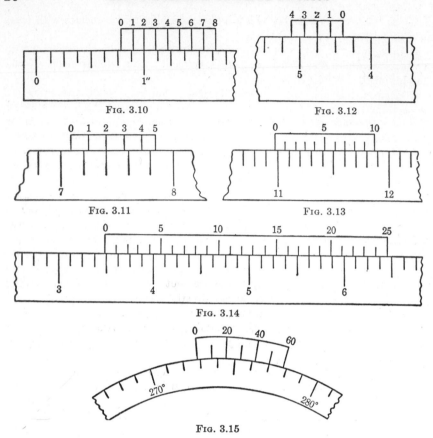

FIG. 3.10

FIG. 3.12

FIG. 3.11

FIG. 3.13

FIG. 3.14

FIG. 3.15

6. A scale whose smallest division is 1 cm is to be provided with a vernier scale that will enable fifths-of-centimeter divisions to be accurately estimated. How many vernier divisions should be used, and how long will each one be?

7. A micrometer caliper has a zero error such that as the jaws are closed the zero mark on the sleeve turns 0.05 mm past the zero on the main scale. The reading of the instrument is 7.82 mm. What is the corrected reading?

8. Given the following data concerning various vernier and main scales, determine the least count and the indicated reading of each instrument:

Vernier number	Total number of vernier divisions	Reading of main scale		Number of the vernier division coinciding with a main-scale division
		Just before the vernier zero	Just after the vernier zero	
A	10	12 mm	13 mm	6
B	5	7.5 in.	7.6 in.	3
C	50	3.65 in.	3.70 in.	23
D	20	14°20′	14°40′	7
E	4	2⅝ in.	2¾ in.	3

9. Sketch a vernier scale that would read to 0.2 cm if the smallest division of its main scale is 1 cm: (a) in the zero position; (b) indicating a reading of 3.6 cm.

PART II—MECHANICS

CHAPTER 4

VECTORS

All measurable quantities may be classed either as scalar quantities or as vector quantities. *Scalar* quantities are those that have only magnitude, such as 10 sec, 4 gm/cm³, and 50 ohms. *Vector* quantities are those that have direction as well as magnitude, such as a velocity of 20 mi/hr N or a weight of 25 lb vertically downward.

To add scalar quantities, one has merely to make the algebraic addition. When one wishes to add two vector quantities, the process is more difficult because the direction must be considered. The *vector sum* of two vector quantities is the single vector quantity that would produce the same result as the original pair.

The addition of vector quantities is greatly simplified by representing the vector quantity graphically. A *vector* is a line segment whose length represents the magnitude of a vector quantity and whose direction is that of the vector quantity. The sense along the line is indicated by an arrow. For example, a force of 100 lb acting at an angle of 30° above the horizontal may be represented by the line *OA*, Fig. 4.1, which is 5 units long and has the correct direction. Each unit of length thus represents 20 lb.

When vectors do not have the same line of action, their vector sum is not their algebraic sum but a geometric sum. This geometric sum may be determined by

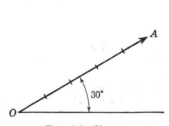

FIG. 4.1   Vector

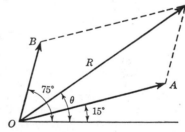

FIG. 4.2   Parallelogram of forces

either graphical or analytical methods. Graphical methods are simple and direct but are limited in precision to that obtainable by drawing instruments. Analytical methods have no such inherent limitations. In this chapter both graphical and analytical methods will be applied to forces, as examples of vector quantities, but the same methods apply to all vector quantities.

The vector sum, or *resultant* of a set of forces, is the single force that will have the same effect, insofar as motion is concerned, as the joint action of the several forces.

Graphical methods are of two kinds, parallelogram and polygon. The parallelogram method is especially suitable when there are only two forces; the polygon method is used when there are more than two.

Consider the two forces represented in Fig. 4.2 by the vectors *OA* and *OB*.

The resultant R is found by constructing a parallelogram having the two vectors as sides and drawing the *concurrent* diagonal. This new vector represents in magnitude and direction the single force that is equivalent to the original pair.

If the resultant of more than two forces is desired, the parallelogram method might be used by finding the resultant of one pair, using this resultant with the next force, and so on. However, for this case the polygon method is quicker.

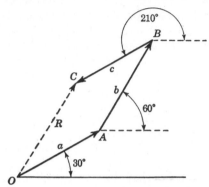

Lay off any one of the vectors, such as a (Fig. 4.3). From the end of the vector OA, construct any one of the other vectors, say b; from the end of AB lay off the next vector c. The resultant is the vector that closes the diagram, OC in Fig. 4.3.

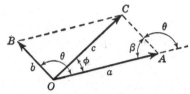

FIG. 4.3   Polygon of forces   FIG. 4.4   Law of cosines and law of sines

Graphical methods for the solution of vector problems are frequently not accurate enough for the purpose in hand. In this case analytical methods are used. When only two vectors must be added, the resultant is readily obtained by the use of the law of cosines and the law of sines. If there are more than two forces the component method is easier.

For the vectors OA and OB of Fig. 4.4, the resultant OC can be calculated if we know the angle θ between the two vectors. From the law of cosines

$$c^2 = a^2 + b^2 - 2ab \cos \beta$$

or
$$c^2 = a^2 + b^2 + 2ab \cos \theta \tag{1}$$

The direction of OC is given by the law of sines

$$\frac{\sin \phi}{\sin \beta} = \frac{b}{c} \tag{2}$$

or, since $\sin \beta = \sin \theta$,

$$\sin \phi = \frac{b}{c} \sin \theta \tag{3}$$

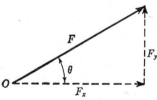

FIG. 4.5   Rectangular components of a vector

Any single force may be replaced by two or more forces whose joint action will produce the same effect as the single force. These various forces are said to be *components* of the single force. The most useful set of components is usually a pair at right angles to each other, Fig. 4.5. The rectangular component of a vector in any direction is simply the projection of the vector on that direction, $F_x = F \cos \theta$ and $F_y = F \sin \theta$.

To find the resultant of a set of forces by the component method, one selects two directions at right angles to each other, commonly vertical and horizontal, calculates the components of the various forces in each direction, adds algebraically each set of components, and computes the resultant by use of the Pythagorean theorem. In Fig. 4.6, X_a, X_b, and X_c are the X components of the forces a, b, and c, respectively, while Y_a, Y_b, and Y_c are the Y components. Then

$$F_x = X_a - X_b - X_c \tag{4}$$
and
$$F_y = Y_a + Y_b - Y_c \tag{5}$$
$$R = \sqrt{F_x{}^2 + F_y{}^2} \tag{6}$$
$$\tan \theta = F_y/F_x \tag{7}$$

The *equilibrant* of a set of forces is the single force that must be combined with the set of forces to maintain the system in equilibrium. The equilibrant (sometimes called antiresultant) must be equal in magnitude but opposite in direction

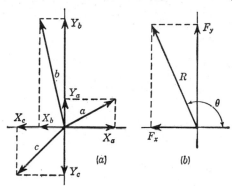

FIG. 4.6   Resultant by component method

to the resultant. If the equilibrant, or the resultant, has zero magnitude the system is in equilibrium and the vector diagram is a closed polygon. A body is in equilibrium, insofar as linear motion is concerned, if the vector sum of all the forces acting on it is zero.

Experiment 4.1

VECTORS: GRAPHICAL METHODS

Object: To study vectors and graphical methods for determining the resultant of several forces.

Method: Each student is assigned a problem. In this problem it is assumed that the magnitude and direction of forces acting on a body are known. The resultant and equilibrant are determined by graphical methods. The accuracy of the result is checked on a force table.

Apparatus: Force table (Fig. 4.7); weight holders; set of slotted masses; ruler; protractor.

The force table is mounted with its surface horizontal. The object on which the forces act is a ring at the center of the table. A pin within the ring prevents motion of the ring before equilibrium is established. Forces are applied by strings that pass over pulleys to weight holders. The direction of each force is read on the circular scale of the force table.

Procedure: 1. Each student is assigned a set of three forces by the instructor. Using two of the three forces, draw vectors representing them, each beginning at the same point. Use as large a scale as possible. Complete the parallelogram and draw the resultant and equilibrant. From the length and direction of the resultant vector determine the magnitude and direction of both the resultant and equilibrant.

FIG. 4.7   Force table

2. Check your result of Step 1 on the force table. Use weights, including the weights of the holders, to represent each of the original forces and the equilibrant. When the ring appears to be in equilibrium, remove the pin, displace the ring somewhat, and tap the table. The ring should return to the central position. Ask the instructor to check all your values. Compare the experimental value of the equilibrant with that determined graphically; express the percentage difference in magnitude and the angular difference in direction.

3. Using the resultant found in Step 1 and the third force, construct a parallelogram to determine the resultant and equilibrant. Test the result on the force table.

4. Using the three original forces, draw to scale a vector polygon. On the diagram draw the resultant and the equilibrant and determine their values. Check the results with those obtained in Step 3.

5. Arrange any four forces on the force table so that they are in equilibrium. Draw a vector diagram to represent these four forces. Comment on the reasons why the diagram is or is not a closed figure.

Experiment 4.2

VECTORS: GRAPHICAL AND ANALYTICAL METHODS

Object: To study vectors and graphical and analytical methods for determining the resultant of several forces.

Method: In a problem assigned each student it is assumed that the magnitude and direction of forces acting on a body are known. The resultant and equilibrant are determined by graphical and by analytical methods. The accuracy of the result is checked on a force table.

Apparatus: Force table (Fig. 4.7); weight holders; set of masses; ruler; protractor.

The force table is mounted with its surface horizontal. The object on which the forces act is a ring at the center of the table. A pin within the ring prevents motion of the ring before equilibrium is established. Forces are applied by strings that pass over pulleys to weight holders. The direction of each force is read on the circular scale of the force table.

Procedure: 1. Each student is assigned three forces by the instructor. Using two of the three forces, determine graphically the resultant and equilibrant by the use of the parallelogram of forces. Use as large a scale as possible in drawing the figure.

2. For the same two forces used in Step 1 compute the magnitude and direction of the resultant by the use of the law of cosines and the law of sines.

3. Check the results on the force table. Set up the two given forces and determine experimentally their equilibrant. The weights of the scale pans must be included in the forces applied to the strings. Reduce the effect of friction as much as possible by displacing the central ring slightly in various directions and observe its return. Unequal tendencies to return to the center after displacements in opposite directions indicate an unbalance that must be corrected. Compare the results found in Steps 1, 2, and 3.

4. Using all three given forces construct a vector polygon to determine the resultant. Use as large a scale as the paper will permit.

5. From the three forces of Step 4 compute the magnitude and direction of the resultant by the component method.

6. Set up the three original forces on the force table and determine as in Step 3 the equilibrant. Compare the results of Steps 4, 5, and 6.

7. Perform other exercises selected by the instructor to illustrate the graphical and analytical methods of finding the resultant and equilibrant of forces.

Review Questions: 1. Define the terms: vector quantity, vector, resultant force, equilibrant force. 2. Describe two graphical methods of finding the resultant of two or more vectors. 3. Explain the difference between vector and algebraic sums. 4. Describe the analytical method of finding the resultant of two vectors. 5. State the law of sines and the law of cosines, illustrating by a sketch. 6. What is meant by a component of a force? rectangular components? 7. Describe the component method used for finding the resultant of several vectors. 8. State the condition necessary for equilibrium insofar as linear motion is concerned.

Questions and Problems: 1. The forces on the force table act on a ring but are said to be concurrent. Explain. If the cords were attached rigidly to the ring would the forces necessarily be concurrent?

2. Indicate whether each of the following is a scalar or vector quantity: speed, velocity, mass, weight, work, torque, and volume.

3. A hammock is supported by two hooks at the same level. A man is seated in the hammock. Under what conditions will the pull on each hook be equal to the weight of the man?

4. A body, weight W, is attached by a string of length L to a hook on a vertical wall. A horizontal force F acting on the body holds it at a distance s from the wall. Derive an equation that gives F in terms of W, L, and s.

5. Three forces of 12, 15, and 20 lb are in equilibrium. If the 12-lb force is directed horizontally to the right, what two configurations in a vertical plane may the other two forces have?

6. A man living on a certain parallel of latitude travels 15 mi SE, then 8.0 mi N, then 5.0 mi NW.   How far will he have to travel SW until he again reaches the original parallel of latitude?   Will he then be at his starting point?

7. A rope 26 ft long is attached to two points A and B, 20 ft apart at the same level.   A load of 250 lb is carried at the middle of the rope.   What force is exerted on A?

8. A 9.0-lb and a 12-lb body are hung from the ends of a string that passes over two fixed frictionless pulleys at the same level.   At a point on the string between the two pulleys a 15-lb body is attached.   What configuration of the string, if any, will permit the system to be in equilibrium?

9. Find the force necessary to draw with uniform speed a 2.0-ton wagon up a grade rising 40 ft in a distance of 200 ft measured along the grade.   What force would be necessary if the force were applied horizontally?   Neglect friction.

10. Find the magnitude and direction of the resultant of the following system of forces: 50 lb at an angle 30° to the right of the vertical, 200 lb at 45° to the left of the vertical, and 150 lb horizontally to the right.

11. Use the component method to find the magnitude and direction of the resultant of the following forces: $A = 2000$ gm-wt at 0°, $B = 1732$ gm-wt at 60°, $C = 1000$ gm-wt at 150°, and $D = 3828$ gm-wt at 225°.

12. A body weighing 100 lb is suspended by a rope.   A second rope attached to the body is drawn aside horizontally until the suspended rope is deflected 30° from the vertical.   Find the stretching force in each rope.

CHAPTER 5

UNIFORMLY ACCELERATED LINEAR MOTION

Bodies in motion seldom travel with uniform speed in a straight line. In any other type of motion there is a change in velocity, that is, there is an acceleration. Many accelerated motions are quite complicated but a few can be analyzed by simple methods. One of these is uniformly accelerated linear motion. A familiar example is the motion of a freely falling object.

The *average speed* $\bar{v}$ of a body is the quotient of the distance s which it traverses and the time t required to travel that distance. In symbols

$$\bar{v} = s/t \tag{1}$$

The cgs unit of speed is the centimeter per second. Many other units are common, such as the foot per second and the mile per hour.

The *instantaneous* speed v of an object is the limit of the distance-time ratio as the time is made vanishingly small. Symbolically

$$v = \lim_{\Delta t \to 0} \frac{\Delta s}{\Delta t} \tag{2}$$

where Δs represents a small increment of distance traversed in the corresponding increment of time Δt.

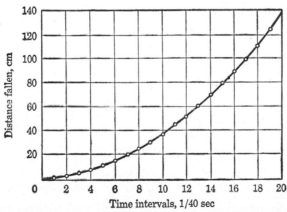

Fig. 5.1 Distance-time relationship for freely falling body

The curve of Fig. 5.1 shows the distance-time relationship of a freely falling body. For any distance-time curve, Eq. (2) states that the instantaneous speed of the body is given by the slope of a line drawn tangent to the curve at the point for the instant in question. If the speed were constant the slope would be the same at every point and the curve would be a straight line. From the shape of

33

the curve of Fig. 5.1 it is evident that the speed of the freely falling body is not constant since the slope of the curve is constantly increasing.

When the velocity of a body varies, its motion is said to be accelerated. *Acceleration* is defined as the time rate of change of velocity. In symbols

$$\bar{a} = \frac{v_2 - v_1}{t} \tag{3}$$

where $\bar{a}$ represents the average acceleration of a body that changes its velocity from v_1 to v_2 in time t. Units of acceleration are obtained by dividing a unit of velocity by a unit of time; thus the cgs unit is the centimeter per second per second and the fps unit is the foot per second per second.

If a body moves in a straight line, making equal changes of speed in equal intervals of time, its acceleration is constant and it is said to be moving with *uniformly accelerated motion*. This is the type of motion produced when a constant net force, parallel to the direction of motion, acts upon the body. The most common example of uniformly accelerated motion is the motion of a freely falling body. In this motion the constant acceleration g is called the *acceleration due to gravity*. The standard value of g is approximately 980 cm/sec² or 32.2 ft/sec².

The relationships between speed, distance, time, and acceleration in uniformly accelerated motion are readily deduced from the definitions of acceleration and average speed. From Eq. (3)

$$v_2 = v_1 + at \tag{4}$$

This equation expresses the dependence of the final speed v_2 upon time t in terms of the initial speed v_1 and the acceleration a. It is the equation of a straight line whose slope is the acceleration a.

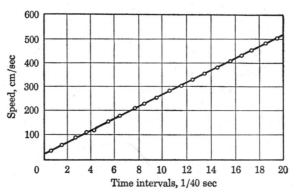

Fig. 5.2   Speed-time relationship for freely falling body

Since the speed-time curve for a body moving with uniform acceleration is a straight line, the slope $\Delta v/\Delta t$, which represents the acceleration, is constant. Such a curve is shown in Fig. 5.2. For a body moving with uniform acceleration the *average* speed during a time interval is identical with the *instantaneous* speed at the middle of that time interval. This special relationship is true only for a linear variation. This fact makes it possible to determine the instantaneous speeds at the middle of successive time intervals by measuring the distance traveled during the successive intervals.   If P_1 and P_2 represent positions of the falling body at the beginning and end of any time interval Δt, then $\Delta s = P_2 - P_1$, and the average speed $\Delta s/\Delta t$ during the interval is the instantaneous speed at the midpoint of the interval

$$v = \Delta s/\Delta t$$

When the speeds thus computed are plotted against time, the resulting curve is a straight line.   The acceleration is obtained by computing the slope of this line.

For *uniformly accelerated* motion the average speed is the arithmetic mean of the initial and final speeds.   Note that this statement holds true only because the acceleration is uniform.   For any other type of accelerated motion the arithmetic average of the speeds does not give the average value.   Thus, for uniformly accelerated motion, Eq. (1) gives

$$s = \bar{v}t = \frac{v_2 + v_1}{2} t \tag{5}$$

When the value of v_2 from Eq. (4) is substituted in Eq. (5), one obtains

$$s = v_1 t + \tfrac{1}{2}at^2 \tag{6}$$

The form of Eq. (6) shows that, for uniformly accelerated motion, the distance-time curve (Fig. 5.1) is a parabola.   The slope of the curve at any point (slope of the tangent line) is the speed at the corresponding instant.

In the study of uniformly accelerated motion it is necessary to measure time intervals or to use known time intervals.   Where the acceleration is relatively high the limitations of space require the use of very short time intervals.   A freely falling body falls 16 ft in the first second.   Thus the time intervals used in studying such motion must be much less than 1 sec.

Experiment 5.1

UNIFORM ACCELERATION ON AN INCLINED PLANE

Object: To study the motion of a body rolling down an inclined plane.

Method: A disk or wheel with a small axle is allowed to roll freely down an incline, starting from rest.   Positions of the disk at the ends of successive equal intervals of time are noted.   By use of the data thus obtained the acceleration is computed.   Distance-time and speed-time curves are plotted and the acceleration is determined from the slope of the speed-time curve.

Apparatus: Incline; disk or wheel; timing device; meterstick.

A suitable incline can be formed by two wooden pieces whose edges are straight and whose length is about 8 ft.   They can be fastened together to form a frame. The angle of the incline can be varied by adjusting the supports.

The disk or wheel may be any uniform metal disk with a small axle and the mass not too near the axle.   A bicycle wheel with cones tightened to make a rigid wheel is quite suitable.

As a timing device a clock with sweep-second hand, a seconds beat clock, or a metronome may be used.

Procedure: 1. Set up the wheel so that it is free to rotate about its axis. Test for balance by allowing it to turn as it will.   If it is perfectly balanced it will remain in any set position. If the wheel is unbalanced attach small objects at appropriate places near the rim until it is balanced. Soft wax may be used to attach the objects.

2. Set up the incline using a small angle of inclination. Place the two wooden pieces sufficiently close to each other that the axle will always remain on both pieces even when the wheel gets near one side. Adjust the supports until the wheel is free to rotate throughout most of the length of the incline.

3. Place the wheel near the top of the incline with its axis accurately perpendicular to the wooden pieces and the wheel centered between them. Release the wheel and measure the time required for the wheel to roll down the incline. If the wheel touches the side on the way down the initial placing of the wheel was not sufficiently accurate. From the time of descent select a suitable time interval to be used in the experiment. Choose a time interval such that there are five to ten such intervals in the total time of descent. Record the chosen time interval.

4. Replace the wheel at the top of the incline and mark its starting position. With the timer in operation release the wheel at the end of one chosen interval. When the wheel is released it is very important that no speed be imparted to it. As the wheel descends mark its position at the end of each interval by chalk or other marker. Repeat the release and check the marks a second time. It may be necessary to make several trials before there is agreement on the positions of the various marks. Measure the distance from the starting point to each of the marks and record the data as in the accompanying table.

POSITIONS AND SPEEDS OF A WHEEL ON AN INCLINE

Time interval	Total distance traversed	Distance traversed during 1 interval	Average speed during 1 interval*	Acceleration
	cm	cm	cm/sec	cm/sec²
0	0.0	....	....	...
..	....	10.0	5.0	5.0
1	10.0	....	....	...
..	....	30.0	15.	5.0
2	40.0	....	....	...
..	....	50.0	25.	5.0
3	90.0	....	....	...

* In the computation of the quantities tabulated in the table it was assumed that the time interval was 2.0 sec.

Computations and Analysis: 1. From the tabulated distance measurements compute the distance traversed by the wheel in each successive time interval and record these values in the third column of the table.

2. By the use of the distance traveled in each time interval and the recorded value of that interval compute the average speed during each interval. Record these values in the fourth column of the table. These average speeds are also the instantaneous speeds at the midpoints of the time intervals.

3. By the use of Eq. (6), $s = v_1 t + \frac{1}{2}at^2$, compute the acceleration, using each of the distances recorded in the second column and the corresponding time. Compute the average value of the acceleration thus obtained.

4. Plot a curve showing the dependence of instantaneous speed upon time. For this purpose use the values of the fourth column and plot each value at the midpoint of the time interval. Carefully interpret this curve. State the significance of its shape and the meanings of the slope and intercepts. Calculate the value of the slope to obtain the acceleration of the wheel. Compare this value with the average value obtained in Part 3.

5. Plot a curve to show the relationship between total distance traversed and elapsed time. Discuss the significance of the shape of the curve and the change of slope.

Experiment 5.2

UNIFORMLY ACCELERATED LINEAR MOTION

Object: To study the motion of a falling body and, in particular, to determine its speed and acceleration.

Method: A vibrating tuning fork of known frequency falls between vertical guides. A stylus attached to one prong of the fork records the vibrations as a sinuous line on a coated surface. From this record, measurements of the distances fallen during successive equal time intervals are made to study the characteristics of the motion and determine the speed and the acceleration.

Apparatus: The acceleration apparatus is shown in Fig. 5.3. The tuning fork F with stylus S attached to one prong is mounted in carriage C. When the carriage is released from its uppermost position by withdrawing the catch H, it falls between the vertical guides G and G' and is caught in the dashpots D and D'. The friction between the carriage and the guides is slight when the apparatus has been properly aligned and leveled. The tuning fork is electrically driven so that constant amplitude of vibration is maintained throughout the time of fall. Consequently the waves traced by the stylus on the coated surface of plate P are of the same amplitude, which facilitates the location of the crests of the wave and therefore the accurate determination of the distances corresponding to the equal time intervals.

The strip of heavy plate glass is held in grooves at its lower and upper ends. The plate may be shifted horizontally and a number of tracings obtained without removing the plate. The surface of the plate on which the records are to be traced may be coated with a thin film of Bon Ami applied with a moist cloth or sponge. A more satisfactory way to obtain a permanent trace is to use a specially coated paper strip held against the glass plate by clamps or Scotch tape.

The auxiliary apparatus includes a 4- to 6-volt storage battery or d-c power supply; a meterstick; a switch; a plumb bob; a small celluloid triangle; and a sharp pointed scriber—needle, pin, or sharp, hard pencil.

Procedure: If the apparatus is already leveled and aligned, the electrical connections made, and if the stylus pressure is satisfactory, some of the procedure 1 to 3 can be omitted. If not, start with Step 1.

1. Remove the glass plate and lay it flat on the table. With the carriage at

the top held by catch H, attach the plumb bob at R and adjust the leveling screws in the base E until the bob hangs over the center of the bolthead between the dashpots. Check the rods for correct alignment and rough spots.

2. Coat the plate as directed by the instructor and place in position. Adjust the stylus until it presses lightly on the coated surface of the plate or the paraffined

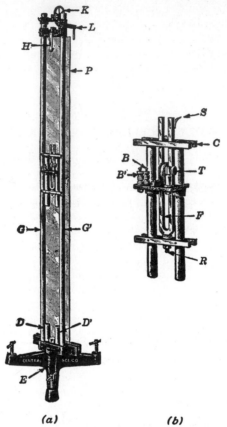

(a) *(b)*

FIG. 5.3 *(a)* Acceleration apparatus. (Wheel K not used in Exp. 5.2.) *(b)* Details of electrically vibrated tuning fork

paper. The pressure should be sufficient to produce a clear trace and permit the crest of each wave to be marked; too much pressure will increase the retarding force due to friction.

3. Connect the voltage source to the binding posts B and B' through the switch; make sure that the fork can fall without interference from the lead wires. Set the fork into vibration by closing the switch and adjusting the small screw T. It may be necessary to give the fork an initial impulse by squeezing the prongs together with the fingers and allowing them to spring apart. Adjust T so that the sparking is a minimum but the vibration is steady. The amplitude of vibration should be several millimeters. *Keep the switch open when the fork is not vibrating and open it after the fork strikes the dashpots.*

4. Make a few preliminary traces to perfect the adjustments and learn the technique.   Have one partner close the switch and adjust the screw T, while the other one holds the prepared plate securely in place at the bottom and slightly tilted backward at the top.   When the fork is vibrating strongly and evenly, quickly bring the plate into position, close the clamp at the top, and take hands off the apparatus.   Immediately release the catch and allow the fork to fall. When it strikes the dashpots, open the switch.   Tilt the plate slightly backward at the top, bring the fork up to position, move the plate slightly to one side, and make another trace close to the first one.   After every two or three runs, check the tightness of electrical connections, contact screw, and the dashpots.   When the fork is adjusted to make a suitable trace, remove the plate to the table and recoat it, or put on a fresh strip of coated paper.   Make two good traces for each pair of students.   Record the frequency stamped on the tuning fork.

5. Place the record strip, or the coated paper, on a flat surface where there is good light.   Beginning near the top, at a place where the trace is distinct and the crests uniform, pin-mark as finely and as accurately as possible the center of the crest of each *fourth* wave.   A small transparent right-angled triangle may be placed transversely across the crest and a fine line drawn with the scriber to bisect the crest and extend about 5 mm above it.

6. Place a meterstick *edgewise* on the trace in such a manner that the graduations are directly touching the trace, and the scale intersects the fine lines at right angles.   Place some even division mark, such as 1.0 cm, on the top line. Leave the meterstick stationary and record the scale readings of the successive positions.   Estimate the readings to fractions of a millimeter, keeping the eye in the same position relative to the scale for each reading.   A typical set of data is shown in the accompanying table.

DATA FOR FALLING TUNING FORK

Frequency of fork 128 vib/sec.   Time interval $\frac{4}{128} = 0.0312$ sec.

No. of time interval	Elapsed time	Scale reading	Distance fallen in 1 interval	Average speed in 1 interval
	sec	cm	cm	cm/sec
0	0	1.00	....	.....
..	......	....	1.56	49.9
1	0.0312	2.56	....	.....
..	......	....	2.78	88.9
2	0.0625	5.34	....	.....
..	......	....	4.18	134.0
3	0.0938	9.52	....	.....

7. Cut off a piece of the coated paper showing the trace to mount in the report.

Computations and Analysis: 1. Compute the time intervals and enter them on the data sheet.   From the observed readings calculate the distance fallen in each successive time interval and tabulate the results as shown in the table. (Note the spacing of the entries.)

2. Calculate the average speed for each time interval and tabulate as shown.

These average speeds are also the instantaneous speeds at the midpoints of the intervals.

3. Plot a graph to show the dependence of the *instantaneous* speed on time for the falling body. Use the tabulated speeds and the time in seconds at the *midpoints* of the time intervals. Place the zero of the abscissa somewhat to the right of the left-hand edge because at zero time (the first marked crest) the body already has a small initial speed.

4. From the slope of the speed-time curve determine the acceleration of the falling body. Compare it with values obtained by other students using the same apparatus and the same adjustment. There is necessarily a certain amount of friction in the apparatus between the stylus and the recording surface and between the frame of the falling tuning fork and the support rods. The observed value of the acceleration a is therefore less than g, the acceleration due to gravity at that location. The values of a for a series of observations made under similar conditions with the same apparatus should, however, check within experimental error.

5. Plot a second curve to show the total distance fallen against total time. Select one of the plotted points on the distance-time curve and draw a tangent to the curve. From the slope of the tangent determine the speed at that instant and compare it with the computed value.

6. Using the data for any two points of the record, compute the initial speed from Eq. (6). Compare this value with the initial velocity determined from the intercept of the speed-time curve.

7. Interpret the graphs by giving conclusions drawn from their shape, their slopes, and their intercepts. Give reasons for all conclusions.

Experiment 5.3

MOTION OF A FREELY FALLING BODY

Object: To study the motion of a freely falling body and, in particular, to measure g, the acceleration due to gravity.

Method: An object is allowed to fall freely and its positions at the ends of successive equal intervals of time are recorded on a coated paper strip by means of electric sparks. From the data thus obtained distance-time and speed-time curves are plotted. The acceleration is determined from the slope of the speed-time curve.

Apparatus: There are two principal units: the fall apparatus and the timing device. Auxiliary apparatus includes a 6-volt storage battery, a switch, a vernier caliper, and a good-quality metric scale.

The fall apparatus, shown in Fig. 5.4 and schematically in Fig. 5.5, provides convenient methods for holding the falling body B suspended by the electromagnet M, for releasing the body at will by opening the switch, and for holding the record strip R of coated paper so that it will receive properly the marks recorded during the fall.

The recording device is a unit that produces a series of electric sparks at equal predetermined intervals of time. The marks that define the positions of the falling body are produced when the sparks jump from the high-potential vertical

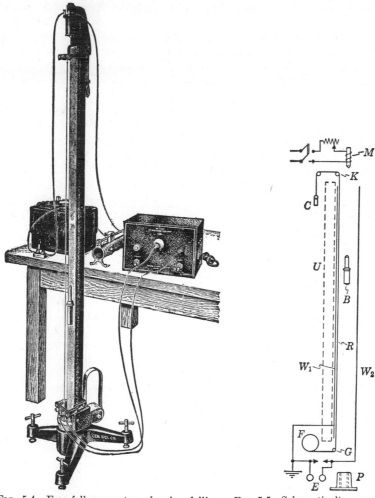

FIG. 5.4   Free-fall apparatus, showing falling body after it has traversed one-half of distance of fall

FIG. 5.5   Schematic diagram of free-fall apparatus, showing electric circuit

wire W_2 to the ridge on B and thence through the coated paper R to a second vertical wire W_1.

The timing mechanism (Fig. 5.6) may be either one of two types. The first consists essentially of a buzzer-type mechanical interrupter vibrating synchronously with the 60-cycle a-c power supply and thus interrupting the current at intervals of exactly $\frac{1}{120}$ sec. This interrupted current induces a high voltage in the secondary of a high-potential transformer. The design of the circuit is such that a spark occurs only at the instant of the peak voltage. The usual a-c power supply is so well regulated that one may assume it to be exactly 60 cycles per second.

The more recent type of timer employs a Thyratron discharge tube, which operates a relay for charging a capacitor. The charges from the capacitor are fed through a high-voltage coil at the rate of 60 per second from a 60-cycle a-c source. Either timer is operated by plugging it into the a-c line and connecting the high-voltage side to the two vertical wires at E (Fig. 5.5).

The fall apparatus must be carefully aligned so that the falling body, throughout its path, will remain uniformly distant from W_1 and W_2 and finally fall accurately into the dashpot P. The dashpot has about an inch of sand in the bottom, and its sides are heavily lined with felt. The prepared record paper, coated on one side with paraffin, is carried in a holder at F. When a record is made, the end of the strip is pulled through the opening at G, thence upward over the wire W_1, and back through the opening at K.

FIG. 5.6 Synchronous spark timer

Procedure: 1. It will be assumed that the fall apparatus has been properly aligned in advance. This important and delicate adjustment should be made only under the personal supervision of the instructor.

2. Energize the electromagnet by closing the switch connecting the storage battery to the binding posts on the electromagnet. When the body hangs motionless, open the switch. The body should fall directly into the dashpot. Use suitable precautions to prevent the body from becoming damaged by striking any object. Leave the switch open except when the body is to be suspended.

3. With the body *not hanging*, draw the coated paper through the opening G at the lower end of the casting, then upward, and back through the opening K. The light coated side must be on the outside. Attach the weighted clip C to the end of the paper to hold it taut.

Make the necessary electrical connections to the timing device. The frame of the apparatus should be grounded.

4. Suspend the body from the electromagnet and lightly touch the body until it hangs motionless. Start the spark timer. Observe that the spark now jumps from the outer wire through the body to the electromagnet and grounded support. A spark will not jump to the inner wire W_1 until the body has been released from the electromagnet. When these conditions have been established, release the body by opening the switch in the electromagnet circuit.

5. When the body has fallen, stop the spark timer and examine the record of the dots on the paper. If any of the spots are missing or are very faint, shift the paper to one side and repeat. Remove the record from the apparatus by drawing it upward. This method puts a fresh strip in place.

6. If the spark timer gives intervals of $\frac{1}{60}$ sec the spots can be used as they appear on the record. Starting near the beginning, but preferably not at the first dot, encircle the spots by using a sharp pencil. If the spark timer has an interval of $\frac{1}{120}$ sec the spots are too close together for accurate measurement and it is preferable to encircle every *third* spot. Number the encircled spots 0, 1, 2, 3, etc.

7. Place the record strip on a flat surface where there is good light. Lay a good-quality metric scale edgewise on the trace in such a manner that the graduations are directly touching the dots. Set some convenient mark, such as 1.00, of the scale at the zero dot. Keep the meterstick stationary and read the position of each dot, estimating the readings to tenths of a millimeter. Record the readings as in the first two columns of the accompanying table. These are the data from which the computations are made.

POSITION AND SPEED OF A BODY DURING FREE FALL

Time intervals	Reading on scale	Distance traversed in 1 time interval	Average speed during 1 time interval
$\frac{1}{10}$ sec	cm	cm	cm/sec
0	1.00	....	....
..	....	0.96	38.4
1	1.96	....	....
..	....	1.58	63.2
2	3.54	....	....
..	....	2.23	89.2
3	5.77	....	....

Computations and Analysis: 1. From the observed readings calculate the distance fallen in each successive time interval and tabulate the results as shown in the third column of the table. (Note the spacing of the entries with respect to the second column.)

2. Calculate the average speed for each time interval and tabulate as in the fourth column of the table. These average speeds are also the instantaneous speeds at the midpoints of the intervals.

3. Plot a graph to show the dependence of the _instantaneous_ speed upon time for the falling body. For this purpose use the tabulated speeds of the fourth column plotted at the _midpoints_ of the time intervals. Interpret this curve carefully by stating the conclusions that can be drawn from consideration of the shape, slope, and intercepts of the curve.

4. Calculate the value of the slope of this graph to obtain the experimental value of _g_. Compare the value thus obtained with the standard value for the locality of the experiment.

5. Plot a graph to show how the total distance fallen depends upon the elapsed time, using the tabulated data. State the conclusions that follow from the shape of this curve and the variation of its slope. What is the physical meaning of the slope at any point? Draw a tangent line at one of the midpoints of an interval and compute its slope. Compare the value thus obtained with the corresponding value in the fourth column of the table.

6. Plot a graph to show the relationship between instantaneous speed and total distance fallen. Take corresponding values of speed and distance at particular times from the first two graphs. Interpret this graph.

Review Questions: 1. Define in words and give the defining equation for the following: average speed; instantaneous speed; average acceleration. 2. Name the cgs and fps units of speed and acceleration. 3. What is meant by uniform velocity? By uniform acceleration? Give examples of each. 4. For what does the symbol g stand, as it is used in these discussions? 5. What sort of motion characterizes an unrestrained body when it is acted upon by a constant external force? 6. Sketch rough curves showing distance-time, speed-time, and speed-distance relationships for a freely falling body. 7. What is the significance of the slope of the distance-time curve at any point? Of the speed-time curve? How would one expect these slopes to vary with time? 8. Show clearly why the *average* speed during a given interval of time is also the actual *instantaneous* speed at the midpoint of the interval for the case of uniform acceleration. 9. Derive and interpret the equation $s = v_1 t + \frac{1}{2}at^2$, assuming as known only fundamental defining equations. 10. Describe the apparatus and technique as used in a selected one of these experiments for the measurements of g.

Questions and Problems: 1. If by some suitable mechanism the falling body had been given an initial downward push instead of being just released, would the resulting observed value of g have been different? Explain.

2. Classify the following as to whether they would introduce *systematic* or *random* errors in this experiment: (a) air friction, (b) estimations of fractional parts of millimeters on the scale, (c) zero error of meterstick, and (d) fluctuations in the frequency of the timing device.

3. If friction be neglected, which of the following statements properly characterizes the motion of a heavy object thrown violently downward from a tall building: (a) uniform speed, (b) uniform deceleration, (c) constant acceleration, (d) uniformly increasing acceleration, and (e) nonuniformly changing acceleration?

4. In Exp. 5.3 when the switch is opened the electromagnet does not instantaneously lose all of its magnetism. What effect, if any, does this have on the results?

5. What would be the appearance of the speed-time curve if the falling body were so light that the effect of air friction could not be neglected?

6. What would be the effect upon the speed-time graph if one of the points of the trace were missing and the omission were unnoticed?

7. In an experiment on acceleration of free fall, three consecutive points were observed at positions 12.82, 18.65, and 25.48 cm, at time intervals of $\frac{1}{31}$ sec. Calculate g from these data.

8. A partial record of a freely falling body showed only three consecutive points. The distance from the first to the second point was 24.55 cm; that from the first to the third point was 58.92 cm. The time interval was $\frac{1}{10}$ sec. Find the acceleration and the instantaneous speed of the body at the time of the first recorded point.

9. How would the observed value of the acceleration be affected if the falling body used were heavier? Explain.

10. In Exp. 5.2 is it appropriate to use the symbol g for the observed acceleration or would the symbol a be better? Why? Could a student justify finding an observed value of the acceleration in this experiment higher than the value of g at the location of the experiment?

11. In an experiment on a falling body it is desired that the total time of fall be about $\frac{1}{3}$ sec. How long should be the trace made by the body?

CHAPTER 6

FORCE AND MOTION

Whenever a net force acts upon a body an acceleration of the body is produced; the acceleration is proportional to the net force, in the direction of the net force, and inversely proportional to the mass of the body. This statement is a form of Newton's *second law of motion*. Symbolically

$$a \propto f \quad \text{and} \quad a \propto 1/m$$

or

$$f \propto ma$$

whence

$$f = kma \tag{1}$$

where f is the net accelerating force, acting on a body of mass m, and a is the resulting acceleration. The factor k is a proportionality constant whose numerical value depends upon the units used in the equation. Units of force and mass are commonly so chosen that k has a value of unity. For *these choices of units and only for these*

$$f = ma \tag{2}$$

Some suitable sets of units are: force in dynes, mass in grams, and acceleration in centimeters per second per second; force in pounds, mass in slugs, and acceleration in feet per second per second; force in newtons, mass in kilograms, and acceleration in meters per second per second.

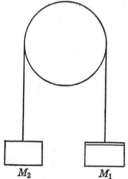

FIG. 6.1 Idealized Atwood's machine

Consider two bodies of equal masses M_1 and M_2, connected by a light flexible cord passing over a practically frictionless light pulley, Fig. 6.1. Since the force M_1g downward on one side of the pulley is equal to the downward force M_2g on the other side of the pulley, the system is in equilibrium and will remain at rest; or, if given an initial impulse, it will continue to move with uniform speed, that is, without acceleration. Now let a small mass m_1 be taken off the body of mass M_1. Since the mass M_2 is now the greater, the system is no longer in equilibrium but is acted upon by a net accelerating force m_1g. The system will therefore move with an acceleration a_1

$$m_1g = Ma_1 \tag{3}$$

where M is the total mass being accelerated ($M = M_1 - m_1 + M_2$).

If a larger body of mass m_2 is removed from M_1 the acceleration will be larger and may be represented by a_2. Then

$$m_2g = M'a_2 \tag{4}$$

where M' is the new total mass being accelerated ($M' = M_1 - m_2 + M_2$).

45

From Eqs. (3) and (4) one may write

$$\frac{m_1 g}{m_2 g} = \frac{M a_1}{M' a_2} \tag{5}$$

A simple method of determining the acceleration of the moving system of Fig. 6.1 involves the measurement of the distance of fall in two *successive equal* intervals of time. Let v_1 be the speed at the beginning of the first interval, v_2 the speed at the end of the first interval, and v_3 the speed at the end of the second interval. Call s_1 the distance fallen in the first time interval t and s_2 the distance fallen during the second interval. Then

$$s_1 = v_1 t + \tfrac{1}{2} a t^2 \tag{6}$$
$$s_2 = v_2 t + \tfrac{1}{2} a t^2 \tag{7}$$

When Eq. (6) is subtracted from Eq. (7), one obtains

$$s_2 - s_1 = v_2 t - v_1 t = (v_2 - v_1)t \tag{8}$$

Divide both sides of Eq. (8) by t^2 to obtain

$$\frac{s_2 - s_1}{t^2} = \frac{v_2 - v_1}{t} \tag{9}$$

But the right-hand member of Eq. (9) is the acceleration a; hence

$$a = \frac{s_2 - s_1}{t^2} \tag{10}$$

From Eq. (10) the acceleration may be calculated from the measured values of s_1, s_2, and t.

Experiment 6.1

NEWTON'S SECOND LAW OF MOTION: SIMPLE ATWOOD'S MACHINE

Object: To study by the use of a simple Atwood's machine Newton's second law of motion; in particular, to show that the accelerating force is proportional to the product of the mass and acceleration of the body acted upon.

Method: Two bodies of equal mass are connected by a strong light thread passing over a light, low-friction pulley. A small mass is removed from one of the two bodies and the distances the system moves in successive equal intervals of time are measured. From these data the acceleration of the system is computed. Similar observations are made for a different value of mass removed. The ratio of the forces is compared with the ratio of the product of the masses and the accelerations.

Apparatus: Simple Atwood's machine; stop watch or stop clock; two weight hangers; light, strong thread; slotted masses; 2-m scale, with four caliper jaws to use as markers.

The simple Atwood's machine consists of a light, low-friction pulley rigidly mounted. The 2-m scale should be set up vertically adjacent to the path of the moving masses, Fig. 6.2. Caliper jaws on the meterstick form convenient markers for the positions of the falling body.

Procedure: 1. Mount the pulley with its axis horizontal and 1.5 to 2 m above the floor. Connect two weight hangers by means of a strong thread. (If the thread is too heavy its weight will introduce a variable and unknown force during the motion of the system.)   The thread should be just long enough for one body to reach the floor when the other is at the pulley.   Load each weight hanger with about 800 gm including the hanger.   Compensate for friction by adding to M_2 sufficient mass that the system will move with uniform speed when M_2 is started *downward*.   This adjustment is possible only for a downward motion of M_2; it cannot move equally well upward, since friction always opposes the motion.

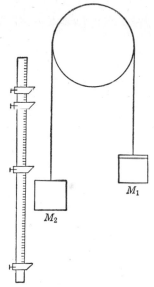

FIG.   6.2   Simple   Atwood's machine

2. Set the first caliper jaw a little below the level of the pulley.   Set the bottom of the weight hanger on that side at the level of this caliper jaw as a starting position.   Keep this marker in the same position throughout the experiment.   Subtract about 20 gm from M_1, release the system, and measure the time of descent.   Stop the moving objects just before one strikes the floor.   Select as the time interval to be used one that is an even number of seconds about one-third the total time of fall.   Record the time interval used.

Set the second caliper jaw a little below the first. Release the moving system at the fixed starting point and start the stop watch as the bottom of the weight hanger passes the second caliper jaw.   By repeated trials adjust the height of the third caliper jaw so that the bottom of the weight hanger passes it just at the end of the first time interval.   Each time stop the moving system before the mass hits the floor.   Similarly adjust the fourth caliper jaw until the bottom of the weight hanger passes it at the end of the second time interval.   Record the position of each caliper jaw.   Record the total mass moved.

3. Repeat Step 2, with twice the subtracted mass used in Step 2.

Computations and Analysis: 1. From the data of Steps 2 and 3 calculate the distances s_1 and s_2 passed over in successive equal time intervals.   By the use of Eq. (10) compute the acceleration in each case.

2. For each accelerating force calculate the product of the corresponding total mass and the acceleration.   Divide subtracted force m_1g by m_2g.   This value is the ratio of the accelerating forces, that is, the left-hand side of Eq. (5).   Calculate also the ratio of the corresponding products of mass and acceleration [right-hand side of Eq. (5)].   Compare the two ratios and discuss any discrepancy.

Experiment 6.2

NEWTON'S SECOND LAW OF MOTION: FALLING-TUNING-FORK ATWOOD'S MACHINE

Object: To study Newton's second law of motion; in particular, to show that accelerating force is proportional to the product of the mass and the acceleration of the body acted upon.

Method: A modified Atwood's machine, consisting essentially of two objects of unequal masses suspended by a cord passing over a light pulley, is used to measure the acceleration of the known masses. The unbalanced forces produce accelerations that are observed by means of traces formed by a vibrating tuning fork constituting a part of one of the accelerated bodies. The acceleration is measured by noting the gain in distance that the fork passes over in successive equal intervals of time. From observations of the accelerations produced by two different known forces acting upon known masses the working equation, set up on the assumption of the validity of Newton's second law of motion, is checked.

Apparatus: Falling-tuning-fork Atwood's machine; 6-volt storage battery; meterstick; DPST switch; plumb bob; trip scales; set of slotted masses.

The modified Atwood's machine is the same as the apparatus used in Exp. 5.2, Fig. 5.3, but with a light wheel K mounted at the top. The tuning fork F with stylus S attached to one prong is mounted in carriage C. The carriage is counter-balanced by the weight of an adjustable bob (not shown in Fig. 5.3) connected to the carriage by a strong cord passing over a light and nearly frictionless pulley K. When a part of the mass of the bob is removed and the carriage is released from its uppermost position by withdrawing the catch H, it falls with an accelerated motion between vertical guides G and G' and is caught in dashpots D and D'. The friction between the carriage and the guides is slight when the apparatus has been properly aligned and leveled. The tuning fork is electrically driven and constant amplitude of vibration is maintained throughout the time of fall. Consequently the waves traced on the coated surface of plate P (Fig. 6.3) are of the same amplitude, which facilitates the location of the crests of the waves and therefore the accurate determination of the distances corresponding to the equal time intervals.

FIG. 6.3   Trace used in determining acceleration

The strip of heavy plate glass is held in grooves at its lower and upper ends. The plate may be shifted horizontally and a number of tracings obtained without removing the plate. The surface of the plate on which the records are to be traced may be coated with a thin film of Bon Ami applied with a moist cloth or sponge. A more satisfactory way to obtain a permanent trace is to use a specially coated paper strip held against the glass plate by clamps or Scotch tape.

Procedure: 1. Remove the glass plate and lay it flat on a table. With the carriage at the top held by the catch H, attach the plumb bob at R and adjust the leveling screws in the base E until the bob hangs over the center of the bolthead between the dashpots. Check the rods for correct alignment and rough spots.

Arrange the wheel and the stop L, which catches the counterpoise at the end of its upward motion, so that the cord does not rub anywhere. Adjust the length of the cord and the position of the stop until the counterpoise is just caught in the stop as the carriage strikes the dashpots.

2. Coat the plate as directed by the instructor and place it in position. Adjust the stylus until it presses lightly on the surface of the plate or the paraffined paper.

The pressure should be sufficient to produce a clear trace and permit the crest of each wave to be marked; too much pressure will increase the retarding force due to friction.

3. Connect the battery to the binding posts B and B' through the switch; make sure that the fork can fall without interference from the wires. Set the fork into vibration by closing the switch and adjusting the small screw T. It may be necessary to give the fork an initial impulse by squeezing the prongs together and allowing them to spring apart. Adjust T so that the sparking is a minimum but the vibration is steady. The amplitude of vibration should be several millimeters. *Keep the switch open when the fork is not vibrating and open it immediately after the carriage strikes the dashpots.*

4. Vary the weight of the counterpoise until the system is so adjusted that the carriage, after being given a slight downward push, will continue to move downward with *uniform* speed. Note that this adjustment is possible only for a *downward* motion of the carriage. In the final adjustment of the balancing weight to produce this condition of equilibrium have the stylus against the coated plate, as its friction is considerable.

5. Having made the adjustments for equilibrium, remove a suitable accelerating weight from the counterpoise. The weight removed should be approximately half the total that is removable; 200 or 300 gm-wt is suggested.

Make a few preliminary traces to perfect the adjustments and learn the technique. Have one partner close the switch and adjust the screw T, while the other holds the prepared plate securely in place at the bottom and slightly tilted backward at the top. When the fork is vibrating strongly and evenly, quickly bring the plate into position, close the clamp at the top, and take hands off the apparatus. Immediately release the catch and allow the carriage to fall. When it strikes the dashpots, open the switch. Tilt the plate slightly backward at the top, bring the fork up to the top position, move the plate slightly to one side, and make another trace close to the first one. After every two or three runs check the tightness of electrical connections, contact screw, and dashpots. When the apparatus is adjusted to make a suitable trace, remove the plate to the table and recoat it, or put on a fresh strip of coated paper. Make a good trace for each student. Record the frequency stamped on the tuning fork. Record the total mass M moved. This value of M will be the sum of the mass of the carriage (stamped on the frame) plus that of the counterpoise *after* the accelerating weight has been removed.

6. Remove from the counterpoise a second weight, equal to the first, to double the accelerating force and, as in Step 5, make a good trace for each student. Record the new total mass M'.

7. Place each record strip, or coated glass, on a flat surface where there is good light. Beginning near the top, at a place where the trace is distinct and the crests uniform, pin-mark as finely and as accurately as possible the middle of one of the crests (Fig. 6.3). A small transparent right-angle triangle may be placed transversely across the crest and a fine line drawn with a scriber or needle to bisect the crest and extend about 5 mm above it. Count and record the total number of wave crests in the entire trace beyond the first one selected and divide the number by two to obtain equal time intervals. Make a line, similar to the one at the first crest, at the crest for the end of the first and second time intervals.

Place a meterstick *edgewise* on the trace in such a manner that the graduations are directly touching the trace, and the scale intersects the fine lines at right angles. Set some even division mark, such as 1.00 cm, on the top line. Record the readings of the scale for each of the three lines, estimating each reading to a fraction of a millimeter.

Computations and Analysis: 1. From the data recorded in Step 7 calculate the distances s_1 and s_2 between the lines for each trace. From the frequency of the fork and the number of waves in each interval compute the time interval t. Compute the acceleration by the use of Eq. (10).

2. Substitute the appropriate values in the working equation, Eq. (5), and determine the percentage difference between the ratio of the two accelerating forces and the corresponding ratio of the products of the total moving mass and the acceleration. Discuss the equality of these ratios as a check of the validity of the working equation.

Review Questions: 1. State Newton's second law of motion. Write the symbolic equation. 2. State some combinations of units that are appropriate to use in Newton's second law to make $k = 1$. 3. Describe the Atwood's machine. 4. Describe the modification of the Atwood's machine to make use of a tuning fork as a timing device.

Questions and Problems: 1. Is the weight of a body the same thing as its mass? Discuss briefly. Is the weight of a body constant at all places on the earth? Is the mass?

2. In each of the following cases, state how the various readings of a spring balance will compare if the balance is attached to the top of an elevator and supports a constant mass. State your reasoning for each case. (*a*) Elevator at rest; (*b*) moving with constant speed upward; (*c*) moving with constant speed downward; (*d*) moving upward with steadily increasing speed; (*e*) moving downward with steadily increasing speed; (*f*) moving downward with acceleration g; (*g*) rising with uniformly diminishing speed.

3. A cord passes over a weightless and frictionless pulley. Masses of 200 and 300 gm are attached to the ends of the cord. Find the distance the masses will move during the fifth second after they are started from rest.

4. A cord passes over a weightless and frictionless pulley. A 50.0-gm mass is attached to each end of the cord. A 2.083-gm mass is placed on the left-hand object and allowed to remain for 3.00 sec, when it is suddenly removed. How far will the masses move in 5.00 sec after starting from rest? How far during the seventh second?

5. Discuss the relative increase in accuracy obtained in Exp. 6.2 by a possible decrease in the error of the following: (*a*) weight of frame and fork; (*b*) weight of balancing mass; (*c*) frequency of fork; (*d*) more accurately calibrated meterstick.

6. An object of mass 8.00 kg is pulled up an inclined plane, making an angle of 30° with the horizontal, by a string that passes over a pulley at the top of the plane and is fastened to a 10.0-kg mass. Neglecting friction, find the acceleration and the tension in the string.

7. An 11.0-lb object is acted on by a constant force that changes its speed from 100 cm/sec to 88.8 cm/sec in 1.667 min. Find the accelerating force. Reduce the acceleration to feet per second per minute.

8. What is the weight of a 3.00-kg mass at a place where a body, starting from rest, falls freely through 44.6 m in 3.00 sec?

9. Derive an equation that expresses the total downward force on a light pulley over which a pair of masses M_1 and M_2 are suspended by a light cord.

CHAPTER 7

FRICTION

Friction is the most common force other than gravity. We soon learn that it requires a force to drag a body along a surface even though the surface is horizontal so that there is no resisting force due to gravity and even though the speed is uniform so that no force is required to produce acceleration. This force is called *frictional force*.

There are two types of frictional force: that when the body is at rest and that when the body is in motion. When the body is at rest the frictional force varies from zero up to a certain limiting value, the force required to start the motion. This force required to start the motion is greater than the force required to maintain constant speed after the body is in motion.

The main factors upon which friction does or does not depend may be summarized by the following approximate statements: (1) the friction between surfaces depends upon the nature of the surfaces and particularly upon the presence of even slight traces of lubricants; (2) the friction is independent of the area of contact; (3) the friction is independent of the speed of one surface over the other; (4) the frictional force is directly proportional to the normal component of the force pressing the surfaces together.

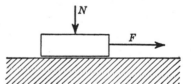

Fig. 7.1 Frictional force F and normal force N

Experience shows that the *tangential* frictional force F necessary to maintain the uniform speed of a body is proportional to the normal force N pressing the two surfaces together, Fig. 7.1. In symbols

$$F \propto N \qquad \text{or} \qquad F = \mu_k N \tag{1}$$

where μ_k is a constant for a particular pair of surfaces and is called the *coefficient of kinetic (sliding) friction*

$$\mu_k = F/N \tag{2}$$

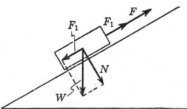

Fig. 7.2 Force on body moving uniformly up incline

It should be noted that F is only the frictional force. If the body is sliding with uniform speed up an inclined plane, Fig. 7.2, an additional force F_1 is necessary to lift the body against gravity and the total force along the plane is $F_1 + F$. Note also that the normal force N is no longer the weight of the body but only a component of the weight. In other cases, such as the brake on a wheel, the normal force has no connection with the weight.

51

In starting (static) friction the *coefficient of starting friction* μ_s is the ratio of the frictional force F_s required to start the body to the normal force N pressing the surfaces together

$$\mu_s = F_s/N \qquad (3)$$

Because coefficients of friction are ratios of forces the coefficients are pure members.

If a body is at rest on an incline and the angle of the incline is slowly increased, the normal component of the weight decreases and the tangential component increases. At some particular angle θ, Fig. 7.3, the component $W \sin \theta$ of the weight parallel to the plane becomes equal to the limiting force F_s of starting friction, and the body begins to slide downward. This angle θ is called the *limiting angle of repose*. For this angle

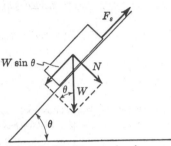

$$\mu_s = \frac{F_s}{N} = \frac{W \sin \theta}{W \cos \theta} = \tan \theta \qquad (4)$$

Fig. 7.3 Limiting angle of repose

Rolling friction is the resistance caused chiefly by the deformation produced where a wheel or cylinder pushes against the surface on which it rolls. The force of rolling friction varies inversely as the radius of the roller, and it is less the more rigid the surfaces. Rolling friction is ordinarily much smaller than sliding friction.

Experiment 7.1

KINETIC AND STATIC FRICTION

Object: To study frictional forces; in particular, to measure the coefficient of kinetic friction and the coefficient of static friction for various surfaces.

Method: Measurement is made of the force necessary to maintain constant speed for a body on a uniform surface and also of the frictional force necessary to start the body moving. The normal force is measured in each case and the coefficient of friction calculated. Normal force, area of contact, and condition of surface are varied to study the influence of each on the frictional force.

Apparatus: Friction board; wooden friction block, with faces of different areas and receptacle for added load; two friction blocks with rubber soles; Hall's carriage; weight hanger; set of slotted masses; glass plate; pulley; meterstick.

The friction board must be flat, smooth, and fine grained. Any nonuniformity in the surface will result in inconsistent observations. The surfaces of the friction blocks must also be uniform on the sides that are used.

Procedure: 1. Measure and record the weight of each of the friction blocks. With the surface of the friction board horizontal place the wooden block on the board with the larger surface in contact with the board. Attach one end of a string to the block and the other to the weight hanger as in Fig. 7.4. Be sure to keep the string parallel to the board. For two different loads on the block measure the minimum force necessary just to *start* the block.

2. With the same block and same loads as in Step 1 turn the block so that it

rests on a side of similar material but smaller area. Again measure the forces necessary to start the block with each of the loads.

3. With the same arrangement as Step 1 measure the force necessary to keep the block in uniform motion. Adjust the load on the hanger so that *after the block is started* it will continue to move at the same speed across the board. Repeat, using the smaller area of contact.

4. Repeat for one load Steps 1 and 3 using the friction block with a rubber sole. Be sure to select the block having a surface free of lubricant.

5. Lay the glass plate over the friction board and measure the sliding frictional force between the rubber-soled block and glass plate for a suitable load.

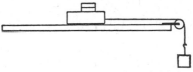

Fig. 7.4 Friction board and friction block

6. Repeat Step 5 first with a film of water on the glass and then with a thin film of oil. For this Step use only the block that has been designated for use with the lubricants. Clean the glass plate and the block carefully after each test.

7. Place the (clean) wooden block on the friction board and elevate one end of the board so that the plane has an angle of about 30°. With a suitable load in the receptacle of the block, determine the force parallel to the plane necessary to just start the block moving up the plane.

8. Remove the weight hanger and, with the block and load of Step 7 on the board, increase the angle slowly until the block just begins to slide. Make several trials to get the average angle. Measure the length and corresponding height of the plane.

9. With the surface of the friction board horizontal place the Hall's carriage on the board. Add a load of 2 kg. As in Step 3 determine the force necessary to maintain uniform motion.

Computations and Analysis: 1. From the data of Steps 1, 2, and 3 calculate the coefficient of friction for each of the observations. Determine the average value of μ_s and that of μ_k. Compare the values of μ_s and μ_k. Does the frictional force depend on the area of contact?

2. From the data of Steps 4 and 5 determine the coefficients of friction of rubber on wood and the coefficient of sliding friction of rubber on glass.

3. From the data of Step 6 determine the coefficient of sliding friction when a lubricant is present. Compare the values obtained with the value for rubber or glass calculated in Part 2.

4. From the data of Step 7 calculate the normal force, the frictional force, and the coefficient of static friction. Compare this value with that computed from the data of Step 1.

5. From the data of Step 8 find the limiting angle of repose and the coefficient of static friction. Compare this value with the corresponding values previously computed.

6. From the data of Step 9 determine the coefficient of rolling friction. How does its value compare with the value of the coefficient of kinetic friction?

Review Questions: 1. Distinguish between static friction and kinetic friction. Which is greater? 2. Define coefficient of friction. Does it have units? 3. How does friction depend upon: (a) area of surfaces in contact; (b) speed of motion; (c) force pressing the surfaces together; (d) nature of the surfaces? 4. What is meant by limiting angle of repose? Deduce the relation between coefficient of friction and limiting angle of repose. 5. Set up and justify an expression for the force necessary to move an object at uniform speed up a rough incline.

Questions and Problems: 1. If a locomotive engineer causes the wheels to spin when starting his train he shuts off the steam and starts over again. Explain reasons.

2. Why would nonuniformity of the friction board or block cause inconsistent results in an experiment to measure the coefficient of friction? In which part of the experiment would this be more important? What effect might prominent grain in the wood have on this experiment? Would the direction of motion affect the friction in this case?

3. A block of mass 1525 gm rests on a horizontal surface. A force of 685 gm-wt is required to start the block into motion but only 512 gm-wt are required to maintain a uniform speed. Compute μ_k and μ_s from these data.

4. A block on an inclined plane just begins to slip if the inclination of the plane is 60°. If the inclination were 30° the block would, after it had been started, continue to slide with uniform speed. What are the coefficients of static and kinetic friction?

5. A block weighing 150 lb is pressed against a vertical wall by a horizontal force of 60 lb. The coefficient of sliding friction is 0.15; the coefficient of static friction is 0.25. (a) How much vertical force is needed to keep the block from falling? (b) What force is needed to move it upward with uniform speed?

6. A 120-lb block is placed on a plane inclined at an angle of 35°. The coefficient of static friction is 0.25; the coefficient of sliding friction is 0.15. What force parallel to the place is necessary (a) to hold the block from sliding down the plane; (b) to pull the block up the plane with uniform speed?

7. The weight on a locomotive drive wheel is 10.0 tons. If the coefficient of kinetic friction between the drive wheel and brake shoe is 0.20 and the coefficient of static friction between the wheel and the rail is 0.30, find the normal brake-shoe force that will produce the maximum retarding force on the locomotive.

CHAPTER 8

WORK AND POWER; MACHINES

Work is defined as the product of a force and the displacement in the direction of the force. In symbols

$$E = Fs \cos \theta \tag{1}$$

where E is the work done, F is the force acting, s is the displacement, and θ is the angle between the force and the displacement. If the displacement occurs in the direction of the applied force, Eq. (1) reduces to the form

$$E = Fs \tag{2}$$

Units of work include the foot-pound, erg (dyne-centimeter), joule (10^7 ergs), and kilowatt-hour.

Power is the time rate of doing work. In symbols

$$P = E/t \tag{3}$$

Common units of power are the horsepower (550 ft-lb/sec) and the watt (1 joule/sec). One horsepower is the equivalent of 746 watts.

A *machine* is a device for applying energy to do work in a way suitable for a given purpose. In the so-called *simple* machine, the energy is supplied by a single applied force, and the machine does useful work against a single resisting force. A compound machine is a combination of two or more simple machines.

The *efficiency* of a machine is the ratio of the useful *work output* of the machine to the total *work input*

$$\text{Eff.} = \frac{\text{useful work output}}{\text{total work input}} = \frac{E_o}{E_i} \tag{4}$$

Efficiency is usually expressed in per cent.

The *mechanical advantage* of a machine involves the forces or distances concerned in the machine. There are two mechanical advantages, an ideal mechanical advantage IMA and an actual mechanical advantage AMA. The *ideal* mechanical advantage is the ratio of the distance s_i moved by the applied force to the distance s_o moved by the opposing force

$$\text{IMA} = s_i/s_o \tag{5}$$

The *actual* mechanical advantage is the ratio of the resisting force F_o to the applied force F_i

$$\text{AMA} = F_o/F_i \tag{6}$$

The useful work output is $E_o = F_o s_o$ and the total work input is $E_i = F_i s_i$. Therefore

$$\text{Eff.} = \frac{E_o}{E_i} = \frac{F_o s_o}{F_i s_i} = \frac{F_o/F_i}{s_i/s_o} = \frac{\text{AMA}}{\text{IMA}} \tag{7}$$

The most common simple machines are the inclined plane, the pulley, and the wheel and axle. Most other machines are combinations of these. The ideal mechanical advantage of each of these simple machines is readily calculated.

The inclined plane, Fig. 8.1, enables one to lift a body of weight W by applying

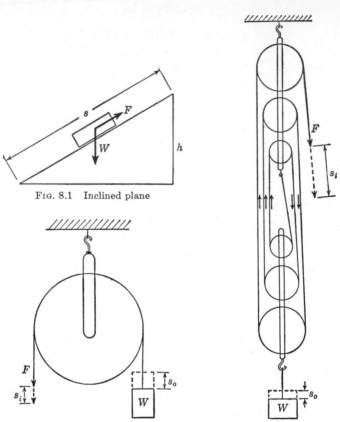

Fig. 8.1   Inclined plane

Fig. 8.2   Single fixed pulley

Fig. 8.3   Block and tackle

a smaller force F along the plane. The IMA is the ratio of the distance s moved along the plane to the distance h that the body is raised

$$IMA = s/h \qquad (8)$$

For a single fixed pulley, Fig. 8.2, the distance s_i moved by the applied force F is equal to the distance s_o moved by the load W. Hence the IMA is 1. For a combination of pulleys the applied force F acts over a larger distance than the load. In the block and tackle of Fig. 8.3, when the load rises a distance s_o each supporting rope must shorten by a distance s_o and the applied force must move a distance $s_i = ns_o$, where n is the number of supporting ropes, here 6.

$$IMA = \frac{s_i}{s_o} = \frac{ns_o}{s_o} = n \qquad (9)$$

In the wheel and axle, Fig. 8.4, the applied force acts through a larger distance than the load.   If the machine turns through an angle θ, $s_o = r\theta$ and $s_i = R\theta$. Hence

$$\text{IMA} = \frac{s_i}{s_o} = \frac{R\theta}{r\theta} = \frac{R}{r} \qquad (10)$$

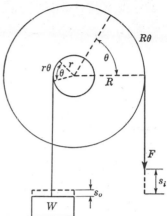

A common method of measuring the power developed by a machine is to use a dynamometer, one form of which is known as a *Prony-brake dynamometer*, wherein the energy output of the machine may be dissipated as heat.   In smaller machines, the brake may conveniently take the form of a strap passing halfway around the pulley and supported by two spring balances, as shown in Fig. 8.5.   The net force against which the pulley turns is given by the difference in the readings of the two balances. The distance which the pulley travels against this force may be obtained from the circumference of the pulley and the number of revolutions that it makes.

FIG. 8.4   Wheel and axle

The working equation may be obtained by substituting in Eq. (3) the observed forces F_1 and F_2, the diameter D of the wheel, and the

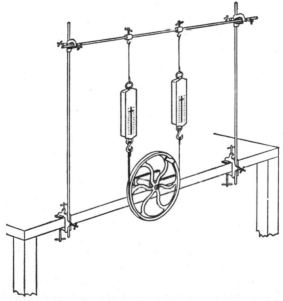

FIG. 8.5   Prony-brake dynamometer arranged for measurement of "manpower"

number of revolutions N made in the time t.   This substitution gives

$$P = \frac{E}{t} = \frac{Fs}{t} = \frac{(F_1 - F_2)\pi DN}{t} \qquad (11)$$

If the power is to be expressed in horsepower, the forces must be in pounds, the diameter in feet, the time in seconds, and the result divided by 550. If the power is to be in watts, the forces must be in dynes (grams-weight multiplied by 980), the diameter in centimeters, the time in seconds, and the result divided by 10^7.

Experiment 8.1

SIMPLE MACHINES

Object: To study and measure the actual and ideal mechanical advantages and efficiencies of various machines.

Method: The forces required to lift given loads by various machines are measured and from these values the actual mechanical advantages are calculated. From the geometrical configurations of the apparatus the ideal mechanical advantages are obtained. The efficiencies are computed from the ratio of AMA to IMA.

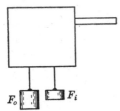

F_o　　F_i

Fig. 8.6　Hidden mechanism

Apparatus: Inclined plane; wooden block; trip scales; set of masses; two triple-tandem pulleys; "hidden mechanism"; meterstick; support rods and clamps.

The hidden mechanism, Fig. 8.6, is a simple machine entirely enclosed within a box with two holes in the bottom. Two cords extend through the holes. When one cord is pulled down the other goes up. To one cord is attached the load and to the other the applied force.

Procedure: 1. Set up the inclined plane at an angle of about 30° with the horizontal. Measure the height and corresponding length of the plane. From these distances compute the IMA of the plane. Measure and record the weight of the wooden block. Place the block on the plane, attach a string to the block, and pass the string parallel to the plane over a pulley to a weight hanger. Load the block by adding one or two 1-kg masses. Add weights to the hanger until the weight F_i of hanger and added weights will just pull the loaded block up the incline with uniform speed after it is started. From F_i and W compute the AMA. From the IMA and the AMA compute the efficiency of the machine.

2. Set up the two triple-tandem pulleys as shown in Fig. 8.3. Add a useful load W of 1200 gm-wt including the weight hanger. Attach a weight hanger to the string and add sufficient weights so that the load moves up with uniform speed after it is started. Record F_i and W. Set a meterstick vertically beside the two weight hangers, move one (F_i) through a distance s_i of about 30 cm, and measure the corresponding distance s_o moved by the load W. Calculate the AMA from the ratio of the forces and the IMA from the ratio of the distances. Compute the efficiency by the use of Eq. (7).

3. Set up the hidden mechanism. Pull *gently* on one cord while grasping the other. Observe whether the second cord moves a greater or a less distance than the first. Attach a load of about 1 kg to the cord that moves the smaller distance. Attach a weight hanger to the other cord and add weights until the load moves up with uniform speed after it is started. Measure corresponding distances moved by the load and applied force. From these data compute AMA, IMA, and efficiency. From the observations you have made try to establish the identity of

the mechanism within the box. Consider what machine could be expected to have the mechanical advantage and efficiency observed.

Experiment 8.2

MECHANICAL ADVANTAGE; WORK; POWER; EFFICIENCY

Object: (1) To study and measure the actual and ideal mechanical advantages and the efficiencies of various machines. (2) To measure the power output of a person under certain conditions by means of a Prony-brake dynamometer.

Method: The forces required to lift given loads by various machines are measured and the actual mechanical advantages are calculated. From the geometrical configurations of the apparatus the ideal mechanical advantages are obtained. The efficiencies are computed from the ratio of the AMA to the IMA. Measurement is made of the power that a person can develop by turning with a hand crank a grooved wheel loaded with a known frictional force.

Apparatus: Inclined plane; two triple-tandem pulleys; trip scales; wooden block; wheel and axle; slotted masses; weight hangers; support rods and clamps; meterstick; Prony-brake apparatus, consisting of a mounted wheel, two 15-kg spring balances, leather belt, and suitable rods and clamps.

Procedure: 1. Set up the inclined plane at an angle of about 30° with the horizontal. Measure the height and corresponding length of the plane. Measure and record the weight of the wooden block. Place the block on the plane, attach a string to the block, and pass the string parallel to the plane over a pulley to a weight hanger. Load the block by adding one or two 1-kg masses. Add weights to the hanger until the weight F_i of hanger and added weights will just pull the loaded block up the incline with uniform speed after it is started.

2. Set up the two triple-tandem pulleys as shown in Fig. 8.3. Add a useful load W of 1200 gm-wt including the weight hanger. Attach a weight hanger to the string and add sufficient weights so that the load will move up with uniform speed after it is started. Record F_i and W. Set a meterstick vertically beside the two weight hangers, move one (F_i) through a distance s_i of about 30 cm, and measure the corresponding distance s_o moved by the load W.

3. Repeat Step 2 for useful loads of 1000, 500, 200, 50, and 0 gm.

4. Repeat Step 2 for arrangements of the pulley system that will make the IMA 5, 4, 3, and 2. Record the observations and results in a table; make a sketch of each arrangement.

FIG. 8.7 Wheel and axle

5. By means of rods and clamps mount the wheel and axle (Fig. 8.7) with its axis horizontal. Find the value of the force F_i applied at the rim of the largest wheel that will just lift a load F_o of 600 gm-wt on the third largest wheel. Measure the diameter of each of these wheels. Measure also the value of s_o when s_i is 20 cm.

6. Combine the wheel and axle with the inclined plane to make a compound machine. Set the inclination of the plane at 30° and adjust the height of the wheel and axle so that the cord from the third largest wheel pulls in a direction up the plane parallel to the plane. Find the load on the largest wheel that will

pull the block and a 2-kg load up the plane with uniform speed after it is started.

7. Arrange the Prony brake as shown in Fig. 8.5. For one observation F_1 may be about 6 kg, and F_2 may be about $1\frac{1}{2}$ kg, both forces being read while the wheel is turning. Turn the pulley as fast as possible, keeping the forces constant. A second observer should determine with a stop watch the time required to turn the wheel through a suitable number of revolutions (25 to 50). Record the necessary data and calculate the power, both in British and metric units. Make all necessary original observations in both systems of units, that is, measure D in both centimeters and in feet and the forces in both grams-weight and pounds.

Computations and Analysis: 1. From the height and length of the plane in Step 1, compute the IMA of the plane. From F_i and W compute the AMA. From the IMA and AMA compute the efficiency of the machine.

2. From the data of Steps 2 and 4 calculate the IMA, AMA, and efficiency of each arrangement of the pulleys.

3. From the data of Steps 2 and 3 calculate the efficiency of the pulley system for each load. Plot a curve to show the relationship between efficiency and useful load. Why is the efficiency zero for zero load? Why does it increase as the load increases? Does it have a maximum value? If so, why?

4. From the data of Step 5 determine the IMA, AMA, and efficiency of the wheel and axle. Compare the ratio of s_i/s_o with that of R/r.

5. From the data of Step 6 determine the IMA, AMA, and efficiency of the compound machine. How is the value of the IMA of the compound machine related to the values for the individual machines? Are the AMA's related in a similar manner? Explain.

6. From the original data of Step 7 calculate the power in horsepower and in watts. Check your results by multiplying the power in horsepower by 746 watts/hp. Compare this value with the original computation of the power in watts.

Review Questions: 1. Define work and power and give the defining equation for each. State cgs and British units appropriate for each. 2. Define ideal mechanical advantage, actual mechanical advantage, and efficiency. What limits, if any, must be imposed on the possible value of each? 3. Derive the expression for the ideal mechanical advantage of each of the following machines: inclined plane, block and tackle, wheel and axle. 4. Derive the working equation for the horsepower developed in using a strap-brake dynamometer. Indicate clearly the unit of each factor in the equation. 5. Set up a similar equation for the dynamometer power output in watts, using cgs units throughout. 6. What is the numerical relation between the horsepower and the watt?

Questions and Problems: **1.** Is it correct to define 1 ft-lb of work as the amount of work required to move a 1-lb body 1 ft? Why?

2. Prove that efficiency = power output/power input.

3. Could any or all of the machines described in this chapter work "backward," that is, could load and applied force be interchanged?

4. A machine is said to be "self-locking" if its efficiency is less than 50%. Prove this statement to be true.

5. Discuss the relative values of the errors introduced into the experiment on man-power by reason of friction in (a) the axle bearings and (b) the strap brake.

6. A hidden mechanism has a load of 25.0 lb which is raised by an applied force of 7.28 lb.   The applied force moves 62.5 in. while the load moves 15.6 in.   Calculate AMA, IMA, and efficiency.

7. A block and tackle that supports a load of 150 lb is 60% efficient.   If a force of 50 lb is needed to raise the load, how are the ropes arranged, and what are the actual and ideal mechanical advantages?   Sketch the arrangement.

8. What are the ideal and actual mechanical advantages of a block and tackle consisting of one movable pulley and a block of two fixed pulleys; the double block is fixed and the system is threaded from the movable pulley?   The useful load is 800 lb and the applied force is 500 lb.   What is the efficiency?

9. The human dynamometer used in an experiment on manpower had a diameter of 1.3 ft.   It was loaded on one side with a force of 22 lb and on the other with 4.0 lb. If 90 revolutions were counted in 45 sec, what was the horsepower output? the output in watts?

10. A strap brake is attached to the pulley of a motor.   The pulley has a radius of 20 cm and exerts a net force of 50 gm-wt on the balances attached to it.   In order that the motor may make 1500 revolutions in 3.0 min, 3080 joules of work must be supplied to it.   Calculate the power output in watts and the efficiency of the motor.

CHAPTER 9

TORQUE; EQUILIBRIUM

The effect of a force in producing a change in the rotation of a body depends not only upon the magnitude of the force but also upon the point at which it is applied and the direction in which it acts. The *torque*, or moment of force, about any axis is the product of the force and the perpendicular distance from the axis to the line of action of the force. The defining equation is

$$L = Fs \tag{1}$$

where L is the torque developed by the force F which acts at a perpendicular distance s from the axis. This distance is frequently referred to as the *lever arm*.

The absolute cgs unit of torque is the centimeter-dyne. The centimeter-gram and pound-foot are gravitational units of torque.

A body is in equilibrium if it has no translational or rotational acceleration. The conditions for equilibrium of a rigid body are (1) the vector sum of all forces must be zero; (2) when the forces are in a single plane, the algebraic sum of the torques about *any axis* must be zero. The algebraic sign of a torque is perfectly arbitrary; for convenience counterclockwise torques are usually called positive and clockwise torques negative.

The *center of gravity* of a body is the point at which all the weight of the body may be considered to act. The weight, then, may always be used as a single force acting downward at the center of gravity.

Experiment 9.1

TORQUE: DEMONSTRATION BALANCE

Object: To study by the use of a demonstration balance the concept of torque and the conditions for a body to be in rotational equilibrium.

Method: The weights (considered unknown) of a pair of scale pans on a demonstration balance are determined by the use of torque equations. Data for these equations are obtained by measuring the lever arms and using one known weight. The weight of the meterstick balance is determined by supporting the system at some place other than the center of gravity, restoring equilibrium by the use of known weights, and writing the equation for the torques acting on the system.

Apparatus: Demonstration balance; vernier caliper; set of masses; beam balance; meterstick, mounted in a horizontal position; an "unknown" mass (about 200 gm).

The demonstration balance, Fig. 9.1, consists of a tripod base B and an upright support rod R; a metal yoke Y attached to the support rod and carrying adjustable knife-edge supports; a metal frame F with two sets of knife-edges, one fixed and the other vertically adjustable, and a sliding collar C to adjust the center of gravity of the system; a meterstick M that slips through the frame to serve as the beam of the balance; two scale pans P_1 and P_2 provided with plumb bobs. The meterstick has another knife-edge K permanently inserted at the 75-cm mark.

Procedure: 1. Arrange the metal frame so that the adjustable knife-edges are about 2 cm above the fixed knife-edges and the sliding collar is near the bottom. Support the beam, without the pans, from the lower knife-edges. Slide the meterstick through the frame until it balances in a horizontal position.

2. Determine the weights of the scale pans by use of torques, as follows: Hang the scale pans from the meterstick, well out toward the ends, and slide one

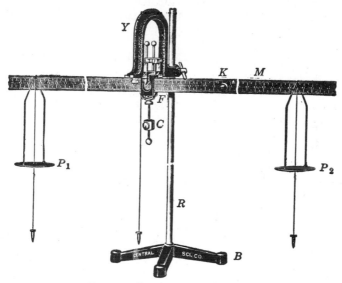

Fig. 9.1 Demonstration balance

of them along until the beam is again balanced in a horizontal position. In this and other adjustments hold the beam lightly while making changes so that it will not fall from its supports nor the pans slide off the beam. Record the position of the knife-edge fulcrum and that of each pan. In all cases make a sketch showing the positions observed.

Using the weights W_1 and W_2 of the pans as unknown, write the equation of torques about the fulcrum as an axis.

Add a 50-gm weight to one pan and, without changing the position of the other, slide the weighted pan toward the fulcrum until the beam is again balanced. Record the new position and write the equation of torques for this setting. Solve the two equations for the unknown weights W_1 and W_2 of the pans. Weigh each pan on the beam balance and compare these values with those obtained by the method of torques.

3. The purpose of this step is to locate exactly the position of the center of gravity. Adjust the frame so that the center of gravity of the system is at the position of the fixed knife-edges. When the system is supported at the lower knife-edges, the beam will, in general, turn about the fulcrum if the center of

gravity is above or to either side of the fulcrum, Fig. 9.2. If the center of gravity is exactly at the point of support, the beam will rest in any orientation without turning.

Set the upper knife-edges about 3 cm above the lower pair and place the collar near the middle of its rod. Remove the pans and balance the beam in a horizontal position. Final delicate adjustment can sometimes be made by rotating the collar around a vertical axis. When this balance is obtained, the center of gravity is in the same vertical line as the fulcrum, Fig. 9.2a. Push one side of the stick downward. If the beam tends to return to the horizontal, the center of gravity is *below* the fulcrum, Fig. 9.2b. If the beam tends to turn still farther downward, the center of gravity is *above* the fulcrum, Fig. 9.2c. Adjust the collar vertically until the beam shows no tendency to move out of any position in which

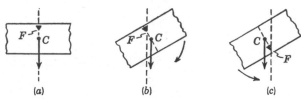

FIG. 9.2   Direction of rotation indicates whether the center of gravity C is above or below the fulcrum F

it is placed. The center of gravity then coincides with the fixed knife edge. *Keep this adjustment for Steps 4 and 5.*

4. In order to determine the weight of the beam, support the bar by knife-edge K at the 75-cm mark. Hang a scale pan near the end of the short section and add weights until the bar balances in a horizontal position. Record the positions of the pan, the fulcrum, and the center of gravity. Write the equation of torques about the fulcrum as an axis and solve for the unknown weight of the beam system. Weigh the bar and frame on the beam balance and compare the value thus measured with that obtained from the torque equation.

5. With the center of gravity still set as adjusted in Step 3, support the system from the *upper* knife-edges. Place the scale pans in the notches near the two ends of the meterstick. Place the "unknown" mass on the left-hand pan and add masses to the right-hand pan until the beam is balanced at an angle of about 20° with the horizontal, Fig. 9.3. Be sure that the angle is small enough that the knife-edges do not rest against the sides of the grooves in which they are set. Measure and record the moment arms s_1 and s_2 by means of the horizontally mounted meterstick. The weight of the beam system acts at C and its moment arm AC is so small that a direct measurement is not sufficiently accurate. By means of a vernier caliper measure the distance FC from the upper to the lower knife edge. Then $AC = FC \sin \theta$. Determine the angle θ either by direct measurement with a protractor or by measuring corresponding distances horizontally and along the meterstick.

Write the equation of torques, using the weights of the beam system and scale pans from the previous steps. Solve for the weight of the unknown. Weigh the unknown on a beam balance and compare this value with that obtained by the equation of torques.

6. The method of double weighing, sometimes used with precision balances,

furnishes a good example of the "law of moments." Hang pans of materially different weights in the notches near the ends of the bar. Slip the bar through the frame until a balance is obtained. This adjustment results in a demonstration balance whose scale pans are unequal in weight and whose lever arms are also unequal. Their respective torques exactly balance each other and are ignored in the following computations. Place an object of unknown weight on one of the

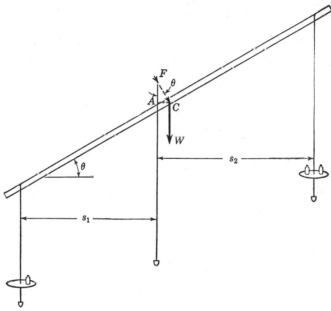

FIG. 9.3   Beam system balanced in a nonhorizontal position

pans, and balance it by placing a known weight on the other. Write the equation of torques, calling the lever arms (assumed unknown) s_1 and s_2. Now place the unknown weight on the other scale pan, and balance again with a known weight. Write another equation of torques, s_1 and s_2 remaining the same as before (but unknown). The solution of these two equations furnishes a value for the unknown weight in terms of the two known weights and nothing else.

7. At the end of the period, remove the meterstick system from its support and leave it on the table.

Experiment 9.2

STATIC EQUILIBRIUM: SIMPLE CRANE

Object: To study the forces and torques involved in the equilibrium of a simple crane boom of negligible weight.

Method: A simple crane boom of negligible weight is loaded at its upper end and held in position by a tie cord. The load, the compressional force in the boom, and the tension in the tie cord are measured. The vector sum of the forces acting at the end of the boom is found and compared with that predicted by the first condition for equilibrium. The torques about the base of the boom are calculated and their sum compared with the zero value predicted by the second condition for equilibrium.

Apparatus: Simple crane; 1-kg weight hanger; kilogram weights; two 15-kg spring balances; heavy-duty low-friction pulley; meterstick with caliper jaws.

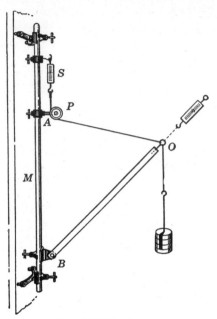

Fig. 9.4   Simple crane

The simple crane (Fig. 9.4) consists of a light but rigid boom (hollow metal tube or wooden rod) with a ring screw O at one end for attaching a tie cord and a simple hinge B at the other for attaching the lower end to the vertical mast M. The tie cord passes around a low-friction pulley P to a spring balance S, used for

Fig. 9.5   Intersection of two segments of tie cord

measuring the tension in the tie cord.   A second spring balance may be attached at O to be used in measuring the compression in the boom.   Since the boom is rigid its length OB is constant.

Procedure: 1. Set up the crane as in Fig. 9.4 so that no two sides of the triangle AOB are equal and no angle is a right angle when the crane is loaded.   Make the point B, at which the lower end of the boom is supported, as nearly as possible vertically below the intersection of the tie cord and the vertical cord at A.

2. Hang a load of about 9 kg, including the weight hanger, from the ring O.   Measure and record the lengths AO, BO, and BA, using the meterstick with caliper jaws.   In the measurement of AO and BA be careful to locate the point A at the intersection of the cords (Fig. 9.5).   Record the load and the tension in the tie cord as indicated by the spring balance.   Measure the compressional force in the boom by a second spring balance at O pulling in a

direction parallel to the boom until the boom is just freed at the hinge.   Record the reading of the spring balance.

3. Repeat Step 2 for other configurations of the crane.

Computations and Analysis: 1. Use the measured lengths AO, BO, and BA to make a scale diagram of the triangle AOB. Choose a scale such that the diagram occupies most of a sheet of paper. On the diagram draw dotted lines to represent the moment arms of each of the forces about B as an axis and about A as an axis. Measure and record the lengths of the moment arms as determined from the lengths of the lines and the scale chosen.

2. Consider the load as the only known force. Write the equation of torques about the axis B and use it to compute the tension in the tie cord. Similarly write the equation of torques about A and use it to calculate the compression in the boom. Compare each of these values with the corresponding force determined experimentally.

3. Draw a vertical vector to represent the load, using a large scale. Measure the angles ABO and OAB of the distance triangle of Part 1. On the vector diagram draw lines making these same angles with the load vector. The intersection of these two lines cuts off the proper lengths of these lines to form the vectors representing the tension of the tie cord and the compression of the boom. Insert arrows in the proper places. From measurement of the lengths of the vectors determine the magnitudes of the two forces. Compare these values with the experimental values and with the values determined in Part 2.

4. On the diagram of Part 1 measure the angle each force makes with the horizontal. Write the equilibrium equations for the horizontal and vertical components of the forces, using only the load as a known force. Solve these equations for the tension of the tie cord and the compression in the boom. Compare these values with those previously determined.

5. Discuss the factors affecting the accuracy of your analysis, especially the effect of omitting the weight of the boom.

Experiment 9.3

STATIC EQUILIBRIUM: SPRING-BOOM CRANE

Object: To study torque and force methods of solving problems involving bodies in static equilibrium, utilizing for this purpose a spring-boom crane.

Method: A spring-boom crane (Fig. 9.6) is so designed that the force in its various parts may be measured.   From the dimensions of the

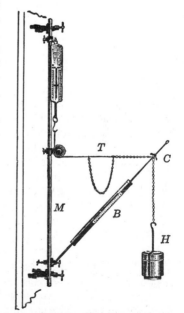

FIG. 9.6   Model crane

crane these forces, for any given load, are calculated by torque or vector polygon methods. Unknown torques and the unknown position and direction of forces are determined by use of the condition for equilibrium.

Apparatus: Spring-boom crane; 15-kg spring balance; weight hanger; kilogram weights; meterstick with caliper jaws; compass; protractor.

The spring-boom crane is made up as follows:

1. A vertical mast M. This consists of a 19-mm rod, 185 cm long, attached to the wall by suitable end supports.

2. An experimental crane boom B. This is a form of compression spring balance having a range of 15 kg. On the sliding rod is a collar C that may be fastened by set screws at any place along the sliding rod. The boom terminates at its lower end in a bent rod. The rod is pivoted in a hole in a right-angle clamp, fastened to the mast, so that the boom can rotate freely in the plane of the mast.

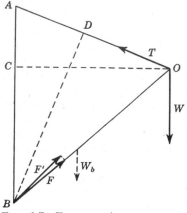

FIG. 9.7 Forces acting on crane boom

3. A tie cord T. This is attached at one end to a 15-kg spring balance, passes under a low-friction pulley and at the other end is attached to one of two stirrups on the movable collar. The second stirrup supports a weight holder H on which the load is placed.

The spring balance is vertically suspended from a clamp attached to the mast. The clamp can be moved up or down to change the effective length of the tie cord or to compensate for the extension of the spring balance.

Procedure: 1. Set up the crane as in Fig. 9.7 in such a manner that when a total load of about 9 kg, including the weight hanger, is suspended from point O the triangle AOB has no right angle and no two sides equal in length. Tap the boom and the spring balance to reduce the effect of friction. Record the reading of each balance.

2. Use a meterstick with caliper jaws to measure AB, BO, and AO, the dimensions of the triangle of the crane. The point A is the intersection of the two segments of the rope passing around the pulley (Fig. 9.5). Point B is the center of the rod on which the crane is pivoted and point O is the center of the rod by which the stirrups are attached to the collar.

3. Change the total load, including the weight hanger, to about 6 kg. Restore the crane to the same dimensions as before by adjusting the position of the collar at O and that of the vertical spring balance. If the pulley at A is not moved, the length AB will remain constant and only AO and BO need be adjusted. Tap the balances and record their readings.

4. Weigh the boom on a suitable balance and record the weight.

5. Repeat Steps 1 to 3 for a different configuration of the crane.

Computations and Analysis: 1. Draw a careful scale diagram of the crane, using the measured lengths AB, BO, and AO. Choose a scale such that the diagram will occupy most of a sheet of paper. Draw dotted lines OC and BD on the diagram to represent the moment arms of each of the measured forces about an axis through B and about an axis through A.

2. The forces acting on the boom are similar to those shown in Fig. 9.7. The magnitude, direction, and point of action of the load W and the tension T of the

tie cord are known.   The magnitude and direction of the weight W_b of the boom are known but the point of application is unknown.   Only the point of application of the force F' that the lower support exerts on the boom is known; the magnitude and the direction are unknown.   The force F along the boom is not the same as the force F'.

Write the torque equation about an axis through B perpendicular to the plane AOB.   Use the forces W and T from Step 1 and the moment arms measured on the diagram of Part 1 of the Analysis.   Include the unknown torque due to the weight of the boom.   Solve the equation for this unknown torque.   From this torque and the measured weight of the boom determine the position of the center of gravity of the boom, assuming that it is on the boom.

3.   Use the data of Step 3 to write torque equations, first about an axis through B, then about an axis through A.   In these equations assume the torque due to the weight of the boom to be that calculated in Part 2, that T and F are not known, and that F is parallel to the boom.   Solve the equations for T and F and compare the values obtained with the experimental values of Step 3.

4.   Use the data of Steps 1 and 4 to draw a vector diagram.   The forces T, W, and W_b are known but F is unknown.   Select a scale such that the diagram will occupy nearly a whole sheet of paper.   Draw vectors, in the proper direction to represent each of the known forces, placing the tail of each vector at the head of the preceding one.   Draw the vector that closes the polygon.   This vector represents the magnitude and direction of the force F'.   Measure the angle ABO on the diagram.   On the force diagram draw a dotted line making an angle equal to ABO with the vertical vector.   This line represents the direction of the boom. Measure the angle between the force F' and the line of the boom.   Draw the projection of F' on the line of the boom.   From the length of the projection determine the compressional force F in the boom and compare it with the experimental value measured in Step 1.

Review Questions: 1. Define torque and write its defining equation.   State some units of torque.   Define lever arm.   2. Explain what is meant by equilibrium. State the conditions necessary to produce equilibrium.   3. Define center of gravity. 4. Show how one may determine the weight of a stick by balancing it with known weights.   5. Explain the fact that the force at the base of a crane boom is not parallel to the boom.

Questions and Problems: 1. Show how one might locate the center of gravity of a thin, uniform sheet of metal by using only a plumb bob.

2. Where is the center of gravity of a doughnut?   Explain reasoning.

3. In order to assist a horse to pull a wagon out of a rut, at what place on the wheel could one apply a force most effectively?   Why?

4. Three unequal forces act upon a body at a point so that the body is in equilibrium.   If the magnitudes of two of the forces are doubled, how must the third force be changed to preserve equilibrium?   Justify your conclusion by diagrams.

5. What configuration of a crane would make W, T, and F equal to each other in magnitude?   Explain your reasoning and illustrate by a vector diagram.

6. Which of the following statements are true?   A body to be in equilibrium (a) must be at rest, (b) must move with uniform acceleration, (c) must have uniform linear and rotary velocity, (d) cannot be acted upon by external forces, and (e) must have the sum of the upward forces acting upon it just equal to its weight.

7. The center of gravity of a 50.0-gm meterstick is located at 51.0 cm and the stick

is supported at 70.0 cm. Where must an 80.0-gm object be hung in order to have equilibrium?

8. A crane boom of negligible weight is 20 ft long. It makes an angle of 30° with a vertical mast. The tie rope, fastened from the upper end of the boom to the wall, makes an angle of 90° with the boom. The maximum tensile force that the rope can safely stand is 5.0 tons. What is the maximum load the crane can support and the thrust in the boom for this load? Solve by both an analytical-vector method and a torque method.

9. In a typical experiment performed as in Step 5 of Exp. 9.1, the following data were recorded: angle made by meterstick, 30°; lever arm of pan 1, 34.3 cm; lever arm of pan 2, 35.0 cm; distance from the fulcrum to the center of gravity, 4.00 cm; mass of pan 1, (at left) 60.2 gm; mass of pan 2, 50.4 gm. Calculate the mass of the meterstick system from these data.

10. A derrick boom 30 ft long weighs 400 lb and is hinged at the bottom to a vertical mast. The boom is held in position by a rope attached at the top, the rope making an angle of 90° with the boom, and the boom making an angle of 30° with the mast at the hinge. If a load of 1.00 ton is carried at the top of the boom, what is the force exerted on the hinge?

11. A crane is constructed with a 200-lb uniform boom B 30 ft long attached to a vertical mast A. The cable C is fastened to the mast at a point 20 ft above the place where B is hinged to A. The boom inclines 30° to the vertical, and C is attached to B at a point 10.0 ft from the upper end of the boom. Find the thrust of B against A and the pull on C when a load of 1.00 kg is attached to the upper end of B. Solve by the graphical method.

CHAPTER 10

ROTATION OF RIGID BODIES.

The *average* angular speed $\bar{\omega}$ of a rotating body is the quotient of the angle θ through which it turns to the time t required to turn through that angle. In symbols

$$\bar{\omega} = \theta/t \tag{1}$$

The *instantaneous* angular speed ω is the limit of the angle-time ratio as the time becomes vanishingly small. Symbolically

$$\omega = \lim_{\Delta t \to 0} \frac{\Delta\theta}{\Delta t} \tag{2}$$

where $\Delta\theta$ represents a small increment of angle turned through in the corresponding increment of time Δt. The unit of angular speed commonly used in physics is the radian per second.

There is a distinction between angular *speed* and angular *velocity* that is similar to the distinction between linear speed and linear velocity. The *direction* of an angular velocity is specified as the direction of its axis of spin; the *sense* of the direction is related to the sense of the rotation as the direction of advance of an ordinary right-handed screw is related to its direction of rotation.

When the angular velocity of a body changes, it has an angular acceleration. The *angular acceleration* α is defined as the time rate of change of angular velocity. In symbols the *average* angular acceleration is given by the defining equation

$$\bar{\alpha} = \frac{\omega_2 - \omega_1}{t} \tag{3}$$

where the angular velocity changes from ω_1 to ω_2 in time t. The common unit of angular acceleration is the radian per second per second.

If the rotating body makes equal changes of angular speed in equal intervals of time, and the axis of rotation remains unchanged, it is said to be moving with *uniformly accelerated* angular motion. This type of motion is produced when a constant net torque, whose axis is the axis of rotation, acts upon the body.

In uniformly accelerated angular motion, average angular acceleration is the same as the instantaneous angular acceleration. In this type of motion, and only in this type, the average angular speed is the arithmetic average of the sum of the initial and final angular speeds.

$$\bar{\omega} = \frac{\omega_1 + \omega_2}{2} \tag{4}$$

From Eq. (1)

$$\theta = \bar{\omega}t \tag{1a}$$

From Eq. (3)

$$\omega_2 = \omega_1 + \alpha t \tag{3a}$$

71

When the values of $\bar{\omega}$ from Eq. (4) and ω_2 from Eq. (3a) are substituted into Eq. (1a), one obtains

$$\theta = \omega_1 t + \tfrac{1}{2}\alpha t^2 \tag{5}$$

Whenever an unbalanced net torque L acts on a body free to rotate, an angular acceleration α is produced that is proportional to the net torque

$$L \propto \alpha \qquad \text{or} \qquad L/\alpha = I \tag{6}$$

The constant I of Eq. (6) represents a quantity analogous to inertia in linear motion and is called moment of inertia (also called rotational inertia). *Moment of inertia* is that property of a body which causes it to resist change in its angular velocity. When Eq. (6) is written in the form

$$L = I\alpha \tag{7}$$

its similarity to the analogous equation for linear motion $f = ma$ becomes more apparent. The relationship represented by Eq. (7) is a form of Newton's second law as applied to rotational motion.

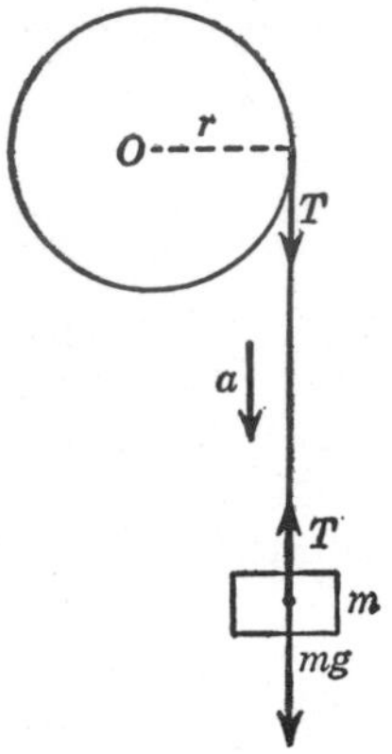

The moment of inertia of a simple geometrical solid can be calculated from equations derived by the use of the integral calculus. For example, the moment of inertia of a uniform solid cylinder of radius R and mass M rotating about its longitudinal axis is given by

$$I = \tfrac{1}{2}MR^2 \tag{8}$$

Fig. 10.1 Forces involved in producing the angular acceleration of a disk

The moment of inertia of an object about any axis may be experimentally determined, no matter how irregular or nonhomogeneous the body may be, by applying a known net torque to the body and measuring the resulting angular acceleration $(I = L/\alpha)$.

If a disk is mounted so that it is free to turn about an axis O and a body of mass m is suspended by a light string passing around the rim of the disk (Fig. 10.1), there will be a torque tending to turn the disk. This torque is given by

$$L_1 = Tr \tag{9}$$

where T is the tension in the string. But the tension T is not equal to the weight of the suspended body because that body has an acceleration a. This acceleration is produced by the net force $mg - T$ and in accordance with Newton's second law of motion

$$mg - T = ma \tag{10}$$

or

$$T = mg - ma \tag{10a}$$

The linear acceleration of the descending body is the same as the linear acceleration of the rim of the disk; it is related to the angular acceleration by the equation

$$a = \alpha r \tag{11}$$

Thus Eq. (10a) becomes

$$T = mg - m\alpha r = m(g - \alpha r)$$

and

$$L_1 = Tr = m(g - \alpha r)r \tag{12}$$

The net torque L is given by $L = L_1 - L_f$, where L_f is the frictional torque.

Therefore

$$I = \frac{L}{\alpha} = \frac{m(g - \alpha r)r - L_f}{\alpha} \tag{13}$$

The moment of inertia of a disk may also be determined from consideration of the energy relationships in the motion of the disk. If the body of mass m falls through a distance h, it loses potential energy mgh. In this process the descending body gains kinetic energy $\frac{1}{2}mv^2$ and the disk gains kinetic energy $\frac{1}{2}I\omega^2$. When frictional torque is neglected, the loss in potential energy is equal to the gain in kinetic energy

$$mgh = \frac{1}{2}mv^2 + \frac{1}{2}I\omega^2 \tag{14}$$

The relationship between the angular motion of the disk and the linear motion of the rim gives $h = r\theta$ (where θ is the angle turned by the disk as the body descends the distance h) and $v = r\omega$;

thus

$$mgr\theta = \frac{1}{2}m(r\omega)^2 + \frac{1}{2}I\omega^2 \tag{15}$$

From measurement of the time t required for the disk to turn through the angle θ it is possible to determine the final angular speed ω. If the body starts from rest the initial angular speed is zero and the final angular speed is ω. Thus, from Eqs. (1) and (4),

$$\bar{\omega} = \frac{0 + \omega}{2} \qquad \text{or} \qquad \omega = 2\bar{\omega} = \frac{2\theta}{t} \tag{16}$$

By the use of Eqs. (15) and (16) the moment of inertia of the disk can be obtained from measurements of the mass m of the descending body, the radius r of the disk, and the angle θ turned through by the disk in the time t.

Experiment 10.1

MOMENT OF INERTIA

Object: To study the angular motion of a body under the action of a constant torque and to determine the moment of inertia of the body.

Method. A constant torque is applied to a metal disk that is free to rotate. The time required for the disk to make a chosen number of rotations starting from rest is measured. The constant angular acceleration is measured and used in the torque equation to calculate the moment of inertia. The final angular speed is also measured and used in the energy equation to make an independent determination of the moment of inertia. These experimental values are compared with the value calculated from the geometric constants and the mass of the disk.

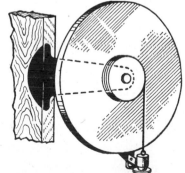

FIG. 10.2 Disk for measurement of moment of inertia

Apparatus: Uniform solid metal disk with small axle, mounted on low-friction, pivot bearings or ball bearings (Fig. 10.2); stop watch or stop clock; set of masses; meterstick; vernier caliper; string; two caliper jaws.

Procedure: 1. Record the mass of the disk and axle as stamped on it; if this value is not given, measure it by the use of a suitable balance. Measure the diameter of the disk with a meterstick provided with caliper jaws. Measure the diameter of the axle with the vernier caliper.

2. Mount the disk in its bearings, adjusting them so that the friction is as low as possible. (Be very careful not to damage these bearings.) Test the balance of the disk. If it is not perfectly balanced, attach small objects near the rim in appropriate places until the disk will remain in any selected position when it is so placed.

3. Attach the string to the axle and wind it carefully in one layer on the axle. In order to neutralize friction, attach a small mass to the string, start the disk in slow rotation, and observe whether the angular speed remains constant. In this observation it is well to time one rotation near the start of the motion and another when the string has nearly all unwound. Adjust the mass attached to the string until the speed is uniform. The torque thus applied is then just equal to the frictional torque. Apply this compensating torque in all the observations.

4. Attach to the string an additional mass of about 100 gm (the most appropriate value depends upon the disk used and should be so chosen that the disk makes 8 or 10 turns before its speed becomes too great for accurate observation). Release the disk and start the stop watch at the same instant. Take care not to impart any speed to the disk in the release. Allow the disk to rotate at least 10 turns and stop the watch at the end of a complete turn. Record the mass of the descending body, the number of turns, and the time elapsed. Repeat these readings several times using the same number of turns each time.

5. Repeat the procedure of Step 4 using a falling body of different mass.

Computations and Analysis: 1. From the data of Steps 4 and 5 compute the average time required for each chosen number of rotations. Calculate the angle, in radians, through which the disk turned.

2. By the use of Eq. (5) compute the angular acceleration of the disk during the measured time interval.

3. By the use of Eq. (13) compute the moment of inertia of the disk. (The radius r must be the radius of the axle, and the mass m is only the added mass.) Compare the values obtained from the two sets of data.

4. Compute the moment of inertia from the geometric measurements. For this computation consider the disk and axle as two cylinders; the disk is a cylinder of radius R and the axle is a cylinder of radius r. Compute the volume of each cylinder, calculate the density of the metal, and from these values compute the masses M_1 and M_2 of the cylinders. Then I is computed from

$$I = \tfrac{1}{2}M_1R^2 + \tfrac{1}{2}M_2r^2$$

Compare the value of I thus obtained with the experimental values.

5. From the data of Steps 4 and 5 compute the average angular speed during the observed motion. Compute the final instantaneous speed by the use of Eq. (16). Use the energy equation, Eq. (15), to compute the moment of inertia of the disk. Compare this experimental value with the other values obtained.

Experiment 10.2

ANGULAR MOTION; MOMENT OF INERTIA

Object: To study angular motion and the concept of moment of inertia; in particular, to determine the effect of a constant torque upon a disk free to rotate, to measure the resulting angular acceleration, and to determine the moment of inertia of the disk.

Method: A constant torque is applied to a disk and the successive positions of the disk are recorded by a spark-timing device of known frequency. The angular speeds are determined from the angular distances and the times. The

FIG. 10.3 Moment-of-inertia apparatus. The disk is shown covered with the coated-paper chart with polar-coordinate rulings. Some spark spots are also shown

angular acceleration is determined from the slope of the graph of angular speed against time. The moment of inertia is obtained from the ratio of the net torque to the angular acceleration. This experimental value is compared with the moment of inertia computed from the geometric constants and the mass of the disk.

Apparatus: Moment-of-inertia apparatus; spark-timing device; set of slotted masses; 50-gm weight hanger; thread; vernier caliper; scales capable of weighing the disk; several coated charts; Scotch tape; meterstick with caliper jaws; C-clamp; pulley; support rods; right-angle clamps.

The moment-of-inertia disk (Fig. 10.3) is mounted on a horizontal axis in precision pivot bearings, so as to turn with very low friction. It is made in three cylindrical steps, of simple geometric form, so that the moment of inertia can be calculated. Upon the ungraduated disk there is fastened with bits of Scotch tape a sheet of paraffin-coated paper with coordinate rulings in degrees. A spark point, mounted on a simple slide with insulating support, is arranged to move across the face of the disk as it rotates. High-potential sparks, passing at regular intervals from the point to the disk, puncture the paper; the heat thus developed melts a bit of the paraffin coating and causes a clearly recognizable spot. The location of these marks upon the printed scale gives the successive positions of the disk at the ends of equal intervals of time.

The accelerating mass is attached by a light string wound around one of the rims of the disk and attached to a pin on its periphery. In order to obtain a greater distance of fall, the string may be passed over a pulley P at some height above the disk as shown in Fig. 10.4.

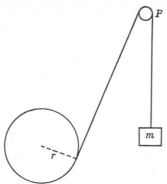

The spark-timing device is the same as that used in the experiment of free fall and described in Exp. 5.3. The two types of this device produce sparks at the rate of 120 or 60 per second.

Procedure: 1. For a preliminary trial, fasten a *used* sensitized chart to the disk by means of a few pieces of Scotch tape touching the edge of the chart and bending over the periphery of the wheel. Replace the disk on its supports, being very careful not to damage the bearings, and tighten the knurled screw. This should hold the disk firmly but with little friction. Wrap the thread several times around the rim; fasten one end to the pin provided for that purpose and the other to the accelerating mass after passing it over the pulley. A fall of the mass of several meters is desirable.

FIG. 10.4 Arrangement for attaching the accelerating mass m over a pulley P to the disk

2. Determine the frictional torque by attaching a small initial mass to the string. Give the disk a *small* initial speed and note whether it continues to rotate *uniformly* at this speed. Adjust this mass m_f until a uniform speed is obtained. The torque thus applied is just equal to the frictional torque. $L_f = m_f gr$.

3. Set the slider so the spark point is near the rim of the wheel. Attach an accelerating mass of about 200 gm. Release the mass without imparting any speed to it and, as it descends, slide the spark point slowly and uniformly toward the axis. Adjust the speed of the slider so that the spark dots are clearly separated for the successive rotations and the point reaches the end of its motion at nearly the same time as the end of the fall. Stop the rotation of the disk before the falling mass hits the floor to avoid damage to mass and floor and to prevent tangling the string.

4. After the method of moving the spark point has been mastered, connect the spark-timing device to the binding posts of the moment-of-inertia apparatus, turn on the switch, release the falling mass, and gradually move the spark point toward the axis of the disk. Just before the accelerating mass reaches the floor, turn off the spark timer and stop the disk. Examine the trace to see that there are no overlapping rings and few missing points.

If the practice trace appears satisfactory, attach a fresh sheet of coated paper and prepare a record trace. Remove the chart and examine the trace. If a few spots are missing the traces can be used, but if numerous points are missing a new record must be made.

5. Because the time interval between the sparks is very short ($\frac{1}{60}$ sec or

$\frac{1}{120}$ sec) there are more spots than are needed. A larger interval, such as $\frac{1}{10}$ sec, is more convenient. Lay the coated paper with the final trace flat on a table under good light. Start with some point near the beginning of the record and mark every *sixth* (or every twelfth) spot by encircling it by means of a sharp pencil. Number the encircled spots 0, 1, 2, 3, etc. Tabulate the data as shown in the accompanying table.

Spot No.	Angular position degrees	Angular distance θ degrees	Average angular speed $\bar{\omega}$ degrees/sec

6. Read the *angular position* of each consecutive *encircled* spot. Record these values in degrees in the second column of the table. For each complete revolution add 360°.

7. Measure the various dimensions of the disk and axle. Draw a diagram similar to Fig. 10.5 and record upon it the various dimensions. Measure and record the mass of the disk, if it is not given with the apparatus.

Computations and Analysis: 1. Subtract each item in the tabulated column of positions from the next successive item. This difference is the *angular distance* passed over in each successive time interval. Tabulate these values in the third column. Divide each of these angular distances by the time interval to obtain the *average angular speed* during the interval. Record these values in the fourth column. This average angular speed during the interval is also the *instantaneous* angular speed at the midpoint of the interval.

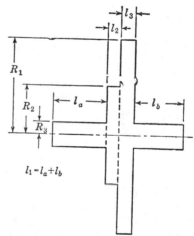

FIG. 10.5 Dimensions of the three-step moment-of-inertia disk

2. Plot a curve showing the relationship between instantaneous angular speed and time. For this purpose use the data of the fourth column plotted at the *midpoints* of the successive intervals. Carefully interpret the shape, slope, and intercepts of this curve. Compute the slope of the curve, which is the angular acceleration of the disk. Reduce the angular acceleration to radians per second per second.

3. By the use of Eq. (13) compute the moment of inertia of the disk. In this equation the radius r is the radius of the part about which the thread was wrapped.

4. Calculate the volume of each of the three cylindrical parts of the disk. Find the total volume and compute the density of the metal. Calculate the mass of each of the cylindrical parts and then compute the moment of inertia of the disk from the relation

$$I = \tfrac{1}{2}M_1R_1{}^2 + \tfrac{1}{2}M_2R_2{}^2 + \tfrac{1}{2}M_3R_3{}^2$$

Compare this computed value with the experimental value previously obtained.

Review Questions: 1. Define, write the defining equations, and state the name of the common cgs unit of each of the following: average angular speed, instantaneous angular speed, average angular acceleration, torque, and moment of inertia. 2. Distinguish between angular speed and angular velocity. 3. How is the direction of angular velocity specified? 4. State the analogies between corresponding quantities of linear and angular motion. 5. Write the equation that expresses the moment of inertia of a uniform circular disk rotating about its major axis. 6. Derive the expression for moment of inertia obtained from consideration of torques. 7. Derive an expression for moment of inertia obtained from consideration of energy. 8. Why is the tension in the cord not equal to the weight of the body attached to it? Set up the proper symbolic expression for the tension in the cord. 9. Describe an experimental method of measuring the angular acceleration of the disk. 10. Derive a measure of the relation between s and θ; between v and ω; between a and α. 11. What is the shape of a curve of angular speed against time for a body rotating under the influence of a constant net torque? What is the significance of the slope of this curve?

Questions and Problems: 1. From the definition of moment of inertia, $I = L/\alpha$, show that the moment of inertia of a particle of mass m, distant r from the axis, is mr^2.

2. From the units of torque and angular acceleration, show that the cgs unit of moment of inertia is the gram-centimeter^2.

3. In the torque and energy equations, why may the mass used to compensate for friction be omitted?

4. In Exp. 10.2, the first spark point does not occur when the angular speed is zero. What effect does this have upon the observed value of α? Explain.

5. If data were taken for angular distances *after* the accelerating force had been removed and the corresponding angular speeds plotted against time, what shape of curve would result? Why?

6. Explain how one could, by a simple experiment involving moment of inertia, determine which of two identically appearing eggs was hard-cooked and which was raw.

7. One end of a string is attached to a 200-gm mass. The other end is wrapped around a disk 20.0 cm in diameter and of mass 3.00 kg. The disk is free to rotate with negligible friction about its horizontal axis. The 200-gm mass starts from rest and descends 1.00 m. Find the following: (a) the loss of potential energy of the system; (b) the angular acceleration of the disk; (c) the gain in kinetic energy of the descending mass; (d) the tension in the string.

8. A flywheel is made of a uniform cylindrical disk of diameter 60 cm and mass 7.0 kg. A *constant* force of 2.45×10^5 dynes is applied tangentially at the rim of the disk. What is the angular speed of the disk 0.50 min after it is started from rest?

9. An object of mass 100 gm is attached to a cord passing over a disk 20 cm in diameter and of mass 3.0 kg, free to rotate about a horizontal axis. What is the speed of the 100-gm object 0.60 sec after it starts from rest?

10. What portion of the total kinetic energy of a rolling solid disk is energy of translation and what portion is energy of rotation?

CHAPTER 11

MOMENTUM; BALLISTICS

The *momentum p* of a body is defined as the product of the mass m of the body and its velocity v. In symbols, the defining equation is

$$p = mv \tag{1}$$

The cgs unit of momentum is the gram-centimeter per second; the fps unit is the slug-foot per second.

It may be shown from Newton's second and third laws of motion that momentum must be conserved in all impacts. The *principle of conservation of momentum* states that the momentum of a body or system of bodies does not change except when an external force is applied.

If an external force acts upon a body, its momentum is changed but, in the process, some other body suffers a change of momentum equal in magnitude but opposite in direction to that of the first body. Suppose an object of mass m_1 moving with velocity v_1 collides with a second object of mass m_2, initially at rest. As a result of the collision the velocity of each object will be changed, that is, each will be accelerated. Call the forces f_1 and f_2 and the accelerations a_1 and a_2, respectively. In accordance with the third law of motion

$$f_1 = -f_2 \tag{2}$$

and from the second law, $f = ma$,

$$m_1 a_1 = -m_2 a_2 \tag{3}$$

These forces necessarily act for the same time Δt. From the definition of acceleration

$$a_1 = \frac{\Delta v_1}{\Delta t} \quad \text{and} \quad a_2 = \frac{\Delta v_2}{\Delta t} \tag{4}$$

where Δv_1 is the *change* in velocity of the first object and Δv_2 is the *change* in velocity of the second object. When these values of the accelerations are substituted in Eq. (3), we obtain

$$m_1 \Delta v_1 = -m_2 \Delta v_2 \tag{5}$$

Since the product of mass and change of velocity represents the change of momentum, it follows that the change of momentum of one object is equal in magnitude but opposite in direction to that of the second object. That is, the total momentum of the system has remained constant during the impact.

In an inelastic impact two objects that collide adhere and have a common velocity after the impact. Consider the two balls of Fig. 11.1. Before the impact one ball has a mass M and a velocity v_1, while the second ball has a mass m and zero velocity. After the impact the combined objects have a common velocity V. In accordance with the law of conservation of momentum, the momentum before the impact must be the same as the momentum after the impact.

$$Mv_1 + m \times 0 = (M + m)V$$

or
$$Mv_1 = (M + m)V \tag{6}$$

Experimental verification of the law of conservation of momentum depends upon ability to measure velocities before and after impact. Where all the velocities have the same direction, measurement of speed is sufficient.

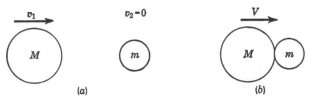

Fig. 11.1   Inelastic collision of two objects: (a) before impact; (b) after impact

Measurements of speed may be direct or indirect. In a direct measurement one measures the distance s traveled and the corresponding time t.

$$\bar{v} = s/t$$

Where the speed is high, special means must be devised for measuring the small interval of time involved.

One method of measuring speed is to project the object horizontally. In the absence of a horizontal force the horizontal speed v remains constant while the body falls vertically with acceleration g. If S_H and S_V represent, respectively, horizontal and vertical distances traveled in equal intervals of time

$$S_V = \tfrac{1}{2}gt^2 \tag{7}$$

and
$$S_H = vt \quad \text{or} \quad v = S_H/t \tag{8}$$

From Eq. (7) $t = \sqrt{2S_V/g}$

and
$$v = \frac{S_H}{\sqrt{2S_V/g}} \tag{9}$$

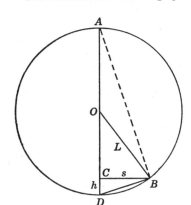

Fig. 11.2   Ballistic pendulum rises to height h when given a horizontal velocity

A second method of determining a speed involves use of a ballistic pendulum. When a pendulum is at rest at its lowest point its kinetic energy is zero and its potential energy is a minimum. If the pendulum of mass M is given a horizontal velocity V by an impact, it will have acquired kinetic energy $\tfrac{1}{2}MV^2$. The pendulum will swing through an arc such that the center of gravity of the pendulum will rise a distance h (Fig. 11.2). In rising this distance the kinetic energy of the pendulum is all converted into potential energy

$$\tfrac{1}{2}MV^2 = Mgh \tag{10}$$

or
$$V = \sqrt{2gh} \tag{11}$$

If the distance h is small it may be computed from the length L of the pendulum and the horizontal distance s moved by the pendulum. In Fig. 11.2, the triangles ABC and BCD are similar. Thus

$$\frac{AC}{CB} = \frac{CB}{CD} \tag{12}$$

or

$$\frac{2L - h}{s} = \frac{s}{h} \tag{13}$$

$$2Lh - h^2 = s^2 \tag{14}$$

Since h is a small quantity, its square is very small in comparison with the other terms of the equation and may be neglected. Making this approximation

$$2Lh = s^2$$

or

$$h = s^2/2L \tag{15}$$

This measurable value of h can be substituted in Eq. (11) to compute the initial speed of the pendulum.

These equations may be employed in experiments to demonstrate the law of the conservation of momentum and to study some of the concepts of ballistics.

Experiment 11.1

CONSERVATION OF MOMENTUM: SPEED OF BULLET

Object: To measure the speed of a rifle bullet and to compare momentums before and after impact.

Method: The speed of a rifle bullet is determined by direct measurement of distance and time. A bullet is fired into a ballistic pendulum and the initial speed of the pendulum is determined from the distance it swings. From these measured speeds and masses, the momentums before and after impact are determined and compared.

Apparatus: Electric motor; rotating shaft carrying two cardboard disks; revolution counter; special protractor; clamp to hold gun; stopping device for bullet; ballistic pendulum; 22-caliber rifle; shells; balance; meterstick.

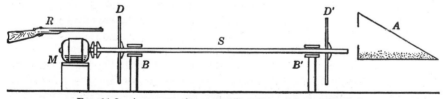

FIG. 11.3 Apparatus for measuring the speed of a rifle bullet

The motor and shaft arrangement is shown in Fig. 11.3. The shaft S, mounted on bearings BB', carries two cardboard disks DD' and is turned by the motor M. The disks are mounted as far apart as the length of the shaft permits. The rifle R is clamped with the barrel parallel to the shaft. When the rifle is fired, the bullet passes through the two disks in succession and to the stop A. The stop should be made of heavy steel with an opening only at the front; the bottom should be covered with a thick layer of sand.

The ballistic pendulum is a heavy block of wood suspended by four string supports, as in Fig. 11.4. The strings should be about 2 m long and slanted outward in the plane perpendicular to the path of the pendulum, to prevent twisting. The meterstick is mounted just below the pendulum. A small cardboard or aluminum rider serves as an indicator of the position of the pendulum.

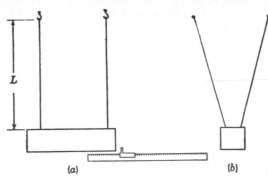

FIG. 11.4   Form of ballistic pendulum

Procedure: 1. Mount the rifle in its clamp with the barrel parallel to the rotating shaft. *Keep the breech open at all times except during the actual firing.* Close the motor switch and allow the system to reach full speed. Fire the rifle, using a .22 short shell.

2. Stop the motor and turn the shaft until the hole in the first disk is visible through the barrel of the rifle. With the disks stationary in this position fire the rifle. Measure the angle θ between the holes in the second disk. Use for this purpose the special protractor which has a semicircular notch at its center to fit the shaft.

3. Repeat the procedure of Steps 1 and 2, using a .22 long-rifle shell.

4. Measure and record the distance between the two disks. With the motor and shaft at full speed attach a revolution counter and allow it to run for 30 sec. Record the initial and final readings of the revolution counter.

5. Mount the rifle in the clamp before the ballistic pendulum with the barrel horizontal and parallel to the plane of the pendulum. Adjust the height of the rifle so that the bullet will strike at about the level of the center of gravity of the wooden block.

6. With the pendulum stationary set the rider on the scale against the end of the block. Read the scale at the end of the rider and record the reading. Insert a .22 short shell and fire the rifle. Read and record the new position of the rider on the scale. Repeat for two other .22 short shells.

7. Repeat Step 6, using .22 long-rifle shells.

8. Remove from its shell a sample of each kind of bullet and measure its mass with a suitable balance. Measure also the mass and length L of the pendulum. (This length should not include one-half the thickness of the block.)

Computations and Analysis: 1. Compute the angular speed ω of the rotating shaft, from the data of Step 4. Compute the time of travel of the bullet from $t = \theta/\omega$. Calculate the speed of each bullet from $v = S/t$, where S is the distance between the two disks.

2. From the data of Step 6, calculate the average distance s that the pendulum moved horizontally. By the use of Eq. (15) find the value of h. By the use of Eq. (11) compute the initial speed V of the pendulum.

3. Calculate the momentum of the .22 short bullet using its mass and the speed determined in Part 1 of the Analysis. Compute the momentum of the pendulum just after the impact. Compare the two values thus obtained.

4. Repeat Parts 2 and 3 using the data for the .22 long-rifle bullet.

5. For each type of bullet, calculate the kinetic energy of the bullet before impact and the kinetic energy of the pendulum after impact. Compare the corresponding energies and explain the discrepancy.

Experiment 11.2

CONSERVATION OF MOMENTUM: BLACKWOOD PENDULUM

Object: To study the elements of projectile motion and the law of conservation of momentum.

Method: A ball is fired horizontally and its range and vertical distance of fall are observed. From these distances and the value of g, the speed is calculated. The ball is fired into a suspended holder, arranged to swing as a ballistic

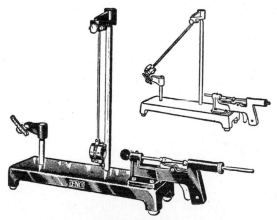

FIG. 11.5 Blackwood ballistic pendulum

pendulum. The speed of the pendulum and ball after impact is computed from the rise of the pendulum. The momentum of the ball before impact and that of the ball and pendulum after impact are calculated and compared.

Apparatus: Blackwood ballistic pendulum, Fig. 11.5; trip scales with masses; metric rule; steel tape measure; level; table clamp; carbon paper; box for catching ball; inclined plane.

The Blackwood pendulum, details of which are shown in Fig. 11.6, is a combination of a ballistic pendulum and a spring gun for propelling the projectile. The ballistic pendulum consists of a massive cylindrical bob C, hollowed out to receive the projectile and suspended by a strong light rod K. The rod is supported at its upper end by steel cone-pivot bearings. The pendulum may be removed from its supporting yoke by unscrewing the shouldered screw L. This screw, when tight against its shoulder, automatically adjusts the bearings so that the pendulum is securely held with little friction.

The projectile is a brass ball B which, when propelled into the pendulum bob, is caught and held by the spring S in such a position that its center of gravity lies in the axis of suspension rod K. The pendulum therefore hangs freely in the same position whether or not it contains the ball. A brass indicator I is attached to the pendulum bob C in such a way that its tip indicates the height of the center of gravity.

When the projectile is caught, the pendulum swings upward and is retained at its highest point by the pawl P, which engages a tooth in the curved rack R. The

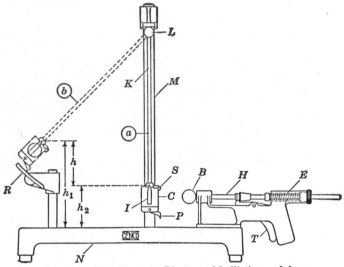

Fig. 11.6   Details of the Blackwood ballistic pendulum

toothed surface of the rack lies on the arc of a circle having its center in the axis of suspension of the pendulum. A scale along the outer edge of the rack provides a means for noting and recording the position of the pendulum after each shot. The ball is drilled so that it may be held on the forward end of the rod H, which is propelled forward by the compressed spring E when the trigger T is pulled.

Procedure: 1. To determine the initial speed of the ball, set up the apparatus as shown in Fig. 11.7. Set the apparatus near one edge of the table. If necessary, wedge it up with cardboard until the base N (Fig. 11.6) is accurately horizontal, as shown by the level. Clamp the frame to the table. Fire the gun and note the approximate place where the ball strikes the floor. To make the gun ready for firing, rest the pendulum on the rack R, put the ball in position on the end of rod H, and holding the base with one hand, pull back on the ball with the other hand until the collar on rod H engages the trigger T. This action compresses the spring E by a definite amount, and the ball is given approximately the same initial speed every time the gun is fired.

2. Place the box at the appropriate location to receive the ball when fired. Lay a piece of paper in the bottom of the box and cover it with carbon paper.

This arrangement automatically marks the spot where the ball strikes. See that paper, box, and gun do not move during the firing. Fire the gun five or six times. Locate the average position of striking and measure the range S_H (Fig. 11.7). Also measure the vertical distance S_V traveled by the ball.

3. To study the collision of the ball with the pendulum, release the pendulum from the rack and allow it to hang freely at rest. Fire the ball into the pendulum and record the position reached by the pendulum. Repeat this procedure five or six times. Record the position each time and calculate the average position. Set the pendulum at this average position and measure the distance h_1 (Fig. 11.6) from the base to the index point for the center of gravity. Release

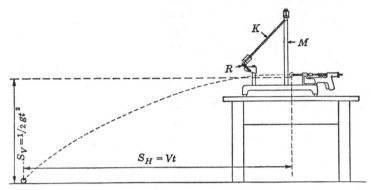

FIG. 11.7 Spring gun arranged for measurement of the initial speed of a projectile from observations of the range and height of fall

the pendulum, allow it to hang in its lowest position, and measure h_2. The difference $h_1 - h_2$ gives h, the distance through which the center of gravity rose after the impact.

4. Remove the pendulum from its support and measure the mass of the pendulum and that of the ball.

5. Elevate the gun so that the ball will be shot at an angle of about 30° above the horizontal. Measure and record the angle. Measure the *vertical* distance between the point of discharge and the level of the receiving surface in the box on the floor.

Assume the initial speed calculated in Part 1 of the Analysis and compute the range of the ball for this position of the gun. Set the box at the calculated position, fire the gun, and compare the calculated and observed ranges.

Computations and Analysis: 1. By use of the data of Step 2 and Eq. (9) calculate the initial speed of the ball.

2. Compute the momentum of the ball just before the collision and that of the ball and pendulum just after the collision. Compare these two quantities. Explain any difference that appears.

3. Compute the kinetic energy of the ball just before the impact and that of the ball and pendulum just after the impact. Compare the two kinetic energies and discuss the result.

Review Questions: 1. Define momentum and write its defining equation. Name the cgs and fps units. 2. Is momentum a vector or a scalar quantity? 3. State the law of conservation of momentum. Show how this law follows as a consequence of

Newton's laws of motion.   4. Derive, in terms of its horizontal range and the vertical distance fallen, an expression for the initial speed of a projectile fired horizontally. 5. What happens to the horizontal component of the velocity of a projectile as time goes on?   To the vertical component of the velocity?   6. Derive the expression for the initial speed of a ballistic pendulum.   What fundamental assumption is made in this derivation?   7. Describe the essential features of the Blackwood ballistic pendulum and the technique of the observations made with it.

Questions and Problems: **1.** Suggest some probable reasons for the difference between observed values of momentum before and after impact.

2. How could one account for an apparent increase in momentum in Exp. 11.2?

3. How might one proceed to determine the efficiency of the spring gun, that is, the ratio of the energy output to the work done on it?

4. In Exp. 11.2, what is the effect upon the equality of the momentums of neglecting the moment of inertia of the pendulum?

5. What happens to the center of gravity of a meteor if the meteor explodes above the earth?

6. On what physical principle or law is based the statement that the horizontal component of the velocity of a projectile remains constant?

7. An 80-gm sphere supported by a wire 150 cm long swings through an arc of 30°. It collides inelastically with a 200-gm block and the two rise together a vertical distance of 1.00 cm.   Show how these data may be used to illustrate the law of conservation of momentum.

8. An 80-gm bullet moving with a speed of 300 m/sec strikes a stationary block of mass 30 kg, suspended as a ballistic pendulum.   How far will the block rise after the inelastic impact?

9. Show that the fractional loss in kinetic energy before and after impact of a ballistic pendulum of mass m struck inelastically by a moving ball of mass M is given by $m/(M + m)$.   Into what form of energy is the lost kinetic energy transformed?

10. In the measurement of v by the free-flight method as in Exp. 11.2, what would be the effect on the accuracy of the measurements if the floor were not truly horizontal?   If the base of the apparatus were not accurately horizontal?

11. When a pendulum is set vibrating it ultimately comes to rest.   What are the various causes of its coming to rest?   Do any such causes affect these momentum experiments?

CHAPTER 12

UNIFORM CIRCULAR MOTION

A body moving with uniform speed in a circle is said to move with *uniform circular motion*. Although the *speed* of the body is constant, its *velocity* is continually changing, for the *direction* of the motion is always changing. Consequently such a body has an acceleration, but this acceleration produces a change only in the direction of the velocity.

Because the acceleration produces a change only in direction it must always be directed at right angles to the direction in which the body is moving; that is, the acceleration is directed toward the center of the circle. The value of this central acceleration a is given by

$$a = v^2/r \tag{1}$$

where v is the speed of the body, or magnitude of the velocity, and r is the radius of the circular path in which it is moving.

A net force is necessary to produce any acceleration whether it involves a change of magnitude or of direction. According to Newton's second law

$$f = ma = m(v^2/r) \tag{2}$$

This net force, directed toward the center of the circle, producing the central acceleration, is called the *centripetal* force. The force is measured in any units suitable for Newton's second law. If the mass is in grams, the speed in centimeters per second, and the radius in centimeters, the force is in dynes.

The name *centrifugal* force is usually given to the equal and reacting force which, by Newton's third law, the *body* exerts away from the center. *Note that the centrifugal force does not act on the moving body.* Both centripetal and centrifugal forces are radial and not tangential.

The equation for centripetal force, Eq. (2), may be written in terms of angular speed by setting $v = \omega r$, whence

$$f = \frac{mv^2}{r} = \frac{m(\omega r)^2}{r} = m\omega^2 r \tag{3}$$

The angular speed ω, in radians per second, is 2π times the angular speed n, in revolutions per second. Therefore Eq. (3) may be written

$$f = m\omega^2 r = m(2\pi n)^2 r = 4\pi^2 n^2 rm \tag{4}$$

A measure of the central force on a rotating object may be obtained from Eq. (4) by measuring the mass, radius of rotation, and angular speed of the rotating body.

87

Experiment 12.1

CENTRIPETAL AND CENTRIFUGAL FORCES

Object: To study the forces involved in the motion of a body traveling with constant speed in a circular path.

Method: A ball is made to swing in a circular path at such speed that it just supports a body of known mass hanging vertically at the center of the path. The central-force equation is checked by measurements of angular speed, mass, and radius of rotation.

Apparatus: Centripetal-force apparatus; meterstick; hooked masses; beam balance; stop watch or stop clock.

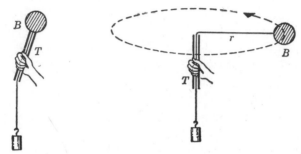

Fig. 12.1   Simple type of centripetal-force apparatus

The centripetal-force apparatus, Fig. 12.1, consists of a rubber ball B attached to a strong light string, which passes through a glass or metal tube T. A hooked mass is attached at the lower end of the string.

Procedure: 1. Measure the mass of the rubber ball. Hang a hooked mass of 200 gm on the other end of the string and hold the tube horizontal with the ball and hooked mass hanging nearly equidistant from the ends of the tube. Because the two masses are unequal, the forces applied to the two ends of the string are unbalanced; the larger mass will go down and the smaller mass will go up. Observe and record the effect.

2. Hold the tube vertical and twirl the ball in a circular path in a horizontal plane. In this motion the centripetal force necessary to constrain the ball in a circular path is supplied, through the string, by the weight of the hooked mass. In reaction the centrifugal force supplied by the rotating ball supports the hooked mass. Adjust the speed of rotation of the ball so that the hooked mass is just supported by the string. The motion should be begun with the tube at arm's length and above the head. When the motion is under control so that the hooked mass is stationary, one observer should count the number N of revolutions in 1 min while the other maintains the constant motion. Next, grasp the string at the bottom of the tube to mark the position of the string while the ball was moving. With the string in this position, measure the distance r from the top of the tube to the center of the ball.

3. For the same hooked mass change the radius of the rotation first to a smaller and then to a larger value. Observe the corresponding changes that must be made to maintain the hooked mass stationary.

4. Repeat Step 2 using a different hooked mass.

Computations and Analysis: 1. From the total number of revolutions N of the ball, the observed time t, and the radius r of revolution calculate the speed v of the ball, $v = 2\pi r N/t$.

2. Compute the centripetal force by the use of Eq. (2). Compare this value of the central force with the weight mg of the hooked mass.

3. Compute the central acceleration by use of Eq. (1).

4. Discuss the observations of Step 3 of the Procedure.

Experiment 12.2

CENTRIPETAL FORCE

Object: To study uniform circular motion; in particular, to compare the observed and calculated values of the centripetal force.

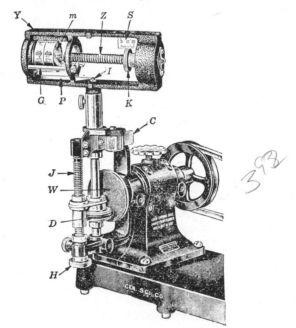

Fig. 12.2   Centripetal-force apparatus mounted on variable-speed rotator

Method: By means of an electrically driven rotator a body of known mass is rotated about a vertical axis in such a way as to produce a definite extension of a spiral spring. From the mass of the body, the speed of rotation, and the radius of the circular path the centripetal force is computed and compared with the gravitational force necessary to produce the same extension of the spring.

Apparatus: Centripetal-force apparatus; variable-speed rotator; stop watch, or clock with sweep seconds hand; weight hanger; slotted masses; vernier caliper; rods and clamps.

The centripetal-force apparatus, shown at the top of Fig. 12.2, consists of a metal frame Y within which is mounted a cylindrical body of mass m attached to a coil spring Z; the entire assembly is rotated about a vertical axis through its center of gravity. The tension of the spring is adjusted by means of a threaded collar K to which the spring is fastened. In some models the position of the collar is indicated by a scale S attached to the frame. Three metal guides G constrain the cylindrical body to move only along the axis of the spring; this axis intersects the axis of rotation at right angles. When the apparatus is rotated about a vertical axis the cylinder moves farther from the axis and produces an extension of the spring. This situation is represented diagrammatically in Fig. 12.3. A specially designed pointer P is loosely pivoted at O and is so shaped that

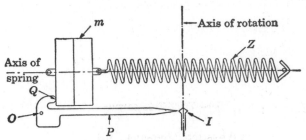

FIG. 12.3   Diagram of speed index

when the cylinder presses against it at Q its tip moves upward through a range of about 5 mm. At the middle of this range is a fixed index I. The index is practically on the axis of rotation, and the position of the pointer can be clearly seen while the apparatus is rotating.

The speed of rotation is controlled by the variable-speed rotator. The speed is determined by the location of the point of contact of the friction disk D with the driving disk W. Turning the milled head H of the screw J carries the friction disk in or out along the radius of the driving disk. The task of counting the number of turns is facilitated by a revolution counter C attached to the frame of the rotator by a steel spring that normally holds the counter disengaged from the rotating spindle. The counter gear is engaged with an identical gear on the spindle by pressing the finger on the end of the spring.

Procedure: 1. With the axis of the rotator vertical make a few preliminary observations to learn the technique of keeping the speed of rotation near the critical value. One observer should pay constant attention to the control of the speed while the other makes the actual time measurement. In order to maintain the critical speed, it is necessary to make continual adjustments either of the position of the friction disk or of the motor speed. The pointer P does not stay in a constant position at the index but oscillates more or less slowly about it. With a little practice it is possible to anticipate the motion so that a compensating motion of the knurled nut is made before the pointer is appreciably away from the index. Thus there will be a slow oscillation that gives an average position at the index.

When the technique of adjusting the rotator has been mastered, record the reading of the revolution counter, engage the gears at the instant the stop watch is started (or as the sweep second hand passes an even minute), and keep the gears meshed for about one minute. Record the new reading of the revolution counter and the time the gears were engaged. Take a series of five such readings.

2. Record the mass m stamped on the cylindrical body attached to the end of the spring and the scale reading of the position of the collar K on the scale S.

3. Remove the centripetal-force apparatus from the rotator and suspend it as in Fig. 12.4. Attach a weight hanger and add masses until the pointer is brought to the index. The cylinder is then in the position in the frame that it occupied during the rotation. Record the total mass sustained by the spring, including that of the weight hanger and that of the cylindrical body. By means of a vernier caliper measure the distance r between the axis of rotation (indicated by the line L_1 scribed on the frame) and the center of gravity of the cylinder (indicated by the line L_2).

4. Change the tension of the spring and repeat the series of observations.

Computations and Analysis: 1. From the data of Step 1 compute the angular speed of rotation n in rotations per second. Use Eq. (4) to calculate the centripetal force.

2. From the mass used to stretch the spring compute the stretching force. Compare this force with the centripetal force previously calculated. Note the percentage difference and try to account for the discrepancy.

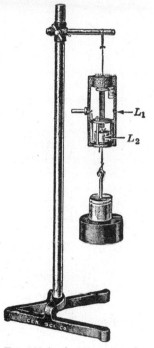

FIG. 12.4 Arrangement for the application of gravitational forces

Review Questions: 1. What is meant by uniform circular motion? 2. How may a body have constant speed and yet be accelerated? 3. Write three expressions for the central force necessary to hold a body in uniform circular motion. State the cgs unit of each factor in the equations. 4. Explain the difference between centripetal and centrifugal forces. Are they radial or tangential? On what body does each act? 5. Sketch the essential features of the apparatus used and describe how it is used to check the expression for centripetal force.

Questions and Problems: 1. Derive the expression for the central acceleration of a particle in uniform circular motion.

2. Could a horizontal axis of rotation be used satisfactorily in an experiment on uniform circular motion? Explain.

3. In data used for computing centripetal force which would be the more serious: a 1% error in observing the time or a 1% error in measuring the radius? Why?

4. In Exp. 12.2 what are the advantages of keeping the same value of r and varying the spring tension for various speeds rather than keeping the spring tension constant and observing the values of r at various speeds?

5. What mass could be lifted by the centrifugal force of an object having a mass of 98 gm if it were revolving in a horizontal circle, 20 cm in diameter, and making 240 rev/min?

6. A body of mass 100 gm rotates on the circumference of a circle 2.00 m in diameter. The speed increases uniformly from 60 to 180 rev/min in 3.00 sec. Calculate the tangential and radial accelerations at the end of the third second and the centripetal force at that same time.

CHAPTER 13

ELASTICITY

Whenever a single force acts upon a body, that body will be accelerated. If other forces act on the body to prevent the acceleration, the body will be distorted, that is, there will be a change in size or a change in shape or both. After the forces are removed, the body may return to its original size or shape.

If after a body is deformed by some force, it returns to its original shape or size as the distorting force is removed, the material is said to be *elastic*. Every substance is elastic to some degree. Every substance fails to be perfectly elastic if the distortion is too great. There is an elastic limit.

Whenever a body is subjected to distorting forces, there is set up within the body an internal *stress*, which is expressed as the force per unit area. Inasmuch as the reacting force is equal to the acting force, we can use the *external* force per unit area as the stress. The fractional distortion produced by the stress is called the *strain*.

Whenever the distortion is an increase (or decrease) in only one dimension, the fractional change in length is called a *tensile* strain and the corresponding stress is a tensile stress. When the distortion is a change in volume, that is, all dimensions are changed by the same fraction, the strain is a *volume* strain. If there is a change only of shape, there is a *shearing* strain.

The greatest stress for which the body will return to its original size and shape is called the *elastic limit*. When a stress greater than the elastic limit is applied to a body, that body acquires a permanent set. For various materials the values of the elastic limit vary over a wide range. Whether a body is *elastic* does not depend upon the value of the elastic limit but upon the exactness in returning to the original size and shape.

Hooke's law applies to any stress and the corresponding strain. For stresses below the elastic limit the stress is proportional to the strain. The ratio of the stress to the strain is called a *modulus of elasticity*.

$$\text{Stress} \propto \text{strain}$$

$$\text{Modulus} = \frac{\text{stress}}{\text{strain}}$$

Experiment 13.1

ELASTICITY OF MATTER; HOOKE'S LAW

Object: (1) To study the elastic behavior of two materials. (2) To study Hooke's law.

Method: The elongations of a spiral spring and of a rubber tube are measured as successively larger loads are added and as the loads are removed. The ability of each body to return to its original length is observed. The relation of elongation to load is determined by means of a graph.

Apparatus: Spring; soft rubber tubing; weight hanger; four 1-kg masses; meterstick; meterstick clamp; table clamp; right-angle clamp; support rods.

The spring and the rubber tube should be very carefully selected to have about the same length and same load capacity. A heavy-duty screen-door spring is satisfactory.

Procedure: 1. Mount the spiral spring so that it hangs vertically. Attach the weight hanger and if necessary add enough mass just to separate the coils. Mount the meterstick vertically in its clamp so that its upper (zero) end is near the bottom of the weight hanger. Record the position of the bottom of the weight hanger with respect to the scale. Add the masses 1 kg at a time and record the resulting positions on the scale. Do not allow the spring to oscillate but hold it in position while the load is being changed and allow it to take its new position gradually.

2. After the masses have all been added, remove them one at a time and record the resulting positions.

3. Repeat Steps 1 and 2 using the rubber tube in place of the spiral spring. Use the same zero load as was used with the spring.

Computations and Analysis: 1. Is the spring perfectly elastic for the loads used, that is, does it return to the initial length after the stress is removed? Is the rubber tube perfectly elastic? Which is more elastic, the spiral spring or the rubber tube?

2. For each load compute the displacements produced for both loading and unloading. Plot two sets of points showing the relationships between displacement and load for the spring during loading and during unloading. Will one curve fit both sets of points? Discuss the shape of the curve or curves. Does the shape indicate that the spring follows Hooke's law? Was the elastic limit exceeded?

3. Repeat the analysis of Part 2 for the rubber tube. Does the curve for unloading coincide with the curve for loading or is it parallel to it? What is the significance of a different location or slope? If the unloading curve does not pass through the origin discuss the significance of that fact.

Experiment 13.2

YOUNG'S MODULUS

Object: (1) To study Hooke's law as applied to a stretched wire. (2) To measure Young's modulus for a specimen of steel.

Method: A series of known loads is applied to stretch a wire whose original length and area of cross section are measured. The corresponding elongations are measured and the values plotted against load. From the slope of the curve the force per unit elongation is calculated and the value of Young's modulus is computed.

Theory: For tensile stresses and strains the modulus of elasticity is called *Young's modulus Y*

$$Y = \frac{\text{tensile stress}}{\text{tensile strain}} = \frac{F/A}{\Delta L/L} \tag{1}$$

where F is the stretching force, A is the area of cross section, ΔL is the change in length, and L is the original length (Fig. 13.1).

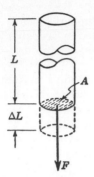

FIG. 13.1 Stress F/A and strain $\Delta L/L$ in stretching a wire

In the study of the elongation of a steel wire by successive loads the principal problem is the measurement of the change in length. The measurement of this small elongation may be made directly by means of a vernier scale as in Fig. 13.2 or by a magnifying device such as the optical lever.

Apparatus: Young's modulus apparatus and accessories; kilogram masses; tape measure; vernier and micrometer calipers; meterstick provided with caliper jaws; extension lamp.

The Young's modulus apparatus may be in either of two forms. The simple form shown in Fig. 13.2 consists of a bracket that supports two wires. At the lower end of the test wire is a scale that moves downward as the wire is stretched by increased loads. The index with the vernier scale is supported by the second wire and remains fixed with constant load.

The second form of apparatus is shown in Fig. 13.3. The wire W whose elongation is to be measured is supported by a chuck C attached to a yoke Y which is clamped to the two support rods R and R'. Near the bottom of the wire is attached a second chuck C' that passes loosely through a hole in a bridge E of adjustable height. The optical lever L consists of a mirror mounted on a three-legged frame. The two front legs rest on the bridge to form an axis of rotation of the mirror; the rear leg rests on the chuck C' and moves up or down with that end of the wire. Several feet from the mirror is a telescope T and vertical scale S. The image in the mirror of the illuminated scale is viewed through the telescope. When the wire is stretched, the angular motion of the mirror causes a different part of the scale to appear at the cross hairs of the telescope. A very small elongation of the wire causes an easily measurable change in reading on the scale.

Consider the diagram of the optical lever and scale of Fig. 13.4. When the mirror is in the position shown by solid lines, light from the point A on the scale S is reflected from the mirror to the telescope. If the wire stretches an amount ΔL the lever turns through an angle θ. By a well-known principle of optics when a mirror is turned through an angle θ the reflected beam is turned through an angle 2θ. Thus for the new position, light from B is reflected to the telescope. From the angles of Fig. 13.4

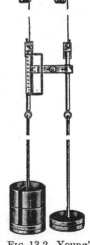

FIG. 13.2 Young's modulus apparatus, using vernier scale

$$\frac{\Delta L}{s} = \frac{d/2}{R} \qquad \text{or} \qquad \Delta L = \frac{sd}{2R} \qquad (2)$$

where d is the difference in the two readings of the scale, s is the distance from the back leg of the lever to the line of the front two legs, and R is the distance from the mirror surface to the scale.

An alternative method of using the optical lever involves evaluation of a con-

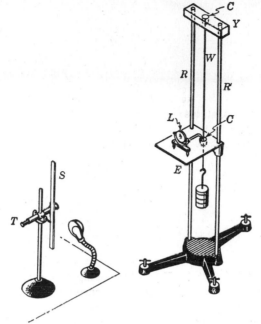

FIG. 13.3   Young's modulus apparatus, using optical lever

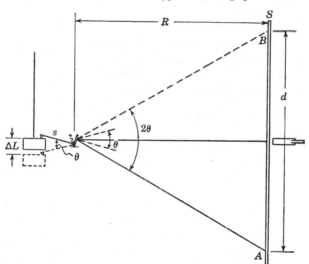

FIG. 13.4   Diagram of optical lever and scale

stant of proportionality on the assumption that the change of reading is proportional to the elongation. A reading is first taken, and then a disk of known thickness t is inserted under the rear leg and a new reading taken. If d' is the difference between the two readings,

$$t = Kd' \quad \text{or} \quad K = t/d' \tag{3}$$

This constant K may be used only for the arrangement of mirror and scale for which it was obtained. When this method is used

$$\Delta L = Kd \qquad (4)$$

Procedure: (If the apparatus with vernier scale is used, omit Steps 2, 5, and 8.) 1. With the wire firmly attached to the upper and lower chucks apply sufficient load to the wire to take up any slack. Adjust the leveling screws in the base until the lower chuck hangs freely within the hole.

2. Set the optical lever in place with the rear leg on the lower chuck in such a position that it will not touch the wire as the chuck descends. Set the telescope and scale 1.5 to 2 m in front of the mirror with the telescope at about the same level as the mirror and with the scale vertical. Focus the *eyepiece* of the telescope by moving it in or out until the cross hairs are most distinct (see Appendix). Illuminate the scale by an extension lamp.

The location of the image of the scale through the telescope frequently gives trouble. Sight along a line *over* (*not through*) the telescope by getting near the mirror and placing the eye so that its image is seen in the mirror. Move back behind the position of the telescope and keep in view the image of the eye. Slide the telescope and scale to this place and adjust it so that, sighting along the telescope, the image of the scale is visible. Then look *through* the telescope and focus it on the scale by using the method of eliminating parallax.

3. With the zero load of Step 1 in place read and record the zero reading of the scale. Add the full maximum load of 8 or 10 kg and note whether the lower chuck still hangs freely. If not, make a more careful adjustment of the leveling screws. Remove the added load and observe whether the reading returns to the zero. If it does not, check the supports and chucks to eliminate slippage.

4. Take a series of scale readings as the load is increased in 1-kg increments up to the maximum. Record the readings as the load is successively reduced.

5. Measure the length s of the optical lever. Press it down upon note paper, draw a fine line through the points made by the front legs, and use the vernier caliper to measure the perpendicular distance from the third leg to this line.

6. Use a micrometer caliper to measure the diameter of the wire. Take readings at 8 or 10 different places on the wire and average the readings.

7. Measure the length of the portion of the wire used, that is, from the bottom of the upper chuck to the point at which the lower chuck is fastened to the wire.

8. Measure the distance R from the scale to the mirror. If the substitution method is used for the optical lever, measure the thickness of a disk with a micrometer caliper; insert it under the rear leg and observe the difference in scale readings.

Computation and Analysis: 1. Subtract the zero reading from each successive scale reading to obtain the difference d. Plot a curve of scale difference against load. Discuss the significance of the shape of the curve in showing the relationship between deflection and load. Compute the slope of the curve to obtain the deflection per kilogram. Use this value to compute the ΔL per kilogram by means of Eq. (2) or (4).

2. Calculate the value of Young's modulus for the material of the wire by substituting proper values in Eq. (1). The $F/\Delta L$ of Eq. (1) may be obtained from the result of Part 1, remembering that $F = mg$. Calculate Y in both dynes

per square centimeter and in pounds per square inch. Compare the calculated value with the value for steel found in tables (see Appendix).

Experiment 13.3

MODULUS OF RIGIDITY

Object: To determine the modulus of rigidity of brass and of steel.

Method: A long rod clamped at one end is twisted through a measured angle by a known torque. From these observations and the length and radius of the rod, the modulus of rigidity of the material of the rod is computed.

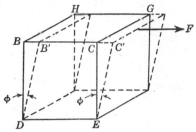

Theory: Consider a cube of material, Fig. 13.5, fixed at its lower face and acted upon by a tangential force F at its upper face. This force causes the consecutive horizontal layers of the cube to be slightly displaced relative to one another. The shearing stress F/A is the ratio of the

Fig. 13.5 Shearing of cubical block

tangential force F to the area A of the upper surface $BHGC$. The strain is measured by the angle ϕ through which each vertical line of the cube is displaced. The shear modulus, or modulus of rigidity, n is given by

$$n = \frac{F/A}{\phi} \tag{5}$$

The angle ϕ expressed in radians is a pure number and the units of n are therefore those of stress; in the cgs system n is in dynes per square centimeter.

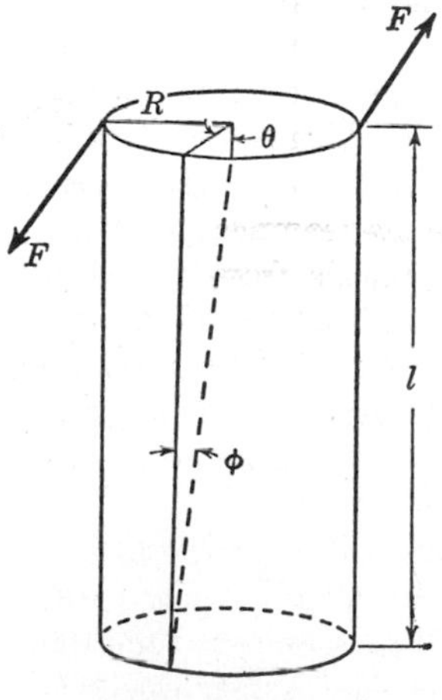

A frequent engineering application of shear is the case in which one end of a solid rod or cylinder is held fast and a torque L is applied at the other end to twist the rod, Fig. 13.6. If each radius of the free end is twisted through an angle θ, each line along the surface is twisted through an angle ϕ. It can be shown that the torque necessary to produce this twist is given by

$$L = \frac{n\pi\theta R^4}{2l} \quad \text{or} \quad n = \frac{2lL}{\pi\theta R^4} \tag{6}$$

where l is the length of the cylinder and R is one-half the external diameter.

Fig. 13.6 Shearing of a cylinder

Apparatus: Torsion apparatus; cylindrical brass rod; two cylindrical steel rods of different diameter; weight hanger; slotted masses; meterstick with caliper jaws; micrometer caliper.

The torsion apparatus, Fig. 13.7, consists of a table clamp, provided with a ball bearing, to which is attached a large wheel. A section of the wheel is graduated in degrees. A vernier arm adjacent to the scale makes it possible to read the position of the wheel to 0.1°. The hub of the wheel contains a socket with a setscrew by means of which the rod is rigidly fastened in a centered position. The other end of the rod is held in a similar socket in a table-clamp tailstock. The wheel has a flat peripheral surface around which passes a steel ribbon carrying a weight hanger. The rods are firmly attached at their ends to brass bushings which fit into the sockets of the table clamps.

Fig. 13.7 Torsion apparatus for determining modulus of rigidity

Procedure: 1. Insert the larger steel rod into the sockets and adjust the clamps until the rod is straight. Tighten the setscrews so that the rod will not slip. Attach the weight hanger and set the vernier so that its zero is near the zero of the scale. Load the hanger until the rod is twisted about 90°. Remove the load and notice whether the wheel returns to the zero reading. If not, one or more of the following conditions may exist: (a) the rod is not firmly clamped at the ends; (b) there is excessive friction; (c) the vernier arm has been moved; (d) the rod is slipping in its bushings. Before proceeding with the experiment correct any condition that prevents the apparatus from returning to its zero reading.

2. Record the zero reading of the scale, add about one-fifth the load necessary to produce a twist of 90°, and record the new reading. Take a series of five such readings with equal increments of load.

3. Repeat Steps 1 and 2 for each of the other rods.

4. By means of the meterstick with caliper jaws measure the length of each rod between the two bushings. Use the micrometer caliper to measure the diameter of each rod at five or six places along the rod. Measure the outside diameter of the wheel by means of the meterstick and caliper jaws.

Computations and Analysis: 1. Compute the angle of twist for each load used. For each rod plot a curve of angle of twist against load. Discuss the significance of the shape of the curves. Calculate the slope of each curve. The reciprocal of the slope is the load per unit angle.

2. From the reciprocal of the slope compute the torque per unit angle L/θ and express this ratio in centimeter-dynes/radian. (Remember that $L = mgr$.) For each rod calculate the rigidity modulus by use of Eq. (6). Compare the calculated values of n for steel and brass with standard values (see Appendix).

3. For each of the two steel rods calculate the ratio L/θ to R^4. Compare the two ratios and explain the significance of the equality.

Review Questions: 1. What is meant when it is said that a body is elastic? 2. State Hooke's law. What is meant by elastic limit? 3. Define modulus of elasticity. What are the cgs and fps units? 4. Define: Young's modulus; modulus of rigidity. 5. Describe the method used for measuring Young's modulus. 6. Explain the use of the optical lever and set up its working equation. 7. Describe the method used to measure the modulus of rigidity.

Questions and Problems: 1. Assuming perfect elasticity, show that Young's modulus for a material is numerically equal to the force that would be necessary to stretch a rod of unit cross section to double its original length.

2. Which will introduce more uncertainty in the determination of Y, an uncertainty of 1.0 mm in the 50.0-cm length of the wire or 0.00010 cm in an elongation of 0.0112 cm?

3. Assuming the optical law that the angle of incidence equals the angle of reflection (for definitions see any elementary textbook), prove that the rotation of a mirror through some angle will rotate the beam of light incident upon it through twice that angle. (For simplicity, start with the light incident normally upon the mirror.)

4. Reduce a modulus of elasticity of 19×10^{11} dynes/cm^2 to pounds per square inch.

5. A wire 60.45 ± 0.05 cm long has an area of cross section of 28.2 ± 0.3 cm^2 and a Young's modulus of $(20.3 \pm 0.2) \times 10^{11}$ dynes/cm^2. Calculate its elongation to the proper number of significant figures when it is stretched by a load of exactly 1 kg.

6. A steel wire 1000 ft long is made up of two parts; one half (500 ft) is $\frac{1}{8}$ in. in diameter and the other half is $\frac{3}{16}$ in. in diameter. The whole is stretched 0.500 in. What is the elongation of each half?

7. In a typical experiment performed to measure Young's modulus of a wire the following data were recorded: length of optical lever, 6.50 cm; diameter of wire, 0.985 mm; length of wire, 90.2 cm; distance from mirror to scale, 400 cm; change in scale reading when load of 3.00 kg is added, 20.5 mm. Determine the value of Young's modulus obtained from these data.

8. A copper wire of $\frac{1}{8}$ in. diameter and a steel wire of equal diameter are placed side by side. Each wire is 1000 ft long and the two wires are jointly subjected to a tensile force of 400 lb. If the values of Y for steel and copper are 30×10^6 and 18×10^6 lb/in.^{2}, respectively, what is the tension in each wire?

9. A 1000-ft length of $\frac{1}{8}$ in. wire is one-half (500 ft) copper, $Y = 18 \times 10^6$ lb/in.^{2}, and one-half steel, $Y = 30 \times 10^6$ lb/in.^{2}. The whole wire is stretched 0.500 ft. What is the elongation of each half?

10. A steel rod is rigidly clamped at one end. At the other end a disk 100 mm in radius is attached. A bearing at the disk end supports the rod. The rod is 100 cm long and 5.00 mm in diameter. What mass must be hung from the circumference of the disk to twist the rod 90°?

11. What must be the diameter of a steel shaft 20 ft long so that the twist will not exceed 2.0° when transmitting a torque of 30,000 lb-ft?

CHAPTER 14

SIMPLE HARMONIC MOTION

In any vibratory motion the net force acting on the moving body is not constant; hence the acceleration of the body varies. The manner in which the acceleration varies characterizes the different types of vibratory motion. The simplest mode of variation is that in which the acceleration is proportional to the displacement. *Simple harmonic motion* (SHM) is the periodic motion in which the acceleration is proportional to the displacement s but opposite in direction. In symbols

$$a = -ks \tag{1}$$

where a is the acceleration and k is a constant.

It can be shown that the period T for SHM is given by

$$T = 2\pi \sqrt{-s/a} \tag{2}$$

Experiment 14.1

SIMPLE HARMONIC MOTION: SPIRAL SPRING

Object: (1) To study Hooke's law of elasticity as applied to a spiral spring. (2) To study simple harmonic motion of translation.

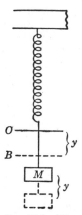

Fɪɢ. 14.1 Displacement of a spring a vertical distance y

Method: 1. The elongation of a spiral spring under the action of successively larger loads is measured. A graph plotted to show the relationship between elongation and load furnishes an example of the linear relation of Hooke's law and enables the experimenter to determine the force constant of the spring.

2. For a series of loads the periods of vibration of a spring are measured. The square of the period is plotted against added mass to show the relationship between mass and period. From the intercept of this curve the fraction of the mass of the spring that is effective in the vibration is determined.

Theory: A common example of SHM is the vibration of a loaded spring that has been displaced from its equilibrium position. When a properly constructed spring is stretched by an applied force, it is found that the elongation of the spring is proportional to the stretching force as long as the elastic limit of the spring is not exceeded. This statement is a form of Hooke's law.

If a force F applied to a spring (Fig. 14.1) produces an elongation y, doubling the force doubles the elongation. The force is proportional to the elongation. In symbols

$$F = Cy \tag{3}$$

100

where C, the *force constant* of the spring, is the force per unit elongation

$$C = F/y \tag{4}$$

The cgs unit of force constant is the dyne per centimeter; the fps unit is the pound per foot. A graph of elongation against force is a straight line for distortions within the limits of perfect elasticity.

When a body of mass M is attached to a spring, such as that shown in Fig. 14.1, the weight Mg causes the spring to be stretched to the position O. Suppose the load is pulled down by an additional force so that the spring is elongated a further distance y to position B. The force necessary to produce this elongation is Cy. If the lower end of the spring is released when in the extended position, the restoring force $-Cy$ produces an acceleration a as given by Newton's second law

$$-Cy = M'a$$
or
$$a = -Cy/M' \tag{5}$$

The force Cy is in a direction opposite to the displacement y as is shown by the negative sign. The mass M' includes not only the load M but also the scale pan, if any, and part of the mass of the spring.

From Eq. (5) it is seen that the acceleration is proportional to the displacement and opposite in direction. Therefore the motion is SHM.

For the vibration of a spring Eq. (5) yields

$$, \quad -y/a = M'/C \tag{6}$$

Therefore, from Eq. (2)

$$T = 2\pi \sqrt{M'/C} \tag{7}$$

where M' is the total mass acted upon by the force of restitution.

The effective mass M' accelerated by the spring is the sum of the mass M of the load, the mass M_p of the scale pan, and some fraction f of the mass M_s of the spring. Theoretically, for a uniform spring, $f = \frac{1}{3}$. Symbolically

$$M' = M + M_p + fM_s \tag{8}$$

Hence from Eq. (7)

$$T = 2\pi \sqrt{\frac{M + M_p + fM_s}{C}} \tag{9}$$

or

$$T^2 = \frac{4\pi^2}{C} (M + M_p + fM_s) \tag{10}$$

Hence the graph of T^2 plotted against M should be a straight line, since all the other factors are constant in a given case. The slope of the line is $4\pi^2/C$; hence, from the observed value of the slope, the spring constant can be computed. When $T^2 = 0$, $M = -(M_p + fM_s)$. Thus, from the negative intercept and the observed values of M_p and M_s, the fraction f may be computed.

Apparatus: The experiment may be performed with either of two alternative sets of apparatus:

1. Medium heavy spiral spring; 50-gm weight hanger; set of slotted masses; stop watch or clock; sensitive triple-beam balance; meterstick; meterstick clamp; right-angle clamp; support rods.

The spiral spring may be any spring that will support a mass of about 500 gm or more. A special spring for this purpose may be used or a door spring or spring from a curtain roller may be substituted.

2. Jolly balance; stop watch or clock; sensitive triple-beam balance; set of accurate brass masses.

The Jolly balance, Fig. 14.2, consists essentially of a vertical helical spring in front of a millimeter scale. The spring may be adjusted vertically and clamped in position. The sliding index is provided with a glass mirror having an etched line against which the image of an indicator disk suspended from the spring is brought into agreement in establishing the zero point and in measuring elongation.

Procedure: 1. Adjust the level of the apparatus so that the scale and spring are both vertical. Hang the pan (or weight hanger) on the spring and if necessary add just enough mass to separate the coils of the spring. The mass of the pan or hanger plus the mass necessary to separate the coils should be recorded as M_p. Read and record the zero position of the indicator. This zero may be either the position of the top of the slider on the scale or the position of the bottom of the weight hanger.

2. Add sufficient load to stretch the spring 5 or 6 cm, adjust the index, and record the added mass and new reading of the scale. Take a series of about eight such readings for uniformly increasing loads. Tabulate the added loads and scale readings.

3. Remove obstructions such as the sliding index. Place the same added load used in Step 1 in the scale pan. Displace the pan vertically a *small* distance and then release it, being careful to avoid any sidewise motion. With a stop watch or clock observe the time of a sufficient number (50 or 100) of *whole* vibrations to give an accurately measurable time interval. In counting the vibrations have each observer count silently and independently. Start counting with *zero* (not one) at the instant the timepiece is started. The period is the total time divided by the number of complete vibrations made during that interval. Take a series of about eight such measurements, using the same increments of mass as in Step 1. The data may be recorded as in the accompanying table.

FIG. 14.2 Jolly balance

Mass in pan gm	Number of whole vibrations	Total time sec	Period sec	Period² sec²
5.00	100	51.8	0.518	0.269

4. Detach the scale pan and indicator (or hanger) and measure its mass by means of the triple-beam balance. If the mass of the spring is not given, measure it also. The spring must be handled very carefully to avoid damage.

Computations and Analysis: 1. Subtract the zero reading of the scale from each succeeding reading of Step 2 to determine the elongations of the spring.

Tabulate the results.   Plot a graph showing the relationship between elongation and added mass.   Discuss the significance of the shape of the curve as a demonstration of Hooke's law.   Find the slope of the curve and its reciprocal.   From the reciprocal of the slope and the acceleration due to gravity, calculate the force constant of the spring.

2. From the data obtained in Step 3, compute and tabulate the periods and the squares of the periods.   Plot a curve showing the relationship between the square of the period and the mass added to the scale pan.   Carefully interpret the significance of the shape, slope, and intercepts of this graph.   Find the slope of the graph and from this value calculate the force constant C of the spring. Compare the value thus obtained with the value previously calculated.   From the intercept compute the value of f, the fraction of the mass of the spring effective in the vibration.   Compare the observed value of f with the theoretical value of one-third.

Experiment 14.2

SIMPLE HARMONIC MOTION: SIMPLE PENDULUM

Object: To study the motion of a simple pendulum as an illustration of simple harmonic motion and to make an experimental determination of the acceleration due to gravity.

Method: The period of a simple pendulum is measured for each of several lengths.   Curves of period against length and square of period against length are plotted.   From the slope of the curve of square of the period against length the value of the acceleration due to gravity is calculated.

Theory: An example of motion that is approximately SHM is the swinging of a simple pendulum.   A *simple pendulum* is a small body, usually a dense sphere, suspended by a cord whose mass is negligible in comparison with that of the sphere and whose length is very much greater than the radius of the sphere.

When the pendulum is displaced from its equilibrium position, the weight of the bob and the tension of the cord produce a resultant force tending to restore the pendulum to the original position.   *If the arc is small*, the motion is approximately SHM and the period depends only upon the length l of the pendulum and the acceleration due to gravity.   In symbols

$$T = 2\pi \sqrt{l/g} \tag{11}$$

This is the fundamental equation of the simple pendulum.   It is valid only for small angles of swing.   For large angles the period is also dependent upon the angle.

When Eq. (11) is squared we obtain

$$T^2 = 4\pi^2 l/g \tag{12}$$

This is the equation of a straight line whose slope is $4\pi^2/g$.   Thus the graph of the square of the period against length may be used to determine g, the acceleration due to gravity.

Apparatus: Simple pendulum; stop watch or clock; meterstick; vernier caliper.

The simple pendulum consists of a metal ball suspended by a light thread from a rigid support (Fig. 14.3).   The support should be sufficiently rigid that no appreciable movement will be imparted to it by the vibration of the pendulum.

Procedure: 1. Measure the diameter of the pendulum bob with a vernier caliper. Make the initial length of the pendulum about 120 cm. Measure the distance from the support to the top of the ball, being careful not to stretch the string in measuring it. Add one-half the diameter of the ball to obtain the length that is tabulated.

2. Start the pendulum vibrating through a small arc, not greater than 5° between extreme displacements. Measure the time required for 50 whole vibrations by means of the stop watch or clock. Count each passage of the bob, *in the same direction*, through the midpoint. Begin the count at *zero* when the timepiece is started. Enter the data as in the accompanying table.

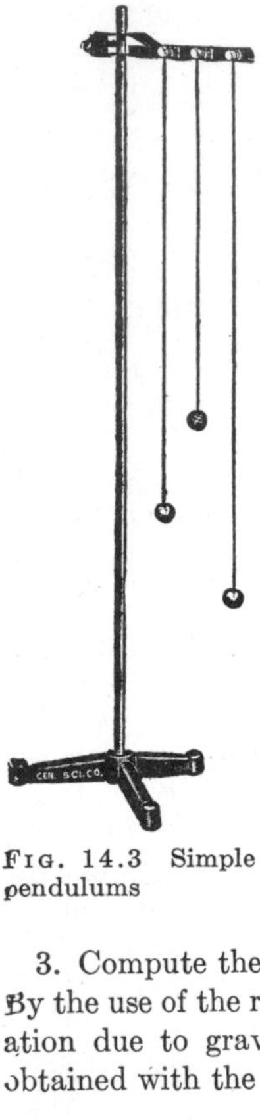

Length	Time of 50 vibrations	Period	Period²
cm	sec	sec	sec²
120	110	2.20	4.84

3. Take a series of six sets of such observations, shortening the length each time by 20 cm. The *angular* displacement must be kept within the limits specified; if the linear displacement is held constant the angular displacement will of course increase.

4. For some convenient length, measure the period when the arc is over 30° and compare it with the period for the 5° arc.

Computations and Analysis: 1. Compute the period and the square of the period for each length of the pendulum and enter the results in the appropriate columns of the table.

2. Plot curves to show (*a*) the relationship between period and length and (*b*) the relationship between the square of the period and length. Discuss the significance of the shape of each curve.

FIG. 14.3 Simple pendulums

3. Compute the slope of the graph of the square of the period against length. By the use of the relationship shown in Eq. (12) compute the value of the acceleration due to gravity from the slope of the graph. Compare the value thus obtained with the standard value for the location of the experiment.

Review Questions: 1. State Hooke's law in general terms; also as applied to a spring. 2. Define force constant of a spring. State its defining equation and the cgs unit. Sketch a curve of elongation against load for a spring. What is the significance of the slope? 4. Define SHM. Cite some examples of SHM. 5. What is a circle of reference? How is it related to SHM? 6. State the general equation for the period of a body executing SHM. 7. State the equation for the period of a vibrating spring. 8. Sketch a curve of period squared against load for a vibrating spring. What is the significance of its slope? of its intercept on the load axis? 9. State the equation for the period of a simple pendulum.

Questions and Problems: 1. In observing the time required for a rather slowly vibrating spring to make a number of oscillations, would it be more accurate to record times when the bob is moving through its equilibrium position or when it is passing one of its extreme positions? Why?

2. How does the period of a vertically oscillating spring vary with each of the following factors: mass of bob; amplitude of vibration; force constant of spring; acceleration due to gravity?

3. Show that both sides of Eq. (10) have the same dimensions.

4. Devise a method for the measurement of g from observations made upon a vertically oscillating spring.

5. Two simple pendulums of exactly the same length have bobs of the same size, one of wood, the other of steel. How will their motions compare? Explain.

6. A block of iron having a mass of 100 gm causes a spring to stretch 39.2 cm. If the spring with this mass attached is set into vertical oscillation with an amplitude of 15.0 cm, what is the period of the motion? Neglect the mass of the spring. What is the maximum speed of the block?

7. A mass of 294 gm is suspended by a band of such elasticity that an additional load of 15.0 gm will stretch it 2.00 cm more. It is extended 2.00 cm and released. Find the period, displacement, velocity, and acceleration 6.20 sec after it passes upward through its equilibrium position.

8. In an experiment on linear SHM, similar to Exp. 14.1, it was observed that an additional load of 12.0 gm would cause the spring to stretch 1.50 cm. The mass of the spring was 4.80 gm and a load of 18.0 gm was attached and set into vibration with an amplitude of 3.00 cm. Calculate (a) the period of the motion and (b) the maximum speed of the load. Assume the factor f to be one-third.

9. Explain how a simple pendulum might be used to assist in geophysical explorations for locating oil.

10. Show why the amplitude of vibration does not appear in the equations for the period of various kinds of SHM.

11. How would the period of a simple pendulum be changed if the pendulum were moved from sea level to (a) the top of a high mountain? (b) to the moon? (c) to the sun? Explain.

12. Compute the length of a seconds pendulum, that is, one whose whole period is 2.00 sec, at a place where the acceleration due to gravity is 980 cm/sec².

13. A simple pendulum 99 cm long requires 9.50 min to make 300 complete vibrations. How long would a simple seconds pendulum be at that place?

14. What is the frequency of a simple pendulum which is 50.00 in. long at a place where $g = 32.16$ ft/sec.²

PART III—FLUID PHYSICS, HEAT, AND SOUND

CHAPTER 15

HYDROSTATIC PRESSURE; ARCHIMEDES' PRINCIPLE

The mass per unit volume of any substance is called the *density* of that substance. The defining equation for density d is

$$d = m/V \qquad (1)$$

where m is the mass contained in the volume V. Some common units of density are gram per cubic centimeter, slug per cubic foot, and kilogram per cubic meter.

The weight per unit volume of a material is called the *weight-density D*. In symbols

$$D = W/V \qquad (2)$$

where W is the weight of the material in volume V. Some units of weight-density are dyne per cubic centimeter, pound per cubic foot, and newton per cubic meter. Because $W = mg$, we have the relationship between density and weight-density

$$D = dg \qquad (3)$$

The *specific gravity* of a substance is the ratio of its density to that of some standard substance. In symbols

$$\text{Sp. gr.} = \frac{d}{d_s} = \frac{D}{D_s} \qquad (4)$$

where the subscript s refers to the standard substance. For solids and liquids the most common standard of comparison is water at 4°C, although water at 15°C is sometimes used; for gases a common standard is air under standard conditions of temperature and pressure, although hydrogen is sometimes used.

Since the numerators and denominators of Eq. (4) have the same units, the quotient, the specific gravity, is dimensionless. The numerical value is thus independent of the system of units used.

The density or the weight-density of any substance may be obtained from the specific gravity by rearranging Eq. (4)

$$d = (\text{sp. gr.})d_s \qquad (5)$$

or
$$D = (\text{sp. gr.})D_s \qquad (6)$$

Because the density of water at 4°C is almost exactly one gram per cubic centimeter, the density of a substance in the cgs system is *numerically* equal to its specific gravity.

Hydrostatic Pressure. *Pressure P* is defined as force per unit area. In symbols

$$P = F/A \qquad (7)$$

where force F acts over an area A. Common units of pressure are the dyne per square centimeter and the pound per square inch.

One type of pressure commonly encountered is that caused by the weight of a

106

fluid. This type of pressure is called *hydrostatic pressure*. The pressure due to a column of liquid is dependent upon the depth h in the liquid and its weight-density D. In symbols

$$P = hD = hdg \qquad (8)$$

Archimedes' Principle. Whenever a body is partly or wholly immersed in a fluid, the hydrostatic pressure of the fluid is greater at the bottom of the body than at the top. Consider a block of rectangular cross section A, Fig. 15.1, immersed in a liquid of weight-density D. On the vertical faces, the liquid exerts horizontal forces, which are balanced on all sides. On the top face the liquid exerts a downward force h_1DA and on the bottom it exerts an upward force h_2DA. The net upward force (buoyant force) on the block is

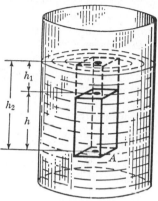

$$h_2DA - h_1DA = hDA \qquad (9)$$

This net upward force is just the weight (volume hA times weight-density D) of the liquid displaced by the block.

It is interesting to note that there is no buoyant force if there is no liquid beneath the body. In Fig. 15.2 the body has been forced to the bottom of the vessel with no liquid beneath. There is a downward force due to the pressure

Fig. 15.1   The net upward force on the block is hDA, the weight of the displaced liquid

of the liquid on the top of the body but no corresponding upward force. Therefore, the liquid merely presses the body against the bottom of the vessel.

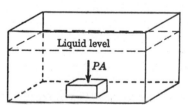

Fig. 15.2   Force on body with no liquid beneath

Archimedes (287–212 B.C.) stated the principle of buoyant forces: *A body partly or wholly immersed in a fluid experiences an upward force equal to the weight of the fluid displaced.*

The principle of Archimedes may be used to determine either (1) the specific gravity of an irregular solid, by immersing it in water, or (2) the specific gravity of an unknown liquid, by immersing a solid, first in water and then in liquid. The measure of specific gravity may be stated.

$$\text{Sp. gr.} = \frac{D}{D_w} = \frac{W_s/V}{W_w/V} = \frac{W_s}{W_w} = \frac{\text{weight of body in air}}{\text{apparent loss of weight in water}} \qquad (10)$$

Experiment 15.1

HYDROSTATIC PRESSURE

Object: (1) To study the variation of pressure with depth in a liquid.   (2) To determine specific gravity and density of liquids by means of hydrostatic pressure.

Method: 1. A cylindrical tube that floats in a liquid is supported by the upward force due to the pressure on the bottom. The pressure is changed by adding weights to the cylinder and the change in depth of the base is noted.

2. Two liquids that do not mix are placed in the arms of a U-tube and the ratio of the densities determined from the ratio of the heights of the free surfaces above the level of the intersurface between the two liquids. For liquids that do mix an inverted U-tube is used.

Apparatus: Hollow uniform cylinder with scale, Fig. 15.3, about 2 in. in diameter and 16 in. long; large hydrometer jar; set of masses; U-tube, Fig. 15.4, inverted U-tube, Fig. 15.5; beakers for liquid samples; oil; distilled water; salt water; meterstick; vernier caliper; hydrometer.

Procedure: 1. Fill the hydrometer jar about half full of water, place the empty cylinder in the water, and add just enough load at the bottom of the tube to hold it upright. Read and record the position of the water surface on the scale of the cylinder. This is the depth h of the bottom of the cylinder for zero load.

2. Add a load of 100 gm to the cylinder and record the new reading of the water surface. Continue to add loads 100 gm at a time until there is an added load of 600 gm. Records the loads and scale readings as in the accompanying table. Measure the diameter of the cylinder by means of a vernier caliper. Weigh the cylinder with its zero load.

ADDED LOADS AND DEPTH OF A CYLINDER IN LIQUID

Added load	Fresh water				Salt water	
	Depth, h	Force, F	Pressure, $P = F/A$	Ratio, P/h	Depth, h'	Ratio, h/h'
gm	cm	dynes	dynes/cm²		cm	
0						
100						
200						
300						
400						
500						
600						

3. Repeat the measurements of depth for the same series of loads using salt water in place of fresh water. Read the specific gravity of the salt solution by means of a hydrometer.

4. If the U-tube is not already filled, pour water into it until it is a little less than half full. Pour oil into one side until the oil level is near the top. Read the level of the surface separating the oil and water, the level of the free surface of the oil, and the level of the free surface of the water.

5. Set up the inverted U-tube, as in Fig. 15.5. Mount two straight glass tubes at least 50 cm long in a vertical position with the lower ends in clean beakers. Connect the upper end of each tube to a Y-tube by rubber tubing.

Connect a rubber tube with a pinch clamp to the third leg of the Y. Fill one beaker with water, the other with salt solution. Open the pinch clamp, draw the liquid about three-fourths the way up the tube, and close the pinch clamp. Measure the level of the liquid in each tube and in each beaker.

Computations and Analysis: 1. From the data of Step 2 calculate the area of cross section of the cylinder. Compute the force F for each load as the combined weight mg of the cylinder, zero load, and added load. Enter these values in the third column of the table. Calculate the pressures from the relation $P = F/A$ and tabulate in the fourth column.

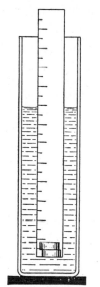

Fig. 15.3 Apparatus for measuring liquid pressure

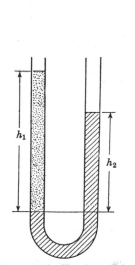

Fig. 15.4 U-tube. Columns of unequal length produce equal pressures

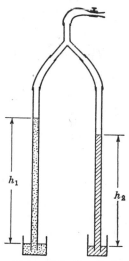

Fig. 15.5 Inverted U-tube

Compute the ratios of P to h. Is this ratio constant? What is the significance of the constancy of P/h? What is the physical significance of the ratio P/h [Eq. (8)]? Plot a graph of P against h. Discuss the significance of the shape of the curve.

2. Compare the depths h' recorded for salt water with the depths h for fresh water. Compute the ratio of h/h' for each observation. What is the significance of this ratio? Compare these ratios with the reading of the hydrometer.

3. From the data of Step 4 set up the equation stating the fact that the hydrostatic pressure due to the column of oil above the common level is equal to the pressure due to the water column above that level. From this equation compute the specific gravity of the oil.

4. From the data of Step 5 compute the length of each liquid column. Write the equation of equal hydrostatic pressures and determine the specific gravity of the salt solution. Compare the value thus obtained with the hydrometer reading and also with the value computed in Part 2 of the Analysis.

Experiment 15.2

DENSITY OF SOLIDS AND LIQUIDS BY DIRECT MEASUREMENT

Object: To determine the density of some solids and liquids by direct measurement of mass and volume.

Method: Several regular solids are accurately measured along all necessary dimensions and their volumes are computed. These solids are weighed on an accurate balance and their densities are computed. Water and several other liquids at known temperatures are weighed in a pycnometer and the densities of the liquids determined. Weighings are made of particles of a solid in air and when taking up part of the volume of the container when the rest is filled with water. The density of the irregular solid is computed from these data.

Apparatus: Cylinders, cubes, and spheres of metal and wood; vernier and micrometer calipers; triple-beam balance; 50°C thermometer; distilled water and other liquids, such as alcohol, acetone, carbon tetrachloride; pycnometer; particles of an irregular solid.

(If carbon tetrachloride is used, be careful not to inhale the fumes or get any in open cuts.)

Fig. 15.6 Pycnometer or specific-gravity bottle for determining the density of liquids

The pycnometer, or specific-gravity bottle, shown in Fig. 15.6, is a small flask with a glass stopper. A capillary opening runs along the length of the stopper and makes it possible to fill the pycnometer completely without leaving a bubble of air in the flask.

Procedure: 1. Record the zero readings of the micrometer caliper and the vernier caliper. Measure the small dimensions of the regular solids with the micrometer caliper; take several readings at different positions and estimate all readings to the closest possible fraction. Measure the large dimensions with the vernier caliper. Record the shape and composition of the solid and the serial number if one is stamped on it.

2. Level the triple-beam balance; set the knife-edge on the agate bearings; adjust the counterpoise until the indicator reads zero. Weigh each of the measured objects on the balance and record the mass to the nearest possible value. In these readings it is not necessary to wait for the balance beam to come to rest. Adjust the riders so that the index swings equally above and below zero.

3. Examine the pycnometer and, if necessary, carefully clean and dry it. Rinse it with distilled water, then with alcohol or ether, and dry it. Record the exact mass of the empty pycnometer and stopper. Fill it with distilled water, replace the stopper, and carefully remove any excess water that runs out. Weigh the pycnometer and the water. Record the temperature of the water. Avoid expansion of the pycnometer due to heat from the hand by picking it up with one or two layers of paper between the fingers and the neck of the bottle. The results of the experiment will have precision only if the temperature of the bottle and the water is the same and if the pycnometer is used at the same temperature throughout the experiment.

4. Pour out the water, dry the pycnometer, and reweigh with other designated liquids. Record the mass of the pycnometer and liquids and the temperature of the liquids.

5. Weigh the pycnometer partly filled with particles of an irregular solid. Weigh the pycnometer with water filling the interstices between the pieces of solid and the remaining volume. Shake the pycnometer to ensure that no air bubbles are trapped. Record the temperature of the water.

Computations and Analysis: 1. From the data of Steps 1 and 2 compute the volume and the density of each of the regular solids. Express all computed values to the correct number of significant figures. Compare the experimental values with standard values for the substances used.

2. From the data of Step 3 compute the mass of water in the pycnometer. By use of Eq. (1) and the table in the Appendix for the density of water at various temperatures calculate the volume of the pycnometer. Calculate the mass and density of each of the liquids used in Step 4. Compare each density with the standard value at the same temperature.

3. From the data of Step 5 determine the mass of water added to fill the pycnometer. This mass is the difference of the two readings. Use Eq. (1) and the density of water to find the volume of water added. Determine the volume of solid as the difference between the volume of the pycnometer and that of the added water. Compute the density of the solid and compare this value with the standard value.

Experiment 15.3

SPECIFIC GRAVITY BY ARCHIMEDES' PRINCIPLE

FIG. 15.7 Hydrometer for measuring the specific gravity of light liquids

Object: (1) To determine specific gravity and density of two solids, one of which is more dense and the other less dense than water. (2) To measure the specific gravity and density of various liquids by means of the Mohr-Westphal balance. (3) To measure the specific gravity of these liquids by means of a hydrometer.

Method: A body is weighed in air and then immersed in a liquid. The apparent loss in weight of the body when immersed in the liquid is, by Archimedes' principle, equal to the weight of liquid displaced by the body. From these measurements, the specific gravity and the density of either the solid body or the liquid may be determined. The specific gravities of various liquids are measured by means of a hydrometer and by the Mohr-Westphal balance.

Apparatus: Triple-beam balance; Mohr-Westphal balance; hydrometer, Fig. 15.7; tall hydrometer jars; beaker; thread; cylindrical metal specimen; solid irregular specimen more dense than water; solid specimen less dense than water; distilled water; liquid specimens; 50°C thermometer.

The Mohr-Westphal balance, Fig. 15.8, consists of a beam B supported on steel knife-edges E. One half of the beam is decimally graduated and supports a glass plummet P for immersion; the other half of the beam is so counterpoised that the beam will balance when the plummet is in air.   Riders are supplied in four values: 1.0, 0.1, 0.01, and 0.001.   The apparent loss of weight of the plummet when immersed in liquid is measured by placing the riders in the notches on the arm until a balance is obtained.   The plummet is so constructed that when it is immersed in water at 15.5°C the unit rider produces a balance when it is placed on the hook.   The proper value of the specific gravity in the respective decimal places is read from the notch in which the rider rests at equilibrium.   If

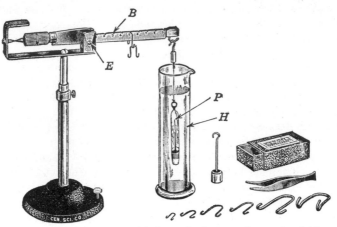

FIG. 15.8   Mohr-Westphal balance, hydrometer jar, and weights

the temperature of the water is not 15.5°C the beam is balanced for water by suitable positions of the weights.   The specific gravity relative to the water used is the ratio of the two readings.

Procedure: 1. Measure with the micrometer caliper the diameter of the metal cylinder at four different places.   Use the vernier caliper to measure the length of the cylinder at four different places.   Weigh the cylinder on the triple-beam balance.

2. Place a beaker of distilled water on the adjustable platform of the triple-beam balance.   Record the temperature of the water.   Suspend the metal cylinder from the hook of the balance by a minimum length of thread.   Arrange the beaker and shelf so that, with the cylinder immersed in the water, none of the weight of the cylinder is supported by the bottom or sides of the beaker.   Record the apparent weight of the cylinder.

3. Repeat the two weighings of Steps 1 and 2 using an irregular solid.

4. Repeat Step 2 with the metal cylinder in a liquid less dense than water and one more dense than water.   Rinse all apparatus in tap water and dry before and after use in each liquid.   When you return a liquid to the stock bottle, be sure that it is the correct one.   Record the temperature of each liquid.

5. Weigh, on the triple-beam balance, the solid specimen less dense than water. Suspend the metal cylinder 2 or 3 cm below the solid specimen and hang both from the hook of the balance.   Adjust the beaker so that the solid specimen is *in air* and the metal cylinder is *in water*.   Record the apparent weight.   Raise the beaker until both solids are immersed and record the new apparent weight.

The difference between these two readings is the apparent loss of weight by the specimen.

6. Pour a liquid from the stock bottle into a tall hydrometer jar. Float the proper hydrometer in the liquid. Read the specific gravity from the hydrometer stem by looking at the scale through the liquid along the under side of the surface. Return the liquid to the proper bottle; rinse and dry the hydrometer and hydrometer jar.

Repeat for a second liquid.

7. Remove the Mohr-Westphal balance from the box and assemble it as shown in Fig. 15.8. Place the leveling screw on the base of the instrument so that it is in position under the arm of the balance. Clean and dry the glass plummet and the small hydrometer jar. Hang the glass plummet by the fine wire from the hook at the end of the balance. With the leveling screw adjust the arm of the balance so that it is horizontal. Partly fill the jar H with distilled water and record its temperature. Immerse the glass plummet in the water and adjust the height of the instrument so that a single strand of fine wire cuts the surface of the water; see that no bubbles adhere to the plummet. Add riders to the beam to restore the balance. If the temperature of the water is 15.5°C a unit rider on the hook (notch 10) should restore balance. At any other temperature another arrangement of riders is necessary. Record the reading of the balance. Empty the jar and dry the plummet and jar. Substitute each of the other liquids in the jar and record the positions of the riders necessary for a balance.

Computations and Analysis: 1. From the data of Step 1 compute the volume and density of the metal cylinder.

2. From the weighings of Steps 1 and 2 compute the apparent loss of weight of the cylinder in water. By use of Eq. (10) compute the specific gravity of the material of the cylinder relative to the water used. From the table in the Appendix determine the density of the water at the observed temperature and compute the density of the metal by use of Eq. (5).

3. In the same manner as in Part 2 calculate the specific gravity and density of the irregular solid specimen. Repeat, using the data of Step 5.

4. From the data of Step 4 calculate the apparent loss of weight of the cylinder in each of the liquids. Use these values with the apparent loss of weight in water previously found to compute the specific gravity of each liquid. Use Eq. (5) to calculate the densities.

5. From the data of Step 7 compute the specific gravity of each of the liquids relative to the water used. This value is the ratio of the reading with the liquid to the reading with water. Compare these values with the specific gravities measured by the hydrometer. From the table in the Appendix determine the density of water at the observed temperature and compute the density of each liquid.

Review Questions: 1. Define density; weight-density; specific gravity. 2. State the observations necessary to measure density by a pycnometer. 3. State Archimedes' principle. 4. Explain how Archimedes' principle may be used to measure the specific gravity of an object lighter than water. 5. Describe the construction and use of the Mohr-Westphal balance; the hydrometer.

Questions and Problems: 1. If the smallest change in mass that the balance will detect is 10 mg, what percentage error does this introduce in Exp. 15.2 in the determination of (a) the density of the heaviest solid and (b) the density of the lightest liquid?

2. What percentage error in the determination of the density of the cylinder in Exp. 15.3 would each of the following introduce: (a) an error of 0.01 gm in the mass? (b) an error of 0.01 mm in the length? (c) an error of 0.01 mm in the diameter?

3. What effect does a rise in temperature have on the density of a solid? the density of most liquids? the density of water at 0°C?

4. State whether each of the following would be apt to produce systematic or random errors in these experiments and explain why: (a) the corners of the cylinder are slightly rounded; (b) the end faces of the cylinder are parallel to each other but are not quite perpendicular to the sides; (c) the temperature of a liquid was 2°C above the temperature of the water when the volume of the pycnometer was determined.

5. Would you be more likely to obtain closer agreement between experimental and standard values with a solid, like copper, which is an element, than one like brass, which is an alloy? With a relatively stable liquid, such as kerosene, or one that takes up water from the air, like glycerin?

6. If, in Exp. 15.1, twice the mass of the granular solid employed had been used, how much would this have increased the precision of the density determination, assuming that all weighings were made with an accuracy of 0.01 gm?

7. A piece of copper whose density is 8.93 gm/cm³ weighs 180 gm in air and 162 gm when submerged in a certain liquid. What is the density of the liquid?

8. A piece of glass of unknown density loses 43.71 gm when weighed in water and 80.36 gm when weighed in concentrated sulphuric acid. What is the specific gravity of the acid?

9. If in the determination of the density of a granular solid, half a cubic centimeter of air is left in the pycnometer when weighed with the solid, will the resulting value of the density be too large or too small? Explain.

10. When placed in a pycnometer, 20 gm of salt displaces 7.6 gm of kerosene. If the density of kerosene is 0.83 gm/cm³, find the volume and density of the salt.

CHAPTER 16

EXPANSION OF GASES

Within a confined gas the actual pressure P may be thought of as consisting of the atmospheric or barometric pressure B plus an added pressure p. The algebraic sign of the added pressure depends upon whether the actual pressure is above and below atmospheric pressure. In symbols

$$P = B + p \qquad (1)$$

where p is positive for pressures greater than atmospheric and negative for those less than atmospheric.

Boyle's Law. The relationship between the pressure exerted by a confined gas and its volume is expressed by Boyle's law, namely: *The temperature remaining constant, the volume V occupied by a given mass of gas is inversely proportional to the pressure P to which it is subjected.* In symbols

$$V \propto \frac{1}{P} \quad \text{or} \quad V = k \times \frac{1}{P}$$

whence
$$PV = k \qquad (2)$$

It is apparent that Eq. (2) is an equation of the second degree, since the left-hand side is the product of two variable quantities. When the pressure is plotted

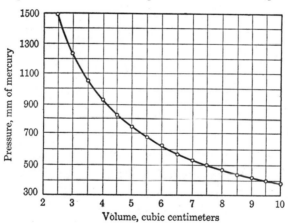

Fig. 16.1   Pressure-volume relationship for gas at constant temperature

as a function of the volume, as in Fig. 16.1, an equilateral hyperbola is obtained.
From Eqs. (1) and (2)

$$(B + p)V = k \qquad (3)$$

or
$$B + p = k \times \frac{1}{V} \qquad (3a)$$

whence
$$p = k \times \frac{1}{V} - B \qquad (4)$$

115

Since B is constant for any given case, Eq. (4) is the equation of a straight line when the variables are added pressure and reciprocal of volume. Such a curve is shown in Fig. 16.2. The slope of the curve is k and the intercept on the p axis is $-B$.

Charles's Law. When the temperature of a confined gas is changed, the gas will change in volume if the pressure is kept constant, or it will exert a different

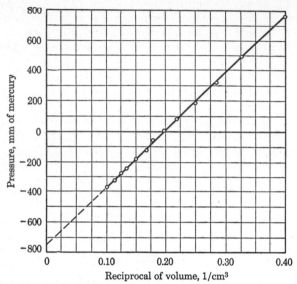

Fig. 16.2 Variation of reciprocal of the volume of a fixed mass of gas at constant temperature, as a function of the pressure above and below atmospheric

pressure if the volume is kept constant. Here we will be concerned only with the pressure change when the volume is kept the same.

Pioneer workers in this field were the Frenchmen Jacques Charles and L. J. Gay-Lussac. The law of pressure change was independently discovered by the two men and is variously known by each of their names. Here the more common usage is followed by referring to it as *Charles's law: The pressure of a gas kept at constant volume increases linearly as the temperature of the gas rises.* In symbols

$$P_t = P_0(1 + \gamma t) \qquad (5)$$

where P_t is the pressure at the temperature t and P_0 is the pressure at some standard initial temperature, usually taken at 0°C. The quantity γ is called the *temperature coefficient of pressure change.* Its defining equation is

$$\gamma = \frac{P_t - P_0}{P_0 t} \qquad (6)$$

or, in words, the temperature coefficient of pressure change of a gas at constant volume is *the fractional change in pressure per unit temperature change, the initial pressure being measured at 0°C.*

The relationship between pressure and temperature is shown graphically in

Fig. 16.3. The slope of the curve divided by P_0 is the coefficient of pressure increase.

It is a fact of great interest that the experimental value of γ for most gases turns out to be approximately $\frac{1}{273}$ per °C. This fact means that for every centigrade degree change in temperature above or below 0°C, the pressure changes by $\frac{1}{273}$ the pressure that the gas exerts at 0°C (the volume being kept constant). Hence if the temperature were lowered by 273°C, the change of pressure would be $273 \times \frac{1}{273}$ of P_0 or the *change* of pressure would equal the initial pressure at 0°C and the final pressure would be zero! This irreducible minimum of temperature is called *absolute zero*, that is, the temperature of an ideal gas at which molecular activity ceases and the pressure consequently is zero. Its value may be roughly checked by experiment by extrapolating (projecting beyond the measured values) the observed pressure-temperature curve until it intersects the

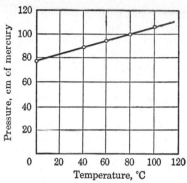

Fig. 16.3 Pressure against temperature, in degrees centigrade, for dry air at constant volume

axis of zero pressure, as in Fig. 16.4. This should occur at a place where $t = -273°C = 0°K$. (Degrees Kelvin are the units of temperature on the absolute scale of temperature.)

A careful distinction should be drawn between a *linear relation* and a *direct proportion* in the present and many similar cases. The pressure here *varies*

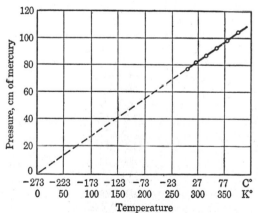

Fig. 16.4 Pressure against absolute temperature, in degrees Kelvin, for dry air at constant volume

linearly with the temperature *in degrees centigrade*, as indicated by Eq. (5) and Fig. 16.3. It is also true that the pressure is *directly proportional* to the temperature *in degrees Kelvin* (absolute), but the pressure is *not* directly proportional to the temperature expressed on centigrade or Fahrenheit scales. This lack of proportionality may be more clearly seen by substituting the value $\frac{1}{273}$ for γ in Eq. (5) and obtaining

$$P_t = P_0\left(1 + \frac{t}{273}\right) = P_0\left(\frac{273 + t}{273}\right) = \frac{P_0}{273}\,T = cT \qquad (7)$$

where T is the temperature in degrees Kelvin and c is (numerically) a constant for any given case. From Eq. (7) and Fig. 16.4 it is apparent that the pressure is directly proportional to the temperature only when the temperature is measured on the absolute scale.

Experiment 16.1

BOYLE'S LAW

Object: To study Boyle's law, at moderate pressures above and below atmospheric, by both analytical and graphical methods.

Method: A fixed mass of air confined in a glass tube is kept at room temperature and subjected to various pressures, ranging from half to double atmospheric pressure. A series of corresponding pressures and volumes are observed and Boyle's law is checked by noting the constancy of their products. The data are plotted in several graphical forms, the interpretation of which also indicates the validity of Boyle's law.

Apparatus: Boyle's-law apparatus; barometer.

The Boyle's-law apparatus is shown in Fig. 16.5. The mass of air on which the measurements are made is confined in the calibrated glass tube T, which has a stopcock K at its upper end. The stopcock tube T and another glass tube T' form the opposite ends of what might be designated an adjustable U-tube. The glass tubes are connected through suitable metal couplings by means of heavy, flexible tubing and this adjustable U-tube is filled with the proper amount of mercury. By means of the metal couplings C and C' the glass tubes are supported in clamps, which can be moved vertically on the support rods G and held in any desired position. Midway between the two support rods is the vertical millimeter scale B, graduated on metal. By means of this scale and a suitable reading device R, the height of mercury column in either of the two glass tubes may be ascertained within 0.1 mm.

The reading device consists of a sleeve that slides freely along the graduated square tube and to which is attached the mirror M and the vernier scale V. A fine horizontal line etched on the mirror permits setting without parallax on the mercury columns, and after the setting has been made the reading is taken by means of the vernier scale.

The air in the closed tube must be dry. It may have been permanently dried in advance. If so the closed tube should never be opened. If not, introduce dry air into the tube by connecting a drying tube to the stopcock tube and running the mercury up and down in the tube, pumping dry air in and out and thereby removing water vapor. Finally, set the level of the mercury where the volume of entrapped air is about one-half the total volume of the closed tube when the level of the mercury is the same in both tubes. Close the stopcock and *keep it closed* throughout the progress of the experiment. (While adjusting the stopcock, steady the closed tube with the hand; otherwise the tube may be snapped off.) Fasten the stopcock with a rubber band. Keep the stopcock well lubricated with stopcock grease.

When the tubes are turned in toward the mirror, great care must be observed not to strike them against the top of the support.

Procedure: 1. Read and record the barometeric pressure (see directions in the Appendix). This reading should be repeated at the end of the experimental

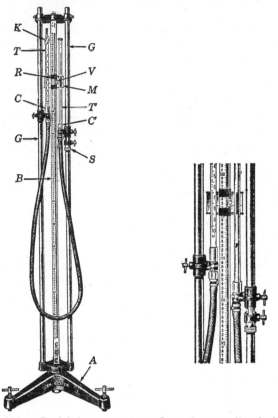

Fɪɢ. 16.5   Boyle's-law apparatus.   Inset shows reading device

work to determine whether the barometric pressure has changed during the experiment.

2. Place the Boyle's-law apparatus in good light where the scale may easily be read.   Bring the mercury in the two tubes to nearly the same level; then open the stopcock.   Level the apparatus by adjusting the leveling screws in the base until the mercury levels in the two tubes coincide with the horizontal line on the indicator.   Test the system for leaks by lowering the open tube or raising the closed tube to decrease the pressure as far as the apparatus will permit.   Allow it to remain in this condition for a few minutes and note whether there is any change in the mercury levels.   If they remain constant, proceed with the experiment.

3. With the pressure in the closed tube as low as the apparatus will permit take readings of the volume of the air and the levels of the tops of the two mercury

surfaces. In tabulating the data, record the following: (a) R_o, the reading of the mercury level in the open tube; (b) R_c, the reading of the level in the closed tube; (c) the "added" pressure p, that is, $R_o - R_c$; (d) the actual pressure $B + p$; (e) V, the volume; (f) $1/V$; (g) the product PV; (h) percentage difference between the observed PV and the mean of all the values.

Take a series of 10 or 12 readings at various pressures ranging from the lowest to the highest attainable. In varying the pressure set the mirror index on the closed tube at $\frac{1}{2}$ cm³ reduction in volume and raise the open tube until the pressure is properly adjusted to give the desired volume. When the open tube is near the top of the frame, the closed tube may be lowered to produce the same effect as further raising of the open tube.

Make all changes slowly, to avoid changing the temperature, and wait a short time before taking the readings. Do not handle the closed tube after the preliminary adjustments are completed. Why? While changing the positions be careful to avoid spilling the mercury out of the open tube. When your observations are complete, bring the tubes back to the position of equal level.

Computations and Analysis: 1. From the data of Step 3 compute and tabulate the following: the added pressure p, the volume V, the reciprocal of the volume $1/V$, the products of PV and their mean value, and the percentage variation of the PV's from the mean. What is the physical significance of the constancy of the various values of PV?

2. Plot a graph of pressure P against volume V, beginning each scale at zero. What is the apparent shape of the curve? What is the significance of this shape?

3. Plot a curve of added pressure p against reciprocal of volume $1/V$. Choose the axis of p near the center of the page and be sure that $-p$ extends as far as 770 mm below the axis. In laying off the scale for $1/V$, begin with $1/V = 0$ at the origin. Discuss the significance of the shape of the curve. Determine the barometric pressure by extrapolating the observed part of the curve (use a dotted line) back to the intercept where $1/V = 0$. Compare this pressure-intercept value of B with the value observed on the barometer.

Experiment 16.2

BOYLE'S AND CHARLES'S LAWS

Object: To study Boyle's law and Charles's law, as applied to air at moderate temperatures and pressures.

Method: To study Boyle's law, a fixed mass of air confined in a glass tube is kept at room temperature and subjected to various pressures, ranging from half to double atmospheric pressure. A series of corresponding pressures and volumes are observed and Boyle's law is checked by noting the constancy of their products. The data are plotted in several graphical forms the interpretation of which also indicates the validity of Boyle's law.

Charles's law for the expansion of gases is studied by the use of a simple form of constant-volume air thermometer. A fixed volume of dry air is subjected to certain measured temperatures and the corresponding pressures are observed. From the resultant pressure-temperature curve the temperature coefficient of pressure increase at constant volume is determined. By extrapolation of this curve the value of "absolute zero" is approximately measured.

Apparatus: Boyle's- and Charles's-law apparatus; barometer; 100°C thermometer; steam generator; Bunsen burner; two metal vessels for water and ice; pinch clamp; rubber tubing.

The Boyle's- and Charles's-law apparatus, Fig. 16.6, is designed so that it may be used to study independently either Boyle's law or Charles's law. When Boyle's law is being studied, the metal-bulb reservoir shown at the left is tightly sealed off by closing a needle valve provided for that purpose and hence this portion of the apparatus may thereafter be ignored. The Boyle's-law apparatus proper consists of two vertical glass tubes, one open at the top and the other closed by a stopcock, held at the lower end by stuffing boxes in an iron reservoir. The reservoir is mounted on a tripod base and is provided with a screw-operated diaphragm for varying the height of the mercury in the tubes and so changing the pressure and volume of the gas confined in the closed tube. The large, milled-head screw has a small pitch and it presses against the corrugated steel diaphragm which forms one side of the mercury reservoir. Readings of the mercury levels to measure the corresponding pressures and volumes are taken from a metric scale mounted vertically between the tubes. A sliding glass cursor provided with a horizontally etched hair line makes it conveniently possible to read the mercury levels with satisfactory precision. The air admitted to the closed tube should be carefully freed from moisture by a method described later. The apparatus may be leveled by having the tripod adjusted on the apparatus so that the leg having the leveling screw is in the plane of the glass tubes. A slight rotation of the screw will then suffice to level the apparatus.

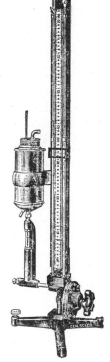

The Charles's-law portion of the apparatus consists essentially of a metal bulb to contain the gas (air) under investigation, connected by a short section of glass capillary tubing to the mercury reservoir and manometer previously described. The volume of the gas is kept constant by adjusting the mercury level with the screw until the level is always

Fig. 16.6 Boyle's- and Charles's-law apparatus

brought to a fixed line etched on the glass tubing beneath the gas bulb. A copper jacket surrounds the bulb proper; it is provided with tubulures so that heating water may be added or removed as desired to control the temperature of the confined air. The water may be brought to and maintained at any desired temperature by an electric immersion heater, or hot or cold water may be introduced from outside.

The reading of the pressure has been facilitated by a horizontal line placed on the meterstick at the side of the open tube; the line is at the same height as the etched line on the tubing of the gas bulb. When the mercury level on the closed tube is adjusted to this etched line, the actual pressure on the gas is merely the barometric pressure plus the difference between the height of the mercury at the top of the open tube and the height of the index line.

Since Boyle's and Charles's laws do not hold for vapors it is essential that the air introduced into the gas bulb be perfectly dry.

I. Boyle's Law

Procedure: *Never allow the level of the mercury in the tubes to come below the lower end of the meterstick, as to do so will often allow air from the reservoir to enter the closed tube and thus to vitiate the results of the experiment. During the experiment, never adjust or open the stopcock on the closed tube, as to do so will change the mass of air, admit moist air, and make necessary a complete drying of a new mass of air.*

Before beginning the experiment the instructor ordinarily will have filled the closed tube with air which is free from moisture. If not, introduce dry air into the tube by connecting a drying tube to the stopcock tube and running the mercury up and down in the tube, pumping dry air in and out and thereby removing water vapor. Finally, tightly close the stopcock tube at a place where the volume of the entrapped air is about one-half the total volume of the closed tube when the level of the mercury is the same in both tubes.

1. Close tightly the needle valve which seals off the Charles's-law gas bulb from the Boyle's-law apparatus. Place the apparatus in good light where the scale may easily be read.

Test for leaks by turning the milled head until the mercury is near the top of the open tube and observing it for a few minutes to see that the level remains constant. Check this also by having the mercury in the open tube near the bottom of the tube. To be sure that no air bubbles are present, turn the adjusting wheel back and forth several times to move the mercury in the open tube from the top to the bottom, meanwhile watching both tubes for bubbles.

As the volume of the closed tube is not calibrated directly in cubic centimeters, the volume of the gas will be measured here in terms of the *length* of the tube above the mercury since the volume is proportional to the length for a uniform bore. Since it is impossible to seal the glass stopcock at the end of the tube and still have a uniform bore right up to the barrel of the stopcock, an etched line on the tube is placed at such a point that the volume in the capillary between that line and the stopcock barrel represents exactly the volume of 1 cm of uniform capillary bore. Hence the corrected scale reading for the top of the enclosed air column is obtained by adding 1.00 cm to the reading of the scale just opposite the etched line on the tube below the stopcock. The corrected value will be designated R_t.

2. Set the barometer and record the reading (see Appendix). Repeat this reading at the end of the experimental work to determine whether the barometric pressure has changed during the course of the experiment.

3. When the apparatus is properly adjusted take a series of 10 or 12 readings of corresponding pressures and volumes, choosing the values so that the open-tube readings vary by about 70-mm intervals over the entire scale. After the pressure has been changed allow the air to come to room temperature before taking the reading. Take one reading when the mercury levels are the same in the open and closed tubes. Tabulate the following: (*a*) R_o, the reading of the open-tube mercury level; (*b*) R_c, the reading of the closed-tube mercury level; (*c*) p, the added pressure; (*d*) P, the actual pressure, or $B + p$; (*e*) V, the volume or $R_t - R_c$; (*f*) $1/V$; (*g*) PV; (*h*) percentage variation of PV from the average values of all PV's.

Computations and Analysis: 1. Calculate the added pressure p for each setting. Note that when the closed-tube readings exceed those on the open tube the confined gas is below atmospheric pressure and the values of p are negative. Compute and tabulate the actual pressure P, the volume V, the reciprocal of the volume $1/V$, the products PV of pressure and volume, and the percentage variation of the PV's from the mean. What is the significance of the constancy of PV?

2. Plot a curve of added pressure p against reciprocal of volume $1/V$. Choose the axis of p near the middle of the page and be sure that $-p$ extends as far as 770 mm below the axis. In laying off the scale for $1/V$ begin with $1/V = 0$ at the origin. Carefully interpret the significance of the curve. Determine the barometric pressure by extrapolating the observed portion of the curve (use dotted line) back to the intercept where $1/V = 0$. Compare this pressure-intercept value of B with that observed by means of the barometer.

II. Charles's Law

Warning! Whenever the temperature is changed rapidly it is necessary to guard against quick changes in the volume of the gas. As steam enters the jacket, the mercury in the gas tube will descend and may go below the lower end of the meterstick and admit air into the bulb. This action should be prevented by raising the pressure to keep the level of the mercury on the gas-bulb tube at the index line. Throughout the entire experiment one observer should continually watch the pressure and keep adjusting the screw to maintain the mercury levels at the desired values. When ice is added, or hot water removed, or steam discontinued, the pressure in the bulb will greatly decrease and the mercury might run into the closed bulb if the adjusting screw were not manipulated to lower the level of the mercury in the closed tube.

Procedure: 1. Attach a short rubber tube to the tubulure at the base of the jacket and close it by means of a pinch clamp. Fill the jacket with a mixture of chipped ice and water. After equilibrium is attained, adjust the pressure until the mercury is brought to the line etched on the glass in the open part of the metal tubing. Record the level I of this index. Record the level M of the mercury in the open tube.

2. Replace the ice and water by water at about room temperature. The ice may be melted and the temperature raised by means of an electric heater or by introducing steam from the steam generator. Adjust the screw to bring the mercury back to the index and record the new level in the open tube. Repeat this process at approximately 20° C intervals until the water is near the boiling temperature. A final convenient value is the one in which steam is passed to the upper tubulure from the steam generator. Open the pinch clamp at the lower tubulure and provide a suitable vessel to catch the condensed water vapor. The temperature of the steam is best obtained from the table of boiling points against pressure, found in the Appendix.

Computations and Analysis: 1. For each temperature compute and tabulate the actual pressure from the relationship

$$P = B + (M - I)$$

2. Plot a curve to show the variation of pressure with temperature in degrees centigrade. Explain clearly the significance of the shape of this curve. Deter-

mine the slope of the curve, choosing points near the ends, and divide the slope by P_0 to obtain the value of γ, the temperature coefficient of pressure increase. Find the percentage difference between the experimental value and the standard value, 0.003663 per °C.

Plot a curve from the observed data to show the variation of pressure with temperature, starting the pressure scale at zero and the temperature scale with -273°C (0°K). Include on the temperature scale the values in both degrees centigrade and degrees Kelvin. Extrapolate by a dotted line the observed data to the axis of zero pressure. Compare this temperature intercept with the standard value for absolute zero. Explain the significance of this curve.

Review Questions: 1. State Boyle's law. Is it rigorously exact? Give the law in the form of symbolic equations. 2. What is the form of a curve of P against V? Of a curve of "added" pressure (above or below barometric) against $1/V$? State the significance of the p intercept of the latter curve. 3. State Charles's law. By what other name is it known? 4. Define the temperature coefficient of pressure variation of a gas at constant volume. State its defining equation. Discuss the significance of its numerical value. How does it vary for different gases? 5. What is "absolute zero"? 6. Sketch curves to show the variation in the pressure of a gas as a function of temperature in degrees centigrade; in degrees Kelvin. Write the equations for these curves. 7. Describe the apparatus and the technique used for studying Boyle's law; for studying Charles's law.

Questions and Problems: 1. Explain fully why it was necessary to wait several minutes after changing the pressure before taking readings.

2. Show by dimensional reasoning that the constant k in the equation $PV = k$ is *not* a mere numerical constant, that is, one having no unit, but that it has the dimensions of work. On what does the numerical value of k depend?

3. Considering the accuracy of your observations of the volume in the Boyle's-law experiment, was it necessary to measure the pressures to tenths of a millimeter in order to secure accurate values of PV? Would more precise readings of the mercury levels have given more accurate values of PV?

4. What would be the effect of performing the Boyle's-law experiment at a higher (but fixed) temperature? What would the curve of p against $1/V$ look like if the room temperature steadily increased during the progress of the experiment?

5. The gas bulb of the Charles's-law apparatus is made of cast iron. What happens to its volume when the temperature is changed from 0°C to 100°C? What effect does this have upon the "constant-volume" assumption? Is the error serious? Why?

6. In the Charles's-law apparatus as used, a portion of the air bulb is connected to the mercury reservoir by a glass capillary tube, the level of the mercury always being brought to a fixed point on this tube. Hence some of the air above this line but below the metal bulb is not subject to the same temperature as the air inside the bulb. Will this introduce a serious error into the results? Why?

7. The level of the mercury in the open arm of a Charles's-law apparatus stood 5 cm below the index when the volume bulb was surrounded by ice water and 20 cm above the index when the bulb was surrounded by steam at normal barometric pressure. Calculate the temperature coefficient of pressure variation at constant volume from these data.

8. The pressure of a gas was measured as 100.0 cm of mercury at 50°C and 114.3 cm of mercury at 100°C, volume constant. What value of absolute zero is obtained from these data?

CHAPTER 17

LINEAR EXPANSION

The change in length per unit length per degree rise in temperature is called the *coefficient of linear expansion*. In symbolic form

$$\alpha = \frac{L_t - L_0}{L_0 \, \Delta t} \tag{1}$$

where α is the coefficient of linear expansion, L_0 and L_t are the initial and final lengths, respectively, and Δt is the change in temperature. Logically the initial temperature should be a fixed standard, such as 0°C; however, because the value of α is very small for solids, the error introduced by using any other initial temperature is not large.

The change in length and the total length are always expressed in the same units; the value of the coefficient is therefore independent of the length unit used but depends on the temperature unit. The value of the coefficient of expansion should be specified as "per degree centigrade" or "per degree Fahrenheit." If ΔL represents the change in length of a metal bar, then

$$\Delta L = \alpha L_0 \, \Delta t \tag{2}$$

The value of α may be found from

$$\alpha = \frac{\Delta L}{L_0 \, \Delta t} \tag{3}$$

Experiment 17.1

COEFFICIENT OF LINEAR EXPANSION

Object: To measure the coefficients of expansion of several metals.

Method: A rod of a common metal is encased in a metal jacket. Its length is measured at room temperature. The change in length is measured when the temperature is raised from room temperature to the temperature of steam. From these observations the coefficient of linear expansion is computed.

Apparatus: Linear-expansion apparatus; steam generator; beaker; 100°C thermometer; rubber tubing; Bunsen burner; rods of various materials; meterstick.

I. *Micrometer-screw Apparatus.* The apparatus shown in Fig. 17.1 is designed for measurement of the increase in length of the rod by means of a micrometer screw. The screw S rests firmly against one end of the rod R in jacket J. Once adjusted and a reading taken of the position of the micrometer screw M at the other end of the rod, screw S must not be moved because it is the datum for the increase in length. An electric circuit is used to indicate the instant that the tip of screw M makes contact with the end of the rod by lighting bulb L (or by the deflection of a voltmeter). From the connection at D, either tap water from faucet F or steam from generator G may be run around the rod. The jacket is held in place by two screws AA', Fig. 17.1a. If these screws are tightened too firmly and the rod fits the corks in the jacket ends too closely, the rod is prevented from expanding and bows inside the jacket. Too low a value of expansion is then obtained at the micrometer screw M.

Procedure: 1. Fill the steam generator two-thirds full of water and light the Bunsen burner under it. Do not let the steam emerge near the apparatus.

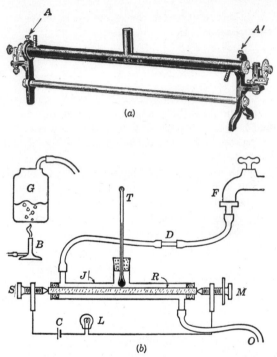

Fig. 17.1   Linear-expansion apparatus, micrometer-screw type

2. Measure the length of the rod carefully with the meterstick and caliper jaws. Record the length of the rod at room temperature and the material of which it is composed. Insert the rod in the jacket through the end corks and place the jacket in the frame. Tighten the top screws just enough to hold the jacket in place with opening O at the bottom. Lead the tubing from O into the sink or into a beaker well below the level of the apparatus.

3. Adjust screw S so that it rests firmly on the center of the flat end of the rod. Check the micrometer scale to be sure that you have the correct units for the main scale and the circular scale. Note whether the values increase or decrease as the screw is turned inward. Turn the screw until contact is established and the bulb just lights. See that the tip strikes the rod squarely. Back off the micrometer screw. Place the thermometer T in the top cork and twist it down with the cork in the opening until the bulb is close to the rod but not touching it. Allow water to flow slowly through the jacket until a steady temperature is reached.

4. Turn the micrometer screw until the bulb just flashes. Record the readings

of the micrometer screw and the thermometer. Turn off the water and drain the jacket. Turn back the micrometer screw about two full turns.

5. Connect the tube from the steam generator to the upper opening; be careful not to scald your hands. Let the steam issue freely through the jacket for about five minutes. Turn the micrometer screw up to contact and record the reading. Back off the screw and repeat in another minute. If the rod is still expanding, repeat at frequent intervals until it reaches its maximum reading. If the steam is passed through too long, the framework expands and the maximum increase of the rod is not measured.

6. Read the barometer and record the reading. Look up in the Appendix, in the table of water-vapor pressures, the corresponding boiling point and use it for the final temperature.

7. Repeat Steps 2 to 5 with a rod of different material.

8. Disconnect the apparatus, empty the steam generator and beaker, mop up the water, and leave everything in neat shape.

II. *Optical-lever Apparatus.* In the optical-lever apparatus (Fig. 17.2) the small increase in length of the rod is measured by means of an optical lever. The test rod R is enclosed in a jacket J. Water or steam may be passed through the jacket to control the temperature. A glass rod G rests on the iron base and serves as a fixed position for the lower end of the test rod. All the expansion of

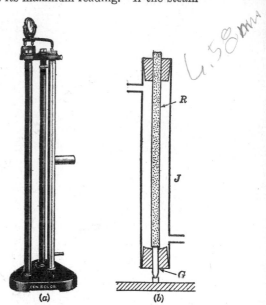

FIG. 17.2 Linear-expansion apparatus, optical-lever type

the rod is upward. The foot of the optical lever rests on the upper end of the test rod and moves upward as the rod expands. The theory and method of operation of the optical lever is described in Exp. 13.2.

Procedure: 9. Measure the length of the metal rod at room temperature.

10. Insert the rod into the jacket. The lower end of the rod must make firm contact with the glass stop at the base. Test for good contact by tapping the upper end of the rod *lightly* with a small coin. A sharp metallic sound indicates good contact but a dull click indicates poor contact. A conical shape of the upper surface of the rubber stopper that holds the glass rod helps get the rod in proper position. The rod must slip readily through the upper rubber stopper; otherwise the rod may bow and fail to show the true expansion.

11. Set the optical lever on the apparatus with the movable leg on the flat end of the rod. Set up and adjust the telescope and scale as directed in Exp. 13.2. Measure and record the distance from the mirror to the scale.

12. Slip the thermometer through the stopper of the middle tubulure and adjust the combination until the stopper is tight and the bulb of the thermometer is near but not touching the rod. Allow water to flow slowly through the jacket until a steady temperature is reached. Record the reading on the scale of the optical

lever and the reading of the thermometer.   Turn off the water and drain the jacket.

13. Connect the tube from the steam generator to the upper opening; be careful not to scald your hands.   Let the steam issue freely through the jacket for 5 min. Observe the reading of the scale of the optical lever.   Allow the steam to flow for another minute and again observe the scale reading.   If the rod is still expanding observe the scale frequently until the reading becomes stationary.   Record this reading.

14. Read the barometer and record the reading.   Look up in the Appendix in the table of water-vapor pressures the corresponding boiling point and use it for the final temperature.

15. Repeat Steps 9 to 13 with a rod of different material.

16. Disconnect the apparatus, empty the steam generator and beaker, mop up any spilled water, and leave everything in neat shape.

Computations and Analysis: 1. From the difference in micrometer-screw readings, or the difference in scale readings of the optical lever, determine the change in length of each rod.

2. From the initial and final temperatures, record the temperature difference.

3. Calculate the coefficient of linear expansion by the use of Eq. (3).   Compare this value with a standard value, taken over about the same temperature range (see Appendix).   Discuss reasons for any difference.

Review Questions: 1. Define the coefficient of linear expansion.   State its defining equation.   2. What are the common metric and British units of the coefficient of linear expansion?   How are they related to each other?   3. Describe a method of determining the coefficient of linear expansion for a metal rod.

Questions and Problems: 1. What percentage error was introduced in this experiment by substituting the length of the rod at room temperature for the length at 0°C?   Did this introduce a significant error into the experiment?

2. Could the method of this experiment be used for a glass rod?

3. A glass baking dish has a coefficient of linear expansion of 8.0×10^{-6} per °C. When the dish is taken from ice water and put into an oven at 160°C, it expands 0.0386 cm.   How long was the dish before putting it in the oven?

4. What must be the length of an iron bar in order that it may expand 1.00 mm on being heated from 0 to 100°C?

5. The steel parts of the San Francisco Bay bridge are about 5.0 mi long.   On a hot day when the temperature is 35°C, how much longer is the bridge than on a day when the temperature is at 0°C?

CHAPTER 18

VAPORS; HUMIDITY

Vaporization is the change of a substance into the gaseous state. Molecules are continually leaving the surface of a liquid and forming a vapor; the vapor molecules are continually striking the liquid surface and entering it again. In a closed vessel an equilibrium state is reached in which the number of molecules returning to the liquid is equal to the number leaving it in the same time. The vapor is then said to be *saturated*.

The molecules of vapor above a liquid exert a pressure. The pressure of the saturated vapor is characteristic of the substance and the temperature but independent of the volume of the vapor.

When there is a mixture of gases, such as in air, each gas or vapor contributes its partial pressure to the whole. The contribution of a vapor cannot ordinarily be greater than the saturated vapor pressure at the temperature of the mixture. When the partial pressure of the vapor is the saturated vapor pressure the mixture is said to be *saturated*.

Water is at all times present in the atmosphere in the solid, liquid, or vapor form. If the air is not saturated, it can be made so either by adding more water vapor or by reducing the temperature until the vapor already present will produce saturation. The temperature to which the air must be cooled, at constant pressure, to produce saturation is called the *dew point*.

The mass of water vapor per unit volume of air is called the *absolute humidity*, and it is expressed in grains per cubic foot (7000 gr = 1 lb) or in grams per cubic meter. The mass of water vapor per unit mass of air is called the *specific humidity*. The ratio of the actual vapor pressure to the saturated vapor pressure at that temperature is called the *relative humidity* and is expressed as a percentage. At the dew point the relative humidity is 100 per cent.

The relative humidity of the air at a certain location and a definite time may be determined by two methods. The *dew-point method* depends on the fact that, because of local cooling, condensation will appear on a surface cooler than the dew point of the atmosphere. The pressure of a vapor is proportional to its density and the pressure of saturated water vapor depends only on the temperature. If the pressure of saturated water vapor at room temperature t is p and the pressure of saturated water vapor at dew-point temperature t_d is p_d, then the relative humidity r is given by

$$r = p_d/p \tag{1}$$

The *wet-and-dry-bulb method* depends on the fact that the rate of evaporation into an atmosphere is a function of the amount of water vapor already present. A current of air should pass over the bulbs at a rate of at least 3 m/sec. From empirical data there has been established the equation

$$e = e' - 0.000367 P(t - t') \left(1 + \frac{t' - 32}{1571} \right) \tag{2}$$

129

in which t and t' are the temperatures of the dry-bulb and the wet-bulb thermometers, respectively, expressed in Fahrenheit degrees, P is the barometric pressure in inches of mercury, e' is the pressure of saturated water vapor at the temperature t' of the wet bulb, and e is the pressure of the water vapor corresponding to the observed temperature t. The Weather Bureau has computed values of e for all feasible values of depression of the wet bulb $t - t'$ and arranged them into tables from which either relative humidity or dew point may be obtained for all the usual values of barometric pressure.

Experiment 18.1

HYGROMETRY

Object: To determine the dew point, vapor pressure, absolute humidity, and relative humidity by several methods.

Fig. 18.1   A hygrodeik

Method: Various types of dew-point and wet-bulb hygrometers are used to determine the pressure of water vapor in the room. From tables of saturated vapor pressure and relative humidity, values are obtained for the dew point, relative and absolute humidity, and mass of water vapor present in the air.

Apparatus: Several forms of hygrometers; ether; 50°C thermometer; polished calorimeter cup; cloth; hygrometric tables.

Hygrometers may be divided into three classes: (1) those in which the actual weight of water present in a measured volume of air is chemically determined; (2) those in which the dew point is determined directly; and (3) those in which the difference in readings of a dry-bulb and a wet-bulb thermometer is utilized.

Figure 18.1 shows a hygrodeik in which wet- and dry-bulb thermometers are mounted along a chart upon which are plotted curved lines. The place where the curved lines from each thermometer cross gives the relative humidity. The dew point and the absolute humidity may then be obtained from tables. Figure 18.2 shows a Mason's hygrometer for which a current of air is passed over a bulb kept moist by a wick. The depression of the wet bulb with respect to the adjacent dry bulb is used to obtain the dew point and the humidity. The sling psychrometer, Fig. 18.3, works on the same principle and the values of dew point and humidity are obtained from the same tables. The bulb is moistened and the two thermometers are swung through the air to produce the lowering of the wet-bulb reading by evaporation.

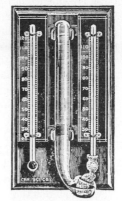

Fig. 18.2 Mason's hygrometer

The dew point may be directly obtained by reading the temperature of cold water in a metal cup on the outer surface of which a light mist forms. In the Alluard hygrometer, Fig. 18.4, a volatile liquid is placed in a polished metal vessel and air is slowly bubbled through the liquid. When the evaporation of the liquid has lowered the temperature of the liquid to the dew point, mois-

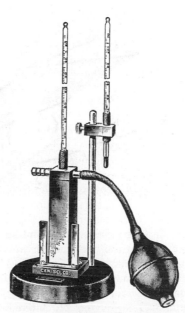

Fig. 18.3 Sling psychrometer

Fig. 18.4 Alluard hygrometer

ture condenses on the polished nickel. From the temperature of the dew point and the temperature of the room, the relative humidity may be calculated from vapor-pressure tables.

Procedure: 1 Observe and record the room temperature near the position at which you are working. Observe and record the barometric pressure.

2. Polish with a dry cloth a small section of the metal calorimeter cup until it is dry and shiny. Fill the cup about two-thirds full of water and stir it with a thermometer. Avoid touching the cup with your hands or breathing directly on it. Gradually add small pieces of ice to the water and carefully observe the temperature at which a thin film of mist begins to form on the outside shiny surface. If the mixture goes down to zero with no observable condensation, add salt to the ice and continue; pour out water as the cup becomes filled. Record the temperature at which the dew *first* appears. Remove any excess ice and continue to stir the water with the thermometer while it is being warmed by the surroundings. If the dew point is high, you might hasten the process by adding a little warm water. Record the temperature at which the dew just disappears. Repeat the procedure for two more readings of the dew point and average the four readings.

3. Fill the Alluard apparatus about three-fourths full of ether and mount a thermometer so that its bulb is immersed in the liquid. Place the apparatus in a favorable light for noting the first appearance of dew on the polished surface. Wipe the polished surface dry and shiny with a cloth. Carefully bubble air through the liquid and note the temperature at which the first trace of dew appears. Allow the liquid to warm up and record the temperature at which the dew disappears. Make a second complete determination and average the values.

4. Soak the jacket of the cloth-covered bulb of the sling psychrometer in water. Make sure that you have plenty of room to operate the instrument without touching any object and whirl it rapidly for 20 or 30 sec. As soon as the motion is stopped, record the readings of both thermometers. Keep the bulb wet and continue the procedure until the readings of the two thermometers have reached steady values. Record them.

5. If time permits, make psychrometric observations in the hall and out of doors. The determination of the wet-bulb reading, and the relative humidity, when the atmospheric temperature is below freezing is an especially difficult procedure.

6. Create for several minutes a steady current of air past a hygrodeik or a Mason's hygrometer and record the thermometer readings, or relative humidity.

7. Measure and record the approximate dimensions of the room.

Computations and Analysis: 1. Using the average experimental dew point, find the corresponding pressure of water vapor at this temperature from the table in the Appendix. From the same table find the pressure of saturated water vapor at the observed room temperature. Calculate the relative humidity.

2. From the sling psychrometer readings, calculate the depression of the wet bulb, $t - t'$. From the hygrometric tables find the relative humidity and the dew point for this value.

3. Compute and record the relative humidity from the readings on the stationary hygrometer.

4. Compute the mean value of relative humidity from Parts 1, 2, and 3. Calculate the percentage departure of the individual values from the mean. Discuss the relative accuracy of the different methods and explain the discrepancies.

5. Compute the volume of the laboratory from its measured dimensions.

From the average value of the relative humidity compute the approximate weight in pounds of the water present in the room in the form of vapor.

Review Questions: 1. Define: saturation, dew point, vapor pressure. 2. What is meant by a saturated vapor? 3. Differentiate between relative, absolute, and specific humidity. 4. Describe several different hygrometers and explain their use.

Questions and Problems: 1. Explain why the dew point is not a fixed temperature in the sense that the freezing point of water is.

2. Discuss the technical accuracy of the bromide that "it is not the heat, but the humidity" that causes discomfort.

3. What are the factors which determine the pressure of the saturated water vapor in the atmosphere?

4. Is it possible for the relative humidity to be greater than 100%?

5. In a room where the temperature is 24°C, an experiment shows the dew point to be 12°C. What is the relative humidity?

6. In which case does the air hold more water vapor: (a) temperature 32°F, dew point 32°F, or (b) temperature 80°F, dew point 50°F? What is the relative humidity in each case?

7. Air at a temperature of 22°C and relative humidity 70% is cooled to 18°C. What is then its relative humidity?

8. The relative humidity in a certain room is 60% at 20°C. (a) Calculate the relative humidity if the temperature drops to 15°C. (b) What is then the dew point?

9. Under what conditions will the wet-bulb and the dry-bulb thermometers read alike?

10. Why must the air be circulated by a fan or other device around the bulbs of stationary hygrometers?

11. The following readings were obtained with a sling psychrometer: dry-bulb reading 76°F, wet-bulb reading 63°F, barometer 29.82 in. of mercury. What is the relative humidity? the dew point?

12. Suppose that Exp. 18.1 were performed in a perfectly dry room (vapor pressure zero). What would be the reading of the wet-bulb thermometer when the dry-bulb thermometer reads 20°C?

13. What would have to be the temperature of a perfectly dry room to have the value of the wet-bulb thermometer read 0°C?

CHAPTER 19

CALORIMETRY

The form of energy that molecules of matter possess because of their motion is known as *heat*. Not to be confused with this concept is *temperature* which is that property of matter which determines the direction of flow of heat between an object and its surroundings.

The unit of heat in the metric system is the calorie. A *calorie* is the heat necessary to raise the temperature of one gram of water one degree centigrade. The corresponding unit of heat in the British system is the British thermal unit (Btu). The Btu is defined as the heat necessary to raise the temperature of one pound of water one degree Fahrenheit. One Btu is equivalent to about 252 cal. These units are not quite constant throughout the range from freezing to boiling; they may be regarded as average values.

Calorimetry is the theory and art of measuring quantities of heat. The most common quantities of heat are (1) the heat required to raise the temperature of unit mass of a substance, in the same state, one degree; (2) the heat required to change unit mass from the solid to the liquid state, with no change in temperature; and (3) the heat required to change unit mass from the liquid to the vapor state, with no change in temperature.

The *specific heat* of a substance is defined as the heat per unit mass per degree change in temperature

$$S = H/m\,\Delta t \tag{1}$$

where S is the specific heat, H is the heat change in the material of mass m, and Δt is the change in temperature. The units are expressed either as cal/gm °C or Btu/lb °F. Because of the manner of defining the calorie and the Btu, the numerical values of specific heat are the same in both systems.

The *heat of fusion* of a substance is defined as the heat per unit mass required to change the substance from the solid to the liquid state at the melting temperature. In symbols

$$L_f = H/m \tag{2}$$

The heat of fusion L_f is expressed in Btu per pound or in calories per gram. The numerical value in the metric system is $\frac{5}{9}$ that in the British system. It is equivalent to the amount of heat given up when a unit mass changes from the liquid to the solid state at the temperature of solidification.

The *heat of vaporization* of a substance is defined as the heat per unit mass required to change the substance from the liquid to the vapor state, without change of temperature. In symbols

$$L_v = H/m \tag{3}$$

The heat of vaporization is expressed in Btu per pound or in calories per gram. It is equivalent to the *heat of condensation*, the heat given up when unit mass of vapor is changed to the liquid state.

Heat quantities are often determined by the *method of mixtures*. This method makes use of the principle that when a heat interchange takes place between two bodies initially at different temperatures, the quantity of heat lost by the warmer body is equal to the quantity of heat gained by the cooler body.

Determinations of heat quantities are carried out with all possible precautions to prevent loss or gain of heat to or from the surroundings. For this purpose various types of calorimeters are used. A simple form of laboratory calorimeter is shown in Fig. 19.1. It consists of a thin polished vessel K, of high thermal conductivity, held centrally within an outer jacket A by a nonconducting ring support H. Conduction of heat is thus minimized between outer and inner vessel and the dead-air space between the vessels minimizes convection currents. Radiation of heat is lessened by having the vessels light in color and highly polished. A wooden cover L with holes for a stirrer and a thermometer helps prevent convection currents to and from the cup and contents from above.

Fig. 19.1 Simple form of laboratory calorimeter

Specific Heat by Method of Mixtures. A known mass of metal shot at a known high temperature is dropped into a known mass of water at a known low temperature. The resulting equilibrium temperature is noted. The heat absorbed by the water and containing vessel is computed and equated to the heat given up by the hot metal. From this equation the specific heat of the metal may be computed. In equation form

Heat lost by shot = heat gained by water + heat gained by calorimeter
$$S_s m_s(t_1 - t_3) = S_w m_w(t_3 - t_2) + S_c m_c(t_3 - t_2) \qquad (4)$$

where m_s, m_w, and m_c are the masses of the shot, cold water, and calorimeter; S_s, S_w, and S_c are their respective specific heats; t_1 is the initial temperature of the shot, t_2 is the initial temperature of the water and calorimeter, and t_3 is the equilibrium temperature.

A mass of water numerically equal to the thermal capacity of the substance is called the "water equivalent" of the substance. It is obtained by multiplying the mass of the substance by its specific heat.

The heat lost to the surroundings when the calorimeter and contents are above room temperature is approximately balanced by the heat gained from the surroundings when the calorimeter and contents are below room temperature if t_2 is as much below room temperature as t_3 is above it.

Heat of Fusion and Heat of Vaporization. When a substance changes state, it liberates or absorbs great quantities of heat, but the temperature remains constant. In the determination of the heat of fusion by the method of mixtures, ice that has been dried is placed in a known quantity of warm water in a calorimeter. When the ice is all melted, the equilibrium temperature will be several degrees below the initial temperature. The heat-lost–heat-gained equation consists of four terms.

Heat absorbed by ice when melting + heat absorbed by ice water
= heat given up by warm water + heat given up by calorimeter

The student may substitute conventional symbols for the different quantities and solve the equation for L_f. In a similar manner steam may be condensed in water and L_v determined.

Experiment 19.1

SPECIFIC HEAT OF SOLIDS

Object: To measure the specific heats of various solid specimens by the method of mixtures.

Method: A solid specimen in the form of finely divided shot is heated in the inner cup of a double boiler to nearly the boiling point of water. The hot shot is quickly poured into a calorimeter containing a known mass of cold water. From the measured masses and the rise in temperature of the water and the drop in temperature of the shot the specific heat of the specimen is calculated.

Fig. 19.2 Double boiler and steam generator

Apparatus: Double boiler for heating shot, Fig. 19.2; bunsen burner; calorimeter; stirrer for water; thermometers, 50°C and 100°C; stirrer for shot; specimens (in form of shot).

Procedure: 1. Fill the lower part of the double boiler one-third full of water and start it heating with the Bunsen burner. Watch the gage throughout the experiment to be sure that the boiler does not run dry.

2. Select one of the materials and make a rough computation of the amount of shot necessary to raise the temperature of 100 gm of water about 10°C for that type of shot. It will be between 100 and 300 gm. Weigh out the proper amount in the form of *dry* shot, pour it into the cup of the double boiler, and place the cup in the outer vessel. Cover the cup with a cork. The cover should be provided with two holes: one for the 100°C thermometer, and one for the stiff wire stirrer. Heat the shot over the steam and keep it well stirred. Do not use the thermometer to stir the shot. Continue heating and watching the thermometer until the temperature of *all* the shot reaches and maintains a temperature of about 95°C.

3. Weigh the inner calorimeter cup (without the fiber ring). If the stirrer for the water is the same material as the calorimeter, weigh it at the same time. If

the stirrer is of some other material, weigh it separately. Record the material being heated, the mass of the shot, the material of the water stirrer, and the mass of the calorimeter cup and stirrer. Add to the cup about 100 gm of water about 5°C below room temperature and reweigh.

4. Insert the inner calorimeter vessel, containing the stirrer and water, in the outer vessel, with the fiber ring separating them. Place the wooden cover over the calorimeter. Insert the 50°C thermometer and the stirrer through the holes in the cover. Stir gently and when the combination is at a uniform temperature 4 or 5°C below room temperature, record the temperature. Quickly pour the shot into the cold water and close the calorimeter. Be careful not to let the bulb of the thermometer get into the hot shot. Keep the thermometer in the water, stir gently the shot and the water; watch the thermometer and note the highest *equilibrium* temperature of the water. Estimate the reading on the thermometer scale as accurately as possible.

5. Repeat Steps 1 to 4 with a different substance. It should have been weighed out and ready to introduce into the double-boiler cup as soon as the first metal was poured into the water.

6. Pour the wet shot onto paper toweling in a flat pan to dry. Do not mix the wet shot with dry shot. Mop up all spilled water and leave things in good order.

Computations and Analysis: 1. Write the symbolic heat-lost–heat-gained equation for the process that you used. Substitute into the equation the experimental data and solve for the specific heat of the shot. Compare your value with the standard value for the substance used (see Appendix).

2. Repeat Part 1 for the second specimen. If the deviation of either value is too great repeat the determination.

Experiment 19.2

HEATS OF FUSION AND VAPORIZATION OF WATER

Object: To determine the heat of fusion of ice and the heat of vaporization of water by the method of mixtures.

Method: 1. A known mass of ice is dropped into a measured quantity of water in a calorimeter. The heat given up by the calorimeter and contents is equated to the heat absorbed by the ice in melting and being raised to the equilibrium temperature. The heat equation is solved for the heat of fusion of ice.

2. The heat liberated by the condensation of steam is measured by allowing it to condense in cold water in a calorimeter. From the observed temperature change of the water and calorimeter and their known water equivalents, an equation is set up from which the heat of vaporization is computed.

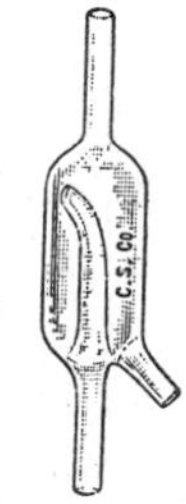

FIG. 19.3 Water trap

Apparatus: Calorimeter (Fig. 19.1); steam generator (Fig. 19.2); water trap (Fig. 19.3); thermometer, 50°C; Bunsen burner; balance and masses; vessels for ice and water; rubber tubing; pinch clamps; barometer; ice; cloths.

Procedure: 1. Record the mass of the inner calorimeter cup and stirrer. If the stirrer is some different material, weigh it separately from the calorimeter cup. Place in the cup about 200 gm of water at approximately 12°C above the room temperature. Record the mass of the water. Put the inner cup into the outer vessel, put on the lid with stirrer and thermometer protruding, and stir gently until equilibrium temperature is reached. Record the temperature.

2. Select one or two lumps of ice with a total mass of approximately 60 gm. Dry the ice carefully and add it to the water quickly just after the temperature of the water has been recorded. Do not splash any water from the cup. Cover and stir carefully; watch the temperature and as soon as all the ice has melted, record the equilibrium temperature.

3. Carefully remove the calorimeter cup and water; shake into the cup all drops of water adhering to the thermometer and stirrer. Weigh the cup and contents to find the actual mass of ice used.

4. The steam generator should have been previously started with the steam hose in a beaker well away from the calorimeter. Insert the steam trap between the steam generator and the hose to be introduced into the water. Lead the short tube with the condensate into a beaker well below the level of the calorimeter cup.

5. Weigh the inner cup with about 200 gm of water about 12°C below room temperature. Record the mass, assemble the calorimeter, and record the temperature when it is steady.

6. Introduce the short tube below the water trap, from which steam is issuing rapidly, into the inner calorimeter cup below the surface of the water. Close off with a pinch clamp the rubber tubing leading to the condensate beaker, but open it to drain if the trap fills half full of water. Watch the temperature of the mixture carefully and, when it is about 10°C above room temperature, remove the steam tube, shaking into the cup any adhered water. Stir the water and watch the temperature. Record the exact maximum temperature.

7. Remove the cup and contents and reweigh. Record the barometer reading.

Computations and Analysis: 1. Record the mass of water and the mass of ice.

2. Set up a heat-lost equal heat-gained equation, and solve for the heat fusion of ice.

3. From the barometer reading and the vapor-pressure tables in the Appendix, record the temperature of the steam. Compute the mass of water and the mass of steam.

4. Set up an equation in which heat lost is equal to heat gained and calculate the heat of condensation, which is equivalent to the heat of vaporization, of the steam.

5. Compare the values you obtained with the standard values of $L_f = 80$ cal/gm and $L_v = 540$ cal/gm. From observed values of uncertainties in temperatures and weighings, estimate the percentage uncertainty in the computed values. List other sources of error in the measured values.

Review Questions: 1. Define: heat; temperature; specific heat; thermal capacity; water equivalent; heat of fusion; heat of vaporization.   2. State the metric and British units for specific heat, heat of fusion, heat of vaporization.   3. Describe the method of mixtures as applied in the determination of specific heat and note the essential precautions.   4. Derive the working equations needed for determining heat of fusion and heat of vaporization.

Questions and Problems: 1. How would the experimental value of the specific heat be affected if some hot water were carried over with the metal in Exp. 19.1?

2. What will be the biggest source of error if too much water is used in the calorimeter into which the shot is poured in Exp. 19.1?

3. A platinum ball of mass 100 gm is removed from a furnace and dropped into 400 gm of water at 0°C. If the equilibrium temperature is 10.1°C and the specific heat of platinum is 0.040 cal/gm °C, what must have been the temperature of the furnace? Neglect the effect of the mass of the calorimeter.

4. In a determination of specific heat, the temperature change of the hot shot was 80 ± 1°C and the temperature change of the water was 4.5 ± 0.2°C. Compute the uncertainty that each of these factors introduces into the determination of specific heat. What error in grams would be permissible in the measurement of 300 gm of shot in order that this uncertainty would be negligible in comparison with the measurement of temperature change?

5. In an experiment on specific heat a copper calorimeter of 100 gm mass containing 200 gm of water was used. The thermometer contained 0.8 cm³ of mercury and 3.0 gm of glass. What error is introduced by neglecting the heat capacity of the thermometer?

6. In an experiment on specific heat the following data were obtained: mass of calorimeter, 200 gm; specific heat of calorimeter, 0.10 cal/gm °C; mass of water, 300 gm; mass of aluminum shot, 500 gm; temperature of material, 99°C; room temperature, 25°C. What should be the initial temperature of the water in order that the final temperature will be as much above room temperature as the initial temperature of the water was below the room temperature?

7. Look up, in some textbook or handbook, data on the specific heat of water. Plot a curve showing its variation from 0 to 100°C.

8. In the measurement of the heat of vaporization of water by the method of mixtures, the following data were obtained: 6.1 gm of steam admitted into a calorimeter of mass 125 gm and containing 130 gm of water at 20°C; the specific heat of the calorimeter, 0.10 cal/gm °C; final temperature of water, 45°C. Determine the value of L_v from these data and compute the percentage error from the standard value.

9. If the ice is wet when placed into the calorimeter so that the added mass consists of 99% ice and 1% water, what percentage error will be introduced into the determination of L_f?

10. Explain the necessity of using a water trap in the determination of L_v and show how its absence would have affected the results.

11. The heat that flows to the thermometer must flow through the thin glass wall of the bulb. How does the lag in the thermometer reading behind that of the water temperature affect the results?

12. Compare the errors introduced by each of the following factors in a determination of L_f: (a) 1 gm of water in a 120-gm block of ice; (b) error of 0.5°C in a temperature change of 21.4°C; (c) error of 0.2 gm in a mass of 100 gm of water; (d) neglecting the 1-gm water equivalent of the thermometer.

CHAPTER 20

HEAT AND WORK

In many familiar processes, such as in automobile mechanical brakes, work is done against friction and heat is produced. When a body falls and strikes a solid surface, the kinetic energy is transformed into heat energy.

J. P. Joule (1843) arranged a set of paddles that were rotated in water by a pair of falling bodies. The churning of the water produced heat that could be calculated from the rise in temperature of the water and its container. The mechanical work performed in driving the paddles was calculated from the weights of the falling bodies and the distance they descended. Joule found that the ratio of the work done to the heat produced was always the same.

The number of units of work W per unit of heat H is called the *mechanical equivalent of heat*. In symbols

$$W = JH \tag{1}$$

where J is a constant independent of the magnitude of W or H but whose value depends on the units in which W and H are expressed. Experimentally determined values for J are 4.18 joules/cal and 778 ft-lb/Btu. The first value is sometimes written 4.18×10^7 ergs/cal.

The principle of conservation of energy states that energy can neither be created nor destroyed but can be transformed from one form into another. The statement of the equivalence of mechanical energy and heat energy is often called the *first law of thermodynamics* and is stated: A *constant ratio exists between mechanical energy and heat energy when either form is converted into the other.* It is this constant ratio that is called the mechanical equivalent of heat, J.

The value of J may be experimentally determined by several methods: (1) a simple method of allowing a quantity of shot to fall repeatedly through a measured height and dividing the work done by the heat produced; (2) by an apparatus designed by H. L. Callendar in which mechanical energy is dissipated by a special brake rubbing on the outside of a rotating brass drum; and (3) by a method developed by J. Puluj and modified by G. F. C. Searle in which a torque is applied to hold a metal cone stationary within a rotating cone, Fig. 20.1.

By reference to Fig. 20.2 it can be seen that by means of the flanged wheel C the weight mg may be made to exert on the friction cone B a torque just sufficient to prevent the cone from turning as cup A is rotated. The same effect would be obtained by holding the cup B fixed and allowing mass m to fall through a certain height and rotate the cup A. The equivalent height is equal to the product of the circumference of the flanged wheel C and the number of rotations of one friction cone with respect to the other. The work done by a falling body of mass m is given by mgh, where h is the distance of fall. The mechanical work done against friction is evidently

$$W = mg\pi dN \tag{2}$$

where g is the acceleration due to gravity, d is the diameter of the flanged wheel at the bottom of the groove, and N is the number of revolutions of the conical cup.

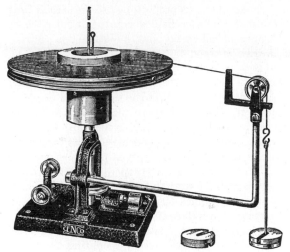

FIG. 20.1 Puluj-Searle apparatus

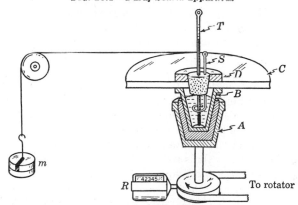

FIG. 20.2 Cutaway view of Puluj-Searle apparatus for measuring heat equivalent of work

The amount of heat necessary to raise the temperature of the cup, friction cone, water, stirrer, and thermometer is

$$H = (m_w + mS + m_e)(t_2 - t_1) \qquad (3)$$

where m_w is the mass of water inside the friction cone; m is the mass of conical cup, friction cone, and stirrer; S is the specific heat of the cup, cone, and stirrer; m_e is the water equivalent of the thermometer; t_2 is the final temperature; and t_1 is the initial temperature.

Experiment 20.1

MECHANICAL EQUIVALENT OF HEAT: SHOT-TUBE METHOD

Object: To make an approximate determination of the mechanical equivalent of heat.

Method: A quantity of lead shot is allowed to fall many times from end to end of a long tube. The mechanical energy is proportional to the total length of fall of the shot; the heat energy is proportional to the specific heat and the rise of temperature of the shot. The mechanical equivalent of heat is, by definition, the ratio between the mechanical energy and the heat.

Apparatus: A tube of wood or heavy pasteboard about 1 m long, provided with a heavy cork stopper for the open end; several pounds of lead shot; thermometer, reading from 0 to 50°C in 0.2°C divisions; meterstick; tin pan.

Procedure: 1. Empty the shot into the tin pan and cool it about 2°C below room temperature. In winter it may be placed outside the window; at other times it may be put on a block of ice. Keep the shot dry at all times.

2. Pour the cold shot into the tube and measure its temperature by means of the thermometer inserted through a stopper at the end of the tube. Replace the thermometer and stopper by a solid stopper. While seated in a chair or on a stool, hold the tube at the center and invert it quickly; repeat this 100 times. The inversion should be so quick that the shot falls from one end to the other rather than sliding along the surface of the tube. Have the end of the tube on the floor when the shot falls. Be careful not to raise or lower the tube while inverting it. (This precaution may be satisfied by mounting the tube in a condenser clamp on a support rod.)

3. After the hundredth inversion, quickly replace the thermometer and record the new temperature.

4. Make two more independent determinations of the rise in temperature produced as in Step 2.

5. Measure with the meterstick the *average* distance through which the shot fell; this distance is measured from the top of the pile at one end to the bottom of the stopper in the other end.

Computations and Analysis: 1. From the specific heat of lead and the rise in temperature calculate the heat developed per gram.

2. From the length of the tube and the number of inversions compute the expended energy per gram. It is not necessary to know the mass of the shot. Why? Compute the mechanical equivalent of heat and compare it with the standard value.

Experiment 20.2

MECHANICAL EQUIVALENT OF HEAT: FRICTION-CONE METHOD

Object: To study the conversion of mechanical energy into heat energy; in particular, to determine the mechanical equivalent of heat.

Method: A measurable amount of mechanical energy is converted into heat energy by the friction of a pair of brass cones one of which is kept fixed by a known torque while the other revolves. The heat energy is computed from the rise of temperature of the cups and the water contained in the inner one, the specific heat of the water and brass, and the mass of the cups and the water. The mechanical work is computed from the measured frictional torque and the number of revolutions made by the outer cup. The mechanical equivalent of heat is the ratio of the work done to the heat generated.

Apparatus: Puluj-Searle mechanical equivalent of heat apparatus; mechanism to drive outer cup, either a variable-speed, friction-drive rotator or a hand crank; stirring rod; wooden disk with grooved edge; iron ring; pulley; weight hanger; set of metric masses; 50°C thermometer; bottle of special lubricating oil for the cones; trip scale and masses; ice; stop watch or clock; clean cloths.

The apparatus used in this experiment is shown in Fig. 20.1 and diagrammatically in Fig. 20.2. It consists essentially of two brass conical cups, or cones, that have been ground carefully so that they fit very snugly together. Be very careful not to damage the surfaces. The conical cup A is mounted in an insulated support that can be rotated either by a hand crank or a motor; the friction cup B fits into cup A. A flanged wheel C is attached to cone B and is weighted by a heavy metal ring D. A cord wound around wheel C supports a mass m. The friction cone B is hollow and contains water, stirred by small stirrer S. The temperature of the water is measured by thermometer T. The number of turns made by the conical cup is recorded on revolution counter R.

Procedure: 1. Weigh the conical cup, the friction cone, and the stirrer, and record the masses. Assemble the apparatus and add water about 10°C below room temperature to the friction cone to fill it nearly to the top of the tapered portion. Record the mass of water added.

2. Make a brief trial run, rotating the cup by hand or with the variable-speed motor, to determine whether the mass m is supported steadily at a constant height while turning at a reasonable speed; if not, change the amount of oil between the friction members until the adjustment is satisfactory. A fair-sized mass (about 150 gm) should be supported by the friction so as to get a rapid rise in temperature in order to minimize the errors due to radiation.

3. When the preliminary trials have been completed, the temperature of the water should still be several degrees below room temperature. If it is not, replace the water and wait several minutes for the cones to reach equilibrium. Record the reading of the revolution counter. Stir the water steadily and record its temperature each 30 sec for 3 min. Then begin to rotate the wheel, keeping mass m at a constant height. Continue to stir the water and record its temperature every 30 sec. When the temperature has risen above room temperature about as much as it started below the temperature of the room, stop turning but continue to stir the water and record its temperature every 30 sec for another 3 min.

4. Record the diameter of the bottom of the groove of the flanged wheel, the value of mass m, and the final reading of the revolution counter. Measure and record the volume of the part of the thermometer immersed in water.

5. Repeat Steps 2 to 4.

Computations and Analysis: 1. Compute the water equivalent of the immersed part of the thermometer; it may be taken as 0.46 times the volume in cubic centimeters of the immersed portion.

2. For each run make a graph of temperature against time. Indicate room temperature by a dotted line. Take as temperature t_1 the point on the curve just before starting to crank; take as temperature t_2 the highest point on the curve.

3. Substitute the proper values into Eqs. (2) and (3) and compute the work W in joules and the heat H in calories. Substitute these values into Eq. (1) and

compute the value of J. Compare this value with the accepted value of 4.18 joules/cal. Compute the percentage error.

Review Questions: 1. Define mechanical equivalent of heat. State the common units in the metric and British systems. 2. Describe a simple method of determining J by falling shot. Derive the working equation. 3. Describe the method of determining J by friction cones. Derive the working equation.

Questions and Problems: 1. Why is it unnecessary to know the weight of the shot used in Exp. 20.1? Could brass or copper shot have been substituted for lead shot? What change would be necessary in the working equation?

2. What are the chief sources of error in the shot-tube method? Could a metal tube be used?

3. How much is the water warmed at Niagara Falls by falling 50 m? What factors might prevent this rise in temperature?

4. Convert into British engineering units the value of J obtained in the friction-cone method of Exp. 20.2.

5. Explain how you can compute work in the friction-cone method when the body that supplies the force remains stationary.

6. A 2000-lb car travels down an incline that makes a constant angle of 10° with the horizontal. The speed of the car remains constant at 20 mi/hr. With your value of J obtained in problem 4 compute the heat developed per second in the brake drums.

7. Why was a uniform speed of rotation not necessary in the friction-cone experiment? What factor must be constant?

8. Would vigorous and long-continued stirring of the water in a calorimeter introduce an error in precision work? Explain.

9. A boy eats 1.0 lb of ice in 10 min. What horsepower is required to melt the ice and raise it to the temperature of the body, 98°F?

10. In a typical experiment to determine the mechanical equivalent of heat by the friction-cone method the following data were taken: mass hung over pulley, 140 gm; radius of disk, 12 cm; mass of cups, 200 gm; specific heat of cups, 0.090 cal/gm °C; mass of water in cups, 30.0 gm; number of revolutions, 2000; rise in temperature, 10.00°C. Calculate the experimental value of J and the percentage error between it and the standard value.

11. Consider the following factors in the friction-cone experiment: flanged wheel not uniform in diameter; cord to pulley not tangent to the circumference of the disk; friction in the pulley supporting mass m; friction members not as cool as the water at the start of turning; a few drops of water spilled during the run; mass m not held at same height during the run. Discuss for each case whether the item will cause an error in the results, whether the error is determinate or indeterminate, and whether the resulting value of J will be larger or smaller because of the error. Justify your answers.

12. Describe a possible method of using a spring balance instead of a hanging mass to counteract the frictional torque.

CHAPTER 21

TRANSVERSE WAVE MOTION

A wave motion is a particular assemblage of harmonic motions. When a particle is vibrating with simple harmonic motion (SHM) its displacement y at any time t is given by

$$y = A \sin \theta = A \sin 2\pi ft = A \sin \frac{2\pi t}{T} \tag{1}$$

where A is the amplitude of the vibration, θ is the phase angle, that is, the angle that expresses the part of the vibration in which the particle is, f is the frequency of the vibration, and T is the period.

If the vibrating particles are those of an elastic medium, each particle vibrates with the same frequency as all the others; but the successive particles are progressively later in phase. If the phase lag is proportional to the distance x from a given point where the phase is zero, the displacement is given by

$$y = A \sin 2\pi \left[\frac{t}{T} - \frac{x}{\lambda} \right] \tag{2}$$

The distance λ is the wavelength, that is, it is the distance between two particles in the same phase; because, when $x = \lambda$, the phase lag $2\pi x/\lambda$ becomes 2π and the particles are again in phase. The relationship of Eq. (2) is the displacement in a transverse wave. The displacements at any one time vary with the distance x and at any one distance x they vary with time t.

Fig. 21.1 Transverse wave in cord. (a) Position of cord at time zero. (b) Position of cord a quarter period later

If a uniform cord of infinite length is subjected to a tension and one end of the cord is given a vibratory motion at right angles to the length of the cord, waves travel along the cord. The motion of any particle of the cord is at right angles to the undisturbed position of the cord and the waves are transverse waves. The waves consist of a regular succession of crests and troughs traveling down the cord. The distance between two successive crests or two successive troughs is one wavelength λ, Fig. 21.1. In a time equal to the period T the crests or troughs travel a distance equal to the wavelength λ. Hence the speed v of the wave is given by

$$v = \frac{\lambda}{T} = f\lambda \tag{3}$$

145

where the frequency f is the number of waves that pass a given point per second. The relationship expressed by Eq. (3) is frequently called *the fundamental equation of wave motion.*

The speed v with which the transverse wave travels along the cord is given by

$$v = \sqrt{\frac{F}{m/l}} \tag{4}$$

where F is the stretching force (or tension) in the cord and m is the mass of a length l of the cord.   In the cgs system of units v is in centimeters per second when F is in dynes, m is in grams, and l is in centimeters.

When two wave trains, of the same frequency and amplitude and having proper phase relations, travel in opposite directions *standing* or *stationary waves* are set up.   When transverse waves are sent along a cord fixed at one end, reflection of the waves takes place at the fixed end.   Two similar sets of waves are traveling in opposite directions along the string.   If the cord is of suitable length, the two sets of oppositely traveling waves produce standing waves.

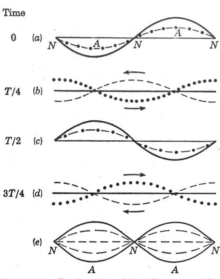

Time

0   (a)

$T/4$ (b)

$T/2$ (c)

$3T/4$ (d)

(e)

A A

FIG. 21.2 Production of standing waves by two progressing waves traveling in opposite directions; curves a, b, c, and d show displacements of the particles due to the two waves and the resultant displacement at fractions of a period 0, $T/4$, $T/2$, and $3T/4$; curve e shows the position of the nodes N and the antinodes A in the stationary waves

The production of standing waves in this manner is shown in Fig. 21.2. In this diagram waves traveling toward the right are represented by the dotted curves and those traveling to the left by dashed curves.   The resultant wave is shown by the heavy solid line.   In a the two waves are coincident and the resultant displacement of the cord at each point is the sum of the two displacements due to the individual waves.   The resultant displacement curve is similar to the individual curves but of double amplitude.   A quarter period later (b), each wave has progressed a quarter wavelength; the displacements at each point are equal in magnitude but opposite in direction and the resultant displacement is everywhere zero.   Figure 21.2c and d show the individual and resultant displacements of the cord at one-half and three-quarters of a period.   The resultant displacements for a whole period are shown in Fig. 21.2e.   It is seen that at certain points, marked N and called *nodes*, there is no displacement of the cord at any time.   At other points marked A and called *antinodes*, the displacement varies from zero to a maximum.   Nodes and antinodes exist only in standing waves.   The distance between a node and an antinode is a quarter wavelength; the distance between two successive nodes or two successive antinodes is a half wavelength.

A cord may be forced to vibrate with any frequency but it will vibrate with maximum amplitude if the frequency, the length of the cord, and the speed in the cord are so related that standing waves are set up. Since both ends of the cord are fixed in position, the ends must be nodes in the standing wave; hence the length L of the cord must be an integral number of half wavelengths. If we combine Eqs. (3) and (4), we obtain

$$f = \frac{1}{\lambda} \sqrt{\frac{F}{m/l}} \tag{5}$$

Since $\qquad\qquad L = N\lambda/2, \qquad \lambda = 2L/N.$

and

$$f = \frac{N}{2L} \sqrt{\frac{F}{m/l}} \tag{6}$$

where for the successive integral values of N the natural frequencies of vibration are given.

From Eq. (5) the relationship between F and λ may be expressed by solving for F

$$F = f^2 \frac{m}{l} \lambda^2 \tag{7}$$

The tension in the cord is directly proportional to the square of the wavelength for a given frequency.

The lowest frequency is produced when the string vibrates in one segment. This minimum frequency is called the *fundamental* frequency of the string. Frequencies greater than the fundamental are called *overtones*. Overtones whose frequencies are integral multiples of the fundamental are called *harmonics*. From the symmetry of a string fixed at both ends it is apparent that all the harmonics are possible.

Experiment 21.1

WAVE MOTION: MELDE'S EXPERIMENT

Object: To investigate by means of the Melde experiment the properties of standing waves in a cord.

Method: An electrically driven tuning fork produces waves in a uniform cord under tension. The tension in the cord is varied to produce stationary waves of different wavelengths in the cord so that it vibrates in one, two, or more loops. From the wavelength of the standing waves and the tension and mass per unit length of the cord, the frequency of the oscillations produced by the tuning fork is calculated.

Apparatus: Electrically driven tuning fork, Fig. 21.3; uniform cord; 6-volt battery; SPST switch; meterstick; pulley; weight hanger; set of weights; analytical balance.

Procedure: 1. Attach one end of the cord to one prong of the tuning fork and pass the other end over the pulley to the weight hanger as in Fig. 21.4. Connect the 6-volt battery through the switch to the binding posts of the tuning fork. Set the fork into vibration by squeezing the prongs together slightly and releasing them. If sparking occurs at the make-and-break mechanism, adjust the contacts to minimize the sparking. With the tuning fork vibrating, adjust the tension of the cord so that the cord vibrates in one segment with maximum amplitude. Measure the length L of the cord from the tuning fork to the pulley. Record the frequency stamped on the fork.

2. Determine as in Step 1 the tensions necessary for the cord to vibrate in two, three, four, five, and six segments, keeping the length constant.

3. With the cord under tension measure and cut off a length l. Determine its mass by means of an analytical balance.

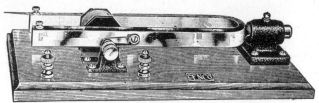

FIG. 21.3 Electrically driven tuning fork

FIG. 21.4 Tuning fork with motion perpendicular to length of cord

FIG. 21.5 Tuning fork with motion parallel to length of cord

4. Apply the tension required for four segments and turn the fork so that its prongs vibrate parallel to the length of the cord, Fig. 21.5. How does this affect the vibration of the cord?

Computations and Analysis: 1. From the data of Step 3 calculate the mass per unit length of the cord.

2. From the data of Steps 1 and 2 calculate the wavelengths for the various tensions. Plot a graph showing the relationship between the tension in the string and the square of the wavelength. What is the significance of the shape of the curve? Compute the slope of the curve. From the slope and Eq. (4) calculate the frequency of the fork. Compare the calculated frequency with the value stamped on the fork.

3. Explain the difference in the vibration observed in Step 4 from that in Step 2.

Experiment 21.2

LAWS OF VIBRATING STRINGS: THE SONOMETER

Object: To study the variation in the fundamental frequency of a vibrating wire with stretching force, length of wire, and mass per unit length.

Method: A piano wire stretched over a sounding box is tuned in unison with tuning fork of known frequency. The tension of the wire is adjusted by means of weights suspended from it and the length of the vibrating segment is adjusted by means of movable bridges. With the tension constant the wire is tuned to several different forks in succession by making adjustments of the length. The relationship between frequency and length is shown by a graph of frequency against the reciprocal of length. A second set of observations is made in which the tension is varied while the length is kept constant. The square of the frequency is plotted against tension. The frequencies computed from measured values of the length, tension, and mass per unit length are compared with the values stamped on the forks.

Apparatus: Sonometer; weight hanger; set of masses; set of tuning forks; rubber hammer; meterstick.

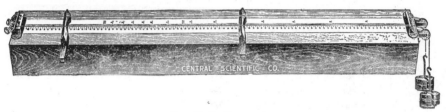

Fɪɢ. 21.6 Sonometer

The sonometer (Fig. 21.6) consists essentially of one or more piano wires stretched over a sounding box. The tension is supplied by means of weights suspended from the wires and the effective length is regulated by movable bridges. In some models the tension is adjusted by means of screws, and the value of the tension is indicated by spring balances attached to the wires.

Procedure: 1. Set the two bridges near the ends of the wire so as to utilize most of the length. Sound the fork of lowest frequency by striking it with the rubber hammer. *Caution: Never strike a tuning fork with a metal or other hard object nor allow the prongs to strike any hard object while vibrating.* Add weights to the weight hanger until the frequency of vibration of the wire is nearly the same as that of the fork. By adjusting one of the bridges, tune the wire exactly in unison with the fork. The inexperienced student will have to practice making this adjustment until he learns to detect a slight difference in pitch. With the ear held close to the fork and the wire, sound the fork, pluck the wire gently, and listen for beats. As the tuning becomes closer the frequency of beats becomes less and when the fork and wire are in unison the beats disappear. The student whose ear is unreliable may facilitate tuning by the following method. Make a rider of a small piece of paper folded in the form of a V and place it on the wire at the midpoint. Sound the fork and place its base firmly in contact with the sounding box. When the wire is in tune with the fork, the forced vibrations set up in the box by the fork will be taken up by the wire and the rider will be displaced.

When the adjustment has been made, measure and record the length of the wire between the bridges. Record the tension of the wire.

2. Repeat the observations of Step 1 for each fork, keeping the tension constant and varying the length to tune the wire to the fork.

3. Keep the length at the lowest value of Step 2 and reduce the tension until the wire is in unison with the fork of second highest frequency. Record the length and tension. Repeat this procedure for each of the other forks.

4. Weigh on a sensitive balance a measured length of the same kind of wire as that used on the sonometer to determine the mass per unit length.

5. (Optional) Use on the sonometer two wires of the same length but with different mass per unit length. Apply a known tension to one wire and adjust the tension in the other until it is in tune with the first one. Record the two tensions. Measure the mass of a known length of wire similar to each sample.

Computations and Analysis: 1. From the data of Steps 1 and 2 plot a curve showing the relationship between the frequency and the reciprocal of the length of the wire. Discuss the significance of the shape of the curve.

2. From the data of Step 3 plot a graph of the square of the frequency against tension. Discuss the significance of the curve.

3. From the data of Steps 1 to 4 compute by the use of Eq. (6) the frequency of the fork used for each observation. Compare the computed frequencies with the values stamped on the forks.

4. (Optional) From the data of Step 5 compute and compare the ratio of the tensions and the ratio of the masses per unit length.

Review Questions: 1. Write the equation that expresses the displacement of a particle executing SHM as a harmonic function of time. 2. Write the equation that expresses the successive displacements of the particles in a transverse wave. 3. Define: wave motion; transverse wave; phase; frequency; period; stationary or standing wave. 4. State the fundamental equation of wave motion. 5. State the necessary conditions for the production of standing waves. 6. Write the equation that expresses the relationship between the frequency of a vibrating string and other factors. 7. Describe the Melde experiment. 8. Describe the sonometer. Explain how it may accurately be tuned to resonance with a given tuning fork. 9. Describe the use of the sonometer for studying the fundamental laws of a vibrating string.

Questions and Problems: **1.** Show that both sides of Eqs. (3), (4), and (5) have the same dimensions.

2. In the Melde experiment, when the cord is adjusted so that it vibrates in one loop, is the end of the cord that is attached to the fork exactly at a node?

3. If the cord had several knots tied at irregular intervals along its length, could standing waves be set up in it? Why?

4. Should distances be measured from the end of the tuning fork or from the first node? Explain.

5. If a system is in resonance with the string vibrating in one loop, will it still be in resonance if the tension is increased by a factor of four? Explain.

6. A certain string 1.00 m long has a mass of 0.375 gm. What stretching force is necessary to tune it to 640 vib/sec? What is the length of the standing wave in the string?

7. A No. 12 galvanized-steel telegraph wire weighing 170 lb/mi is strung on poles 200 ft apart. A transverse wave is started at one pole, travels to the next pole, and is reflected back to the first pole in a total time of 2.20 sec. What is the tension in the wire?

8. A copper wire 1.00 m long vibrates in two segments when under a tension of 250 gm-wt. If the mass per unit length of the wire is 0.0100 gm/cm, what is the frequency of the fundamental note?

CHAPTER 22

SOUND WAVES; RESONANCE

A sound wave consists of a succession of compressions and following rarefractions that travel through an elastic medium. Such a wave will travel through a gas, liquid, or a solid. In each case the speed of the wave depends upon the elastic modulus and the density of the medium. In symbols

$$v = \sqrt{E/d} \tag{1}$$

where v is the speed of the wave, E is the modulus of elasticity, and d is the density of the medium. For solid rods or wires the modulus used is Young's modulus. For extended mediums the bulk modulus is applicable.

The compressions and rarefractions in a sound wave occur too rapidly for the temperature to remain constant or for transfers of energy to take place. Therefore the process is adiabatic. The adiabatic bulk modulus of a gas in the product γP, where γ is the ratio of the specific heat at constant pressure to the specific heat at constant volume and P is the pressure of the gas. Thus

$$v = \sqrt{\gamma P/d} \tag{2}$$

For air $\gamma = 1.40$. If the pressure is in dynes per square centimeter and the density is in grams per cubic centimeter Eq. (2) gives the speed in centimeters per second.

The speed of sound in a gas varies with temperature. From the general gas law, $P/Td =$ constant, where T is the absolute temperature. Therefore the speed of a sound wave is directly proportional to the square root of the absolute temperature

$$\frac{v_1}{v_2} = \sqrt{\frac{T_1}{T_2}} \tag{3}$$

The speed v of a wave is related to the wavelength λ and the frequency f by the relationship

$$v = f\lambda \tag{4}$$

If the wave travels in more than one medium, the frequency stays the same but the wavelength changes as the speed changes.

Every vibrating body has certain natural frequencies of vibration, those for which standing waves are set up within the body. In a cylindrical tube closed at one end a compressional wave traveling down the tube is reflected at the closed end. Standing waves will be set up for those vibrations that require the closed end to be a node and the open end to be an antinode. For a given frequency, the shortest length that satisfies these conditions is $L = \frac{1}{4}\lambda$, Fig. 22.1a. Standing waves of this same frequency will be set up in a tube whose length is $\frac{3}{4}\lambda$, Fig.

22.1*b*; in a tube whose length is $\frac{5}{4}\lambda$, Fig. 22.1*c*; or for any other odd number of quarter wavelengths.

When a body is acted upon by a periodic force that has a frequency equal to a natural frequency of vibration of the body, the amplitude of vibration of the body builds up rapidly. This phenomenon is known as *resonance*. In sound, resonance causes an intensification of the sound from a source. If a sounding tuning fork is held over the opening of a cylindrical tube closed at one end and the length of the tube is varied, the loudness of the sound will be much increased when the length of the vibrating column becomes an odd number of quarter wavelengths of the sound in air.

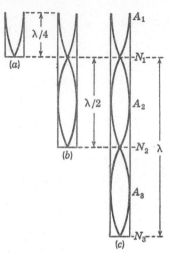

FIG. 22.1   Standing waves in air columns

Because the maximum disturbance does not occur exactly at the mouth of the tube, the length A_1N_1, Fig. 22.1, is not exactly a quarter wavelength. An end correction, equal approximately to 0.6 the radius of the tube, must be added to the measured length.

If a uniform metal rod is clamped at its midpoint and one end is stroked with a rosined cloth, the rod is set into longitudinal vibration. Standing waves are set up in the rod with a node at the midpoint and an antinode at each end. The length of the rod is a half wavelength for the sound wave in the metal.

The vibrations of the metal rod can be transmitted to a gas by means of the arrangement shown in Fig. 22.2. A light disk *D* attached to one end of the rod moves back and forth with the end of the rod and causes the gas to vibrate with the frequency of the rod. When the length of the gas column is properly adjusted,

FIG. 22.2   Diagram of Kundt's-tube apparatus

standing waves are set up in the gas. Cork dust in the bottom of the tube will be swept from the antinodes and will pile up at the nodes. The dust piles will be spaced at intervals of a half wavelength of the sound in the gas. This apparatus was devised by August Kundt and is known as Kundt's tube.

The frequency of the sound wave in the gas is the same as that in the rod. The wavelength λ_m in the metal is $2l_m$ and the wavelength λ_a in the gas is twice the average distance between two nodes. Since

$$f_m = f_a$$
$$\frac{v_m}{\lambda_m} = \frac{v_a}{\lambda_a} \tag{5}$$

From Eq. (5) and measurements of the distances in the Kundt's tube the speed of sound in the metal of the rod may be calculated.

Experiment 22.1

SPEED OF SOUND IN AIR AND IN METAL

Object: To determine the speed of sound in air and that in a metal by resonance methods.

Method: 1. A tuning fork of known frequency is sounded at the mouth of a vertical column of air in a glass tube partly filled with water. The length of the air column is adjusted by regulating the water level until the air column is in resonance with the fork. Several resonance positions are located and from the average distance between resonance positions the wavelength in air is determined. From the wavelength and the known frequency of the fork the speed of sound in air is calculated and compared with the standard value.

17. 3

2. Longitudinal vibrations are set up in the rod of a Kundt's-tube apparatus and the position of the tube is adjusted to obtain standing waves. Distances between nodes are measured and from these measurements the wavelength in air is calculated. The wavelength in the metal is determined from the length of the rod. From these wavelengths and the speed of sound in air previously determined the speed of sound in the metal of the rod is calculated.

Apparatus: Resonance apparatus, Fig. 22.3; two tuning forks of different frequency, for example, 512 and 440; Kundt's-tube apparatus, Fig. 22.4; meterstick; rubber hammer; thermometer; barometer; rosined cloth; C-clamp.

The resonance apparatus, Fig. 22.3, consists of a glass resonance tube supported on an iron stand. The water level is adjusted by regulating the height of a reservoir that slides on the iron standard and is connected to the resonance tube by

Fig. 22.3 Resonance tube

a rubber hose. A metric scale attached to the resonance tube, or mounted beside it, serves to measure the water level. A tuning fork is mounted on the standard with its prongs over the end of the resonance tube.

Fig. 22.4 Kundt's-tube apparatus

Procedure: 1. Mount one of the tuning forks on the resonance apparatus so that the prongs vibrate horizontally above the end of the tube. Raise the water level in the tube until it is near the top. Strike the tuning fork with the rubber hammer. (Never strike the fork with a hard material nor allow the prongs to strike a hard material during the vibration.) Lower the water level until the loudness of the sound is a maximum. Mark the position of this water level by means of a rubber band around the tube. Check the location several times before accepting it as final. In a similar manner locate two other resonance positions. Record the three positions. Record the frequency of the fork.

2. Repeat Step 1 using the second fork.

3. Read the barometer and record the barometric pressure. Read and record the temperature of the air in the resonance tube. Note the diameter of the tube.

4. Check the Kundt's-tube apparatus and, if necessary, adjust the rod so that it is clamped exactly in the middle and the disk does not touch the walls of the tube. Measure and record the length of the metal rod and note the kind of material. See that the cork dust is evenly spread in the tube.

Stroke the rod lengthwise with a lightly rosined cloth, but do not let your hand slip off the end of the rod. Adjust the position of the tube to produce standing waves. The best position of the tube is that for which the dust quickly collects into heaps. If the tube is gently rotated a small amount about its axis so that the dust is on the side of the tube, the dust heaps are more readily formed. Avoid too vigorous stroking of the rod because the rod becomes heated at one end and is no longer uniform. If it does become heated, cool it before proceeding.

When the dust heaps are clear, measure and record the distance between two well-defined nodes as far apart as possible. Record the number of *segments between these nodes*.

5. (Optional) Replace the air in the tube with carbon dioxide by allowing a stream of this gas to pass through the tube for two or three minutes. Repeat Step 4 with the carbon dioxide present.

Computations and Analysis: 1. Calculate the average distance between the resonance points found in Step 1. From this average distance compute the wavelength of the sound in air. From the wavelength and the known frequency of the fork calculate the speed of sound in air. Repeat for the data of Step 2.

Observe that the distance from the top of the tube to the first resonance position is less than half the distance between two resonance points. From the computed wavelength and the position of the first resonance point, calculate the position of the antinode near the top of the tube. Is it a distance six-tenths the tube radius above the top of the tube?

2. From the data of Step 3 find by the use of tables the density of air for the observed pressure and temperature. Express the pressure in dynes per square centimeter. By the use of Eq. (2) compute the speed of sound in air and compare the value thus obtained with the experimental values.

3. From the data of Step 4 calculate the wavelength of sound in air. From the length of the rod compute the wavelength in the metal. Assume the speed of sound in air to be that determined in Part 1 of the Analysis. Use Eq. (5) to calculate the speed of sound in the metal of the rod.

Find in the tables the density of the metal used and from Eq. (1) compute the modulus of elasticity for the material of the rod. Compare this experimental value with that found in tables.

4. (Optional) From the data of Step 5, compute the speed of sound in carbon dioxide, assuming as known the speed in the metal rod. Compare this experimental value with the value found in tables.

Review Questions: 1. What is a sound wave? 2. What is the relationship between the speed of sound in a medium and the properties of that medium? 3. Explain why the modulus of elasticity of a gas in Eq. (2) is γP and not P. 4. How does the speed of sound in a gas vary with temperature? 5. What is resonance? Show that various tube lengths may produce resonance. 6. Describe how wavelength and speed are measured by use of the resonance tube. 7. Explain how the speed of sound in a metal is measured by a Kundt's-tube apparatus.

Questions and Problems: 1. In what ways would the experimental data be altered if the resonance-tube experiment were carried out at a lower temperature?

2. Describe a method of using the resonance tube to determine the frequency of an unknown tuning fork if a fork of known frequency is available.

3. What is the lowest frequency to which a resonance tube 1.00 m long will respond under the conditions of your experiment?

4. At constant temperature what is the effect of increased pressure upon the speed of sound in a gas? Explain by reference to Eq. (2).

5. What happens to the tuning fork while a condensation travels down to the first resonance point and back to the top of the tube? Show how the condensations and rarefactions each build up or reinforce. Trace these constructive interferences through one or two complete cycles.

6. Explain how the relative density of air and helium could be found by using first one and then the other as the gas in the Kundt's tube.

7. Would the dust figures in the Kundt's tube be altered if the glass tube were open? Explain.

8. In the Kundt's-tube apparatus the disk is at an antinode of motion for the vibrating rod but near a node for the vibrating air column. Explain.

9. A tuning fork is held over a resonance tube, and resonance occurs when the surface of the water in the tube is 10.00 cm below the fork. It next occurs when the water is 26.00 cm below the fork. If the velocity of sound is 345 m/sec, calculate the frequency of the fork.

10. In a typical experiment performed with a Kundt's tube, the following data were obtained: temperature, 20°C; length of steel rod, 125 cm; distance between nodes of cork-dust figures, 8.00 cm. What is the frequency of the note emitted? What is the Young's modulus of the steel if its density is 7.85 gm/cm^3?

11. If the water level is at the first position of problem 9, what is the frequency of the tuning fork of next higher frequency that will produce resonance at the same position?

12. Do the data of this experiment furnish any evidence concerning the variation of the speed of sound with frequency or wavelength? Explain.

13. If the rod in the Kundt's tube were clamped at two places, each one-fourth the distance from the ends of the rod, what change should be expected in the cork-dust pattern? Explain.

PART IV—MAGNETISM AND ELECTRICITY

CHAPTER 23

MAGNETIC FIELDS

Much can be learned about the properties of magnets and the nature of magnetic fields by examining their effects on small pieces of magnetic substances, since such materials themselves become magnets and are acted upon by magnetic forces. All magnets may be classified either as *natural* or *artificial* and as *permanent* or *temporary*. Natural magnets are certain iron ores known as magnetite (Fe_3O_4). Familiar examples of permanent artificial magnets are the bar magnet and the horseshoe magnet. Electromagnets are examples of temporary magnets.

The regions near the ends of magnets show magnetic properties most clearly. Such localities are called *magnetic poles*. That end of a freely suspended magnet which tends to turn toward the north is referred to as a north-seeking or N pole; the south-seeking end is called the S pole. Experience shows that like poles repel each other, while unlike poles attract.

The quantitative law for the force between magnetic poles is known as *Coulomb's law*, which, stated symbolically, is

$$F = mm'/\mu s^2 \tag{1}$$

where F is the force which a pole of strength m exerts on another pole of strength m' placed a distance s apart.

The symbol μ is used to represent a quantity known as *permeability*, which is a concept having to do with the magnetic qualities of the medium in which the poles interact. In the electromagnetic system of units μ is arbitrarily assigned a value of unity for empty space. Air has a permeability nearly equal to 1, as do most other "nonmagnetic" materials. Iron and many alloys have values of μ which range up into the thousands.

A *unit pole* (u.p.) is one which exerts a force of one dyne on another unit pole placed one centimeter away from it in empty space.

The region near a magnet or an electric current is known as a *magnetic field*. Such a field may be defined as a region in which a force is exerted on a magnetic pole. The *direction* of a magnetic field is conventionally taken as the direction of the force on an N pole at the point considered. *Magnetic field strength H* is defined as the force per unit N pole that the field exerts on the pole

$$H = F/m \tag{2}$$

The unit of field strength is the *oersted*, which is the strength of a magnetic field in which there is a force of one dyne on a unit pole at the point considered. Field strength is a *vector* quantity.

A small magnetic compass in a magnetic field will align itself in the direction of the field because of the force action of the field on the poles of the compass. If a short compass is moved from place to place in a magnetic field in such a manner that the motion is always in the direction the N pole points, the path traced is called a line of force. Such a *magnetic line of force* is defined as the line that shows at every point the direction of the field. A typical array of lines of force is that around a bar magnet, Fig. 23.1.

When an unmagnetized object of magnetic material, such as soft iron, is placed

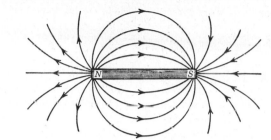

FIG. 23.1 Lines of force near a bar magnet

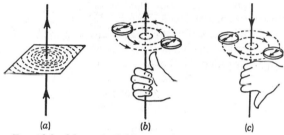

(a)　　　　　　(b)　　　　　　(c)

FIG. 23.2 Magnetic field near a long, straight current

near a magnet, the object becomes temporarily *magnetized by induction*. The portion of the object nearest the N pole of the magnet becomes an S pole. This phenomenon may be used to explain the reason iron filings tend to line themselves up with the lines of force in a magnetic field.

It is a fact of great importance that there is a magnetic field near an electric current. Such fields are responsible for the widespread use of electromagnets. The direction of the magnetic field associated with a current is given by the famous *right-hand rule:* grasp the wire by the right hand with the thumb in the direction of the conventional current; the encircling fingers show the direction of the field. For a long, straight wire the lines of force are concentric circles, with the wire as the center, Fig. 23.2. The field caused by a circular coil of a few superimposed turns is an important case, shown in Fig. 23.3. Probably

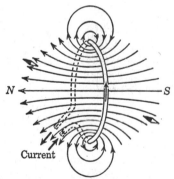

FIG. 23.3 Magnetic field of the current in a circular coil

the most widely used magnetic field is that caused by a helix, or solenoid. This field, Fig. 23.4, somewhat resembles that of a bar magnet. The polarity of an electromagnet consisting of an iron core inside a solenoid is illustrated in Fig. 23.5.

When magnetic fields are superimposed, a resultant field is produced which is the vector sum of the component fields. A compass needle points in the direction of the resultant field. For example, a bar magnet and the earth's field combine to produce the resultant field shown in Fig. 23.6. At certain places, called *neutral points*, the resultant field is zero. At such a place a small compass needle would show no directive tendency. This furnishes a method for measuring the pole strength of a magnet, if the strength of the horizontal component of the earth's

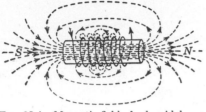

FIG. 23.4   Magnetic field of solenoidal current

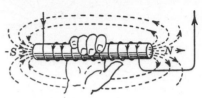

FIG. 23.5   Polarity of electromagnet

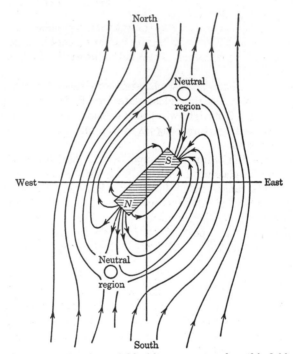

FIG. 23.6   Resultant field of bar magnet and earth's field

field is known. In Fig. 23.7 a long bar magnet is shown in an *N-S* direction with the *N* pole to the north. The field caused by the magnet at a point *P* along the perpendicular bisector of the magnet is directed toward the south, while the earth's field H_E is toward the north. At a certain neutral point the two superimposed fields will neutralize each other. For this point we may write, from the similar triangles,

$$\frac{l}{s} = \frac{H}{m/\mu s^2} \tag{3}$$

Since H, the field caused by the magnet, is equal in magnitude and opposite in direction to the known horizontal component of the earth's field and the distances are measurable, this expression can be used as a working equation for an experimental measurement of pole strength.

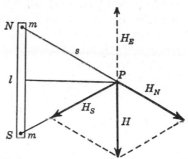

FIG. 23.7 Neutral point in fields of bar magnet and earth

Experiment 23.1

MAPPING MAGNETIC FIELDS

Object: To study the magnetic fields caused by magnets and electric currents.

Method: Iron filings are sprinkled on a glass plate under which are placed various types of magnets. The arrangement of the filings traces out the magnetic fields. The fields caused by electric currents in a straight wire, a circular coil, and a solenoid are similarly mapped. The pole strength of a bar magnet is measured by finding the neutral point when the magnet is parallel to the earth's field whose horizontal component is known.

Apparatus: Natural magnet; two horseshoe magnets; two bar magnets; soft-iron bar; brass bar; soft-iron ring; glass plate; iron filings; frame for bar magnet; small compass; mounted straight wire; coil; solenoid; battery; reversing switch.

Procedure: 1. Roll a natural magnet in iron filings and make a sketch of it. Comment on the results.

2. Sift a very fine film of iron filings upon a glass plate under which is placed a bar magnet. Tap the glass until the iron filings assume a regular shape. Make a sketch showing the arrangement of the filings. Place a small compass at different positions on the glass where the lines are distinct, and note results.

Alternate Procedure. A permanent record of the map of the iron filings can be made by sliding blueprint paper under the glass plate and exposing it to a strong source of light for about 1 min. The paper should then be removed, washed in water, and dried.

3. Repeat Step 2 using the following: (*a*) a horseshoe magnet; (*b*) two unlike poles of bar magnets with the long axis of the magnets in the same line, thus: *S*———*NS*———*N* (space about 1 in. between poles); (*c*) two like poles of bar magnet; (*d*) horseshoe magnet with soft-iron bar about 1 in. from the poles of the magnet. Repeat with brass bar. Comment on results. (*e*) Two horseshoe magnets with unlike poles facing each other and about 4 in. apart; (*f*) repeat (*e*) with a soft-iron ring between the magnets but not touching them. Note especially the shielding effect of the soft iron on the space inside the ring.

4. Examine and sketch the field around a single straight current-carrying conductor. Determine the directions of the lines of force with the compass. Repeat with current reversed. Does the right-hand rule apply in each case? (Use about 20 amp; in this and all other exercises keep the switch closed as briefly as possible.)

5. Study and sketch the field inside a flat coil. Note the effect of reversing the current. Does the right-hand rule apply? Is the field uniform?

6. Repeat Step 5 for a solenoid.

7. Find the pole strength of a long bar magnet by the method of neutral points.

Place the magnet with its axis in the N-S magnetic meridian. Locate the neutral point where the compass needle behaves as if the resultant field is zero, that is, where the needle takes no preferred direction. The region over which this condition is observed is of finite size. Move the compass back and forth along the perpendicular bisector from one place where it just shows a preferred orientation through the neutral region to a second place where it just shows orientation. Midway between these two places is the neutral point. When the neutral point is located, measure its distance to one of the poles. The poles are not localized at the ends of the magnet and a rough estimate of their position should be made by noting where the iron filings in Step 2 seem to converge. Measure the distance l between the poles. Obtain the value of the horizontal component of the earth's magnetic field from the instructor. Calculate the value of m by the use of Eq. (3)

Review Questions: 1. Classify the types of magnets. Give examples of each. 2. What are magnetic poles? State the convention concerning polarity with respect to the geographical meridian. 3. State Coulomb's law of magnetism. What is the significance of the symbol μ? State some approximate values of μ for typical substances. 4. Define unit pole. 5. What is a magnetic field? 6. Define magnetic field strength. Name and define the unit of field strength. Is field strength a vector or a scalar quantity? 7. What are magnetic lines of force? Explain how they may be mapped. Sketch the appearance of the lines of force near various types of magnets. 8. Sketch the lines of force near a straight current, a coil, and a solenoid. State the right-hand rule. 9. What is a neutral point? Explain how this concept may be used to measure pole strength.

Questions and Problems: 1. Explain clearly why iron filings line up in a magnetic field. Would copper filings be equally satisfactory? Why is it desirable to tap the glass plate? Why is it preferable to use very few filings?

2. Describe a method for mapping a magnetic field by the use of a small compass.

3. Do lines of force portray positions where the force on a unit pole is everywhere the same? If not, what do lines of force represent?

4. What is the effect of iron beams and pipes in a laboratory on the assumed standard value of the earth's magnetic field for the particular locality?

5. A magnet is placed in the east-west direction in the earth's magnetic field. Locate any neutral points.

6. When the N pole of a strong magnet is brought near a compass, which pole of the compass is repelled by the magnet? Show why the force may become one of attraction when the magnet and the needle are very close together.

7. A magnet 15 cm long has poles of strength 250 u.p. What is the resultant magnetic field strength at a point 12 cm from each pole?

8. An N pole of strength 300 u.p. is 6.0 cm in air from the S pole of a similar magnet. Calculate the magnetic field strength midway between these poles. What would the field strength be if both poles were N poles?

9. A magnet is placed in the N-S magnetic meridian with the N pole facing north. The magnet has poles of strength 350 u.p., 18 cm apart. Where is the neutral point if the horizontal component of the earth's magnetic field is 0.25 oersted?

CHAPTER 24

ELECTRIC FIELDS AND POTENTIAL

When two objects that have been electrified, or charged, are brought near each other, a force is exerted between them. Bodies of like charge repel each other; bodies of unlike charge attract each other. *Coulomb's law* gives the quantitative relationship between these electrostatic forces: The force F between charges Q and Q' varies directly with each charge, varies inversely with the square of the distance s between the charges, and is a function of the nature of the medium surrounding the charges. Stated in symbols

$$F = QQ'/Ks^2 \tag{1}$$

The factor K is introduced to provide for that property of the medium around the charges which affects this force. This concept is usually called the *dielectric constant*. In the electrostatic system of units (esu) K is arbitrarily assigned a value of 1 esu for empty space. Air has approximately the same dielectric constant as empty space.

The *electrostatic unit of charge* is defined from Coulomb's law as that charge which is acted upon by a force of one dyne when placed one centimeter away from another unit charge in a vacuum. This unit of charge is usually called a *statcoulomb*.

The region in the vicinity of a charged body possesses special properties because of the presence of the charge. For example, another charge brought into this region will experience a force. Such a space is known as an electric field. An *electric field* is defined as any region in which there would be a force upon a charge brought into the region.

The direction of an electric field at any point is defined as the direction of the force upon a positive charge placed at that point.

Electric field intensity is analogous to magnetic field strength in magnetism. *The intensity (or strength) of the electrostatic field* at a point is defined as the force per unit positive charge at that point. The symbolic defining equation is

$$\mathcal{E} = F/q \tag{2}$$

where $\mathcal{E}$ is the electric field intensity and F is the force exerted upon the charge q. Note that field intensity is a vector quantity. The electrostatic unit of electric field intensity is the dyne per statcoulomb.

The direction of electric fields at various points may be represented graphically by lines of force. A *line of force* in an electric field is a line so drawn that a tangent to it at any point shows the direction of the electric field at that point.

The diagram of Fig. 24.1 shows a plane section of the electric field near a pair of charges that are equal in magnitude by opposite in sign. The field at any point is the resultant of the superposition of the two component fields due to the charges at A and B. For example, at point b the direction of the resultant force

161

on a positive charge would be along the vector drawn tangent to the line of force at that point.

Since there is a force on a charge in an electric field, work is usually required to move the charge from one place to another in the field.   The amount of work

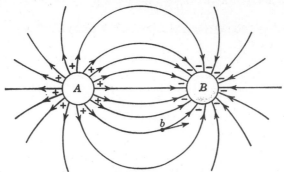

Fig. 24.1   Lines of force representing the electric field near two equal charges of opposite sign

required is related to the difference of electric potential between the points considered.   This *potential difference* V is defined as the work done per unit charge transferred between the points in question.   The defining equation for potential difference is

$$V = W/q \qquad (3)$$

where W is the work done and q the charge moved.   Conventionally q is always understood to be a positive charge.   In the electrostatic system, V is expressed in *statvolts* when W is in ergs and q in statcoulombs.   One statvolt is approximately equal to 300 volts.

Consider the simple electric field illustrated in Fig. 24.2.   The charge $+Q$ produces an electric field with lines of force extending radially outward.   A small test charge $+q$ at any point in the field is acted upon by a force.   Hence work is required to move q between any such points as B and C.   The work done per unit positive charge is a measure of the potential difference between these points.   From the principle of the conservation of energy the work done in moving a charge from one point to another in an electric field is independent

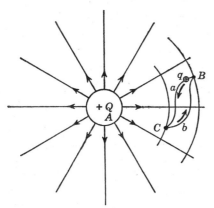

Fig. 24.2   Potential difference in electric field

of the path over which the charge is transported.   Otherwise energy could be created or destroyed by moving a charge from one point such as B in Fig. 24.2 to C by path a, requiring a certain amount of work, and returning by path b, requiring a different amount of work.

Since both work and charge are scalar quantities, potential must be a scalar quantity. The potential near an isolated positive charge is positive in the sense that work must be done by an external source to move a positive test charge from a point outside the field up to any point near the first charge. The potential near an isolated negative charge is negative, because energy is supplied by the field in moving a positive test charge from a point outside the field to any point near the negative charge.

It may be shown that *the absolute potential V of a point near an isolated charge Q* is given by the equation

$$V = Q/Ks \tag{4}$$

where s is the distance from the point in question to the charge that produces the field. The resultant potential of a point situated near a number of charges is the *algebraic sum* of all such component potentials; this may be symbolically stated

$$V = \sum \frac{Q}{Ks} \tag{5}$$

A surface so selected that all points on it have the same potential is called an *equipotential surface*. A line on such a surface is known as an *equipotential line*. Equipotential surfaces in an electric field are always perpendicular to the lines of

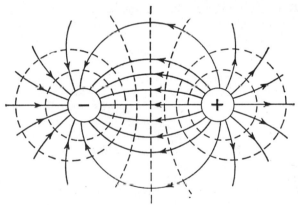

FIG. 24.3 Equipotential lines (shown broken) in the field of two equal charges of unlike sign

force because the line of force shows, by definition, the direction of the force upon a test charge, and there can be no force normal to this direction. Hence no work is done in producing a small displacement of a test charge normal to a line of force, and this normal is therefore an equipotential line. If the equipotential lines can be drawn, the lines of force can be immediately constructed; they are at all points perpendicular to the equipotential lines that they intersect. For example, in Fig. 24.3 are shown the lines of force and the equipotential lines in a plane surface containing two charges of equal magnitude and opposite sign.

When a potential difference is maintained between points on a conducting surface, there is a flow of electricity from the place of higher to the place of lower potential. The lines of flow are the paths followed by the electric charges. These lines of flow are perpendicular to the equipotential surfaces.

That these lines of flow have exactly the same configuration as the lines of force in an electrostatic field must follow from the fact that their configuration is unchanged as the current is reduced. If the current were reduced to zero by increasing the resistance of the plate, the lines of flow would thus become lines of force; the equipotential lines would remain unchanged throughout.

Experiment 24.1

MAPPING EQUIPOTENTIAL LINES

Object: To map the equipotential lines and lines of force in the electric field of two equal charges of unlike sign.

Method: An electric field is produced in a high-resistance material by connecting two electrodes near the edges to a source of electric current. Points of

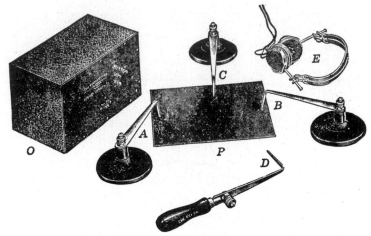

Fig. 24.4 Electric field mapping apparatus

equal potential in this field are discovered by using two additional **movable electrodes** connected to a device that indicates a potential difference when the electrodes are not on an equipotential line. A series of points thus located traces out an equipotential line. The first exploring electrode is then moved and another equipotential line is similarly plotted. After the whole field is thus explored the lines of force are drawn so that they are everywhere normal to the equipotential lines.

Apparatus: Poorly conducting surface; source of potential (10-volt battery or an audio oscillator); four electrodes; galvanometer, or headphones; mapping paper.

Three types of poorly conducting surfaces are described; only one need be used in the experiment.

Type 1. A sheet of graphite-coated paper P (Fig. 24.4) has a high electrical resistance. A battery (or an audio oscillator) is connected to the electrodes A and B. The movable electrodes C and D are connected through a galvanometer (or headphones). The plot of equipotential lines is made directly on the black paper by using chalk marks or pin pricks.

Type 2. A poorly conducting plate is mounted on a wooden base (Fig. 24.5) with a pantograph device for the semiautomatic transfer of the equipotential points from the plate at D to the record sheet at S.   The auxiliary pieces are the same as those shown with the Type 1 apparatus.

Type 3. The poor conductor is a liquid in a glass tray (Fig. 24.6), with the same accessories as used with the Type 1 apparatus.   Ordinary tap water is usually

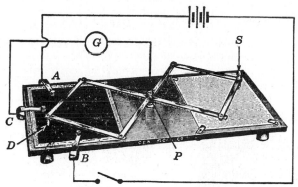

FIG. 24.5   Pantograph device for mapping equipotential lines

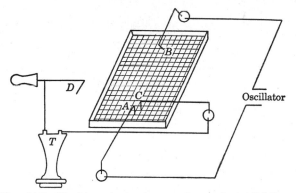

FIG. 24.6   Liquid conductor for mapping equipotential lines

sufficiently conducting for this purpose; if not, a little salt may be added.   A battery gives poor results with Type 3 apparatus; an audio oscillator O (Fig. 24.4) and a set of headphones E are preferable.

The audio oscillator and headphones may be used with any of the three types of apparatus.   A small induction coil or some other source of alternating or intermittent current may be substituted for the audio oscillator.

Procedure: 1. Connect the potential source through a switch to the two fixed electrodes A and B, placed upon and near the end of an imaginary line running lengthwise through the center of the conductor.   Attach the galvanometer (or headphones) to the other electrodes C and D.   (In the pantograph form make the D connection at binding post P; clamp the record paper to the board after inserting a piece of thick cardboard or a number of sheets of paper under the record sheet. Adjust the needle point by means of the locknut so that it just protrudes when pressed down.)

2. Locate first on the record sheet the positions of the electrodes to which the battery is attached. Set the electrode to which the galvanometer is attached at any desired position. Then move the exploring electrode over the conductor until the galvanometer deflection is zero. It is well to move the exploring point until a deflection is obtained first in one direction and then in the opposite direction; move it back and forth across zero until the point of no deflection is located. The point thus located has the same potential as the reference point and should be shown on the student's sheet. Locate in this manner a sufficient number of similar points to determine the locus of the equipotential line. For maximum sensitivity in locating the points, the exploring electrode should be moved back and forth in a direction perpendicular to the approximate line of equipotential points. Pay particular attention to the shape of the lines near the edge of the plate.

3. Having finished one line, move the reference point and repeat the process until the entire field has been mapped.

4. Place a circular disk of metal near one of the main electrodes upon the conductor, weight it down firmly with a heavy object, and again map the field in the modified or distorted condition. In determining these equipotential lines, pay particular attention to the region near the disk. What is found about the potential difference between the various points *on* the disk (not between the disk and a point *outside* the disk)?

5. *Optional Exercises:* Replace the point terminals A and B by rectangular bars and note the parallel type of field thus produced. Then use one rectangular bar for A and the point electrode for B and map this field.

6. Trace the equipotential lines in black ink. Then sketch in, in colored ink, the lines of flow. They should be drawn everywhere perpendicular to the equipotential lines. In the case of the field distorted by the disk, draw the lines of flow rather thickly near the disk, where the lines of flow pass perpendicularly into the disk.

Review Questions: 1. What type of force is observed between like charges? between unlike charges? 2. State Coulomb's law of electrostatics, both in words and in symbols. 3. What is the significance of the factor K in the equation for Coulomb's law? 4. Define the esu of charge. What is it ordinarily called? 5. Explain what is meant by an electric field; by direction of an electric field. 6. Define electric field intensity. What is the esu of field intensity? Is it a vector or a scalar concept? 7. What are lines of force in an electric field? Sketch the lines of force near (*a*) a positive charge; (*b*) a negative charge; (*c*) a positive charge near an equal negative charge. 8. Define potential difference. What is the esu of potential? 9. Show why the PD between two points in an electric field is independent of the path. Is PD a vector or a scalar quantity? 10. State the expression for the electric potential of a point near a single charge; near a number of charges. 11. What is an equipotential surface? an equipotential line? How are equipotential lines related to lines of force?

Questions and Problems. 1. A map of equipotential lines has places in which the lines are much closer together than in other regions. Is the field strong or weak where the lines are close together? Explain.

2. State some similarities between gravitational, magnetic, and electric fields

and the laws of force in these cases. What differences are there in the respective phenomena?

3. Sketch a curve to show how the electric field intensity near a charge varies with distance from the charge.

4. Draw an approximate curve to indicate the variation of electric potential with distance away from a concentrated charge.

5. Sketch the lines of force and the equipotential lines for two nearly equal like charges.

6. A small positively electrified metallic sphere is brought into contact with a similar uncharged sphere and then withdrawn. At a distance of 10 cm the two spheres repel each other with a force of 16 dynes. Find the original charge on the electrified sphere.

7. Three points A, B, and C are on the corners of an equilateral triangle. At A there is a charge of $+300$ esu. At B there is a charge of -300 esu. If the distance from A to B is 10 cm, what is the electric field strength at C? What would be the force on a charge of 50 esu placed at C?

8. Two like charges each of 275 statcoulombs are 15.0 cm apart, in air. What is the field intensity midway between them? What is the potential?

9. Answer question 8 for two unlike charges.

10. How much work is required to move a charge of 33 statcoulombs from a point 50 cm away from a concentrated charge of 825 statcoulombs to a place which is only 35 cm from the charge?

11. A charge of 12.5 statcoulombs is moved 25.0 cm against an electric field that increases at a constant rate from 10.0 to 42.3 dynes/statcoulomb. What work is required to move the charge?

CHAPTER 25

ELECTRIC INSTRUMENTS AND CIRCUITS

Most electrical measurements involve the detection or measurement of electric currents. The basic electric instrument for this purpose is the galvanometer. Ammeters and voltmeters also utilize the galvanometer principle. An elementary acquaintance with the way in which these instruments and other electrical devices are used is helpful, even before one is ready to master the principles involved in their design and operations.

Ohm's Law. Many of the instruments to be considered in this chapter involve a knowledge of the concepts of electric *current, voltage,* and *resistance.* These concepts are related by the famous principle known as *Ohm's law:* In a given metallic conductor the ratio of the voltage to the current is a constant, known as the resistance

$$\frac{\text{Voltage}}{\text{Current}} = \text{resistance}$$

Stated symbolically, using V for voltage, I for current, and R for resistance,

$$V/I = R \tag{1}$$

This equation may be expressed in the alternate forms

$$I = V/R \quad \text{or} \quad V = IR \tag{2}$$

The practical unit of current is the *ampere;* the unit of voltage is the *volt;* the unit of resistance is the *ohm.*

Conventional Wiring Diagrams. Much of the apparatus frequently found in electric circuits is represented by conventional forms in wiring diagrams. Some of those forms are illustrated in the Appendix.

A simple series circuit containing a battery, SPST switch, a rheostat, and an

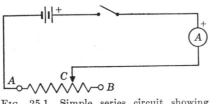

ammeter for measuring current are shown schematically by the conventional wiring diagram of Fig. 25.1. The slide-wire rheostat is a convenient form of resistor for use with comparatively large currents. When the connections are made to the terminal A (or B) and the sliding contact C,

Fig. 25.1 Simple series circuit showing ammeter

the resistance is variable between zero and the value indicated on the plate of the rheostat. When the connections are made to A and B (and not to C), the resistance is fixed at the total value indicated.

As shown in Fig. 25.1 the current in a circuit is measured by placing an ammeter in *series* in the circuit. The positive terminal of the ammeter is the one at which

168

the flow of *conventional* positive electricity enters.   The negative terminal is the one at which the *electron* (negative) flow enters.

Voltmeters are placed in *parallel* with the device whose voltage is to be measured.   For example, in Fig. 25.2 the voltmeter measures the voltage between the terminals *A* and *B* of the resistor.

Resistors. In addition to the slide-wire rheostat previously mentioned, heavy currents are sometimes controlled by *carbon rheostats*.   These devices are made of a series of graphite disks enclosed within insulated iron tubes.   Varying pressures are applied to the disks by a wheel and screw.

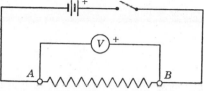

FIG.   25.2   Voltmeter in parallel with resistor

The variable contact resistance offered by the graphite enables the resistance of the rheostat to be adjusted, within moderate limits, by continuous variations.

The most commonly used form of carefully adjusted resistor is the *resistance box*.   The design of the plug-type box is illustrated in Fig. 25.3.   When a plug is removed the electricity must go through the resistance wire, thus inserting the

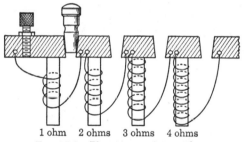

1 ohm   2 ohms   3 ohms   4 ohms

FIG. 25.3   Plug-type resistance box

indicated resistance into the circuit.   In this type of resistance box only small currents (less than 1 amp) may ordinarily be used.   Higher currents cause excessive heating and seriously injure the resistance spools.

In addition to the plug-type resistance box, Fig. 25.4*a*, many forms of dial-type boxes are used, Fig. 25.4*b*.   A familiar form is the dial-decade type, in which one dial varies the resistance in steps of 1 ohm from 1 to 9 ohms, the next dial varies the resistance in steps of 10 ohms from 10 to 90 ohms, etc.

The greatest single source of error and annoyance in electrical experiments is usually found in poor, loose, and variable contacts.   The erratic change in resistance of bad contacts results in variable currents and the introduction of stray resistances that are not included in the values indicated by the resistance box. This difficulty may be minimized in a plug-type resistance box by keeping all the plugs clean and well seated in the holes between the brass blocks.   Plugs should be inserted with a firm downward push, accompanied by a twisting motion. Plugs should be removed with a similar upward twisting pull.   Contactors under binding posts should be very tightly fastened.

Cells and Batteries.   The most familiar type of portable cell is the *dry cell*. A dry cell is an example of a *primary cell*, which is a type that must be replaced or have its elements renewed after a certain amount of use. When several cells are connected together, the combination is called a *battery*. A widely used type of battery is that used in automobiles. Such a battery is made up of three storage cells in series. A *storage* cell is one that can be recharged by a reversed current, from the generator in the case of the automobile. The automobile battery has

(a)   (b)

FIG. 25.4   Precision forms of resistance boxes: (a) plug type; (b) dial-decade type

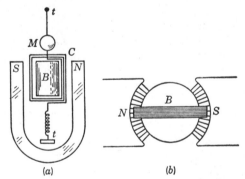

(a)   (b)

FIG. 25.5   Essential features of the d'Arsonval galvanometer

lead plates and the electrolyte is sulphuric acid. A battery used extensively in student laboratories is made up of *Edison* cells, which are lighter and more rugged than the lead-storage cell.

Electric Meters. Most d-c meters utilize the basic design of the d'Arsonval galvanometer, Fig. 25.5. A movable coil C, Fig. 25.5a, is suspended between the poles N and S of a U-shaped magnet by means of a light metallic ribbon. Connections to the coil are made at the terminals marked t. The cylinder of soft iron B and the pole faces are skillfully shaped to produce the radial magnetic field shown in Fig. 25.5b. The deflection of the coil is read by reflecting a beam of light from the mirror M onto a scale or by the use of a telescope and scale, (see Fig. 28.2). The current in the coil produces a magnetic field, which reacts with the field of the magnet to produce the turning torque. The controlling counter torque is furnished by the torsion of the suspension. The return of the coil to its zero position is accomplished by this *control* device.

In order to prevent the coil from oscillating about the zero position when the current is interrupted, or around the final equilibrium position when there is a steady current, some form of reduction of this motion, called *damping*, must be provided. Damping is usually accomplished by the generation of induced currents developed by the relative motion of the conducting frame of the coil and the stationary magnetic field of the magnet. These induced currents tend to stop the motion and thus provide effective damping.

In many portable galvanometers and other such instruments the suspension is in the form of hardened steel pivots in jeweled bearings. In these instruments the control is furnished by spiral springs above and below the coil. Readings are made by a light pointer attached to the coil and moving over a fixed scale (see Fig. 29.1).

The current required to produce a full-scale deflection of a galvanometer is very small. Ammeters used for large currents have essentially the same types of movable coils and magnets as the portable galvanometer with jeweled bearings. Such an ammeter is provided with a low-resistance bypass (shunt) in parallel with the coil so that the shunt carries most of the current. Hence ammeters have very low resistances.

FIG. 25.6   A multirange meter

The voltage across a galvanometer even at full-scale deflection is very small, sometimes only a few thousandths of a volt (millivolt) or more often only a few millionths of a volt (microvolt). Voltmeters consisting of a galvanometer in series with a high-resistance coil, called a multiplier, are used for large voltages. Hence voltmeters have high resistances.

A single multirange instrument is often used instead of a number of separate meters. An instrument of this type, shown in Fig. 25.6, may be used either as a galvanometer, millivoltmeter, voltmeter, milliammeter, or as an ammeter.

Experiment 25.1

CONNECTION OF ELECTRICAL DEVICES IN CIRCUITS

Object: To observe the pertinent facts concerning the connection and use of basic meters and apparatus in electric circuits.

Method: The apparatus is permanently set up and the observer moves to the various locations and notes the essential facts associated with the particular devices to be studied.   Adjustments of the apparatus are made as directed herein or as typed on cards at the locations.

Apparatus: Galvanometers; ammeters; voltmeters; multirange meters; resistors; switches; cells and batteries.

Procedure: 1. Before beginning this experiment the student should carefully note in the Appendix the Conventional Forms for Wiring Diagrams.

2. Follow the instructions given herein and on the cards at the various tables. The tables may be visited in any order in which the apparatus is free, although it is desirable to pass in sequence to Tables 4, 5, and 6.

Table 1. *Types of Resistors.*   Examine a set of resistance spools, observing the different types of wire.   Make a sketch of the block of coils and show connecting wires leading to coil No. 3.   Make a drawing of the rheostat, showing how it could be used as a fixed resistor; as a variable resistor.   Sketch the rheostat as adjusted to give the resistance indicated on the tag.   Return the rheostat slide to the end position after use.   Examine and sketch the carbon rheostat.

Table 2. *Resistance Boxes.*   Inspect a plug-type resistance box that has been disassembled to allow study of the internal construction.   Note the type and arrangement of the resistance spools, the manner in which the plugs are inserted, and the paths of the current.   Make sketches of the various resistance boxes to indicate their external features.   Show on the sketches the plugs or dials arranged to give the value indicated on the tag attached to each box.   Return each box to zero resistance after use.

Table 3. *Galvanometers.*   Study carefully the various galvanometers available; make a sketch of each type to indicate the essential features.   Make a tabulation as in the accompanying table to describe the main characteristics of such instruments as the following: tangent galvanometer, wall-mounted galvanometer, portable galvanometer with ribbon suspension, and galvanometers with jeweled-bearing suspensions.

Manufacturer	General type	Suspension	Control	Damping	Reading device
Cenco	Tangent galv. with fixed coil, movable magnet	Jeweled bearing	Earth's magnetic field	Friction	Fixed pointer

Adjust the plane of the coil of the tangent galvanometer so that it is in the direction of the earth's magnetic field, as indicated by the compass needle.   Note the fact that the pointer is at right angles to the needle.   Observe the deflection when the switch is closed; note the rather poor damping.   Reverse the current and note the effect produced.   Leave the switch closed as briefly as possible.

Demonstrate the sensitivity of the wall-mounted galvanometer by holding the lead wires in your hands and noting the deflection caused by the small emf generated from the voltaic action of the moisture on the hands.   Touch the terminals to your tongue and notice the larger deflection produced.

Note the functions of the clamp and zero-adjusting device on the portable Leeds & Northrup galvanometer.

Table 4. *Connection of Ammeters in Circuits.* Draw a wiring diagram of the circuit to show that the ammeter is connected in *series* in the circuit. Indicate the polarity of the cell and the ammeter terminals. (The red bushing on the cell indicates the positive terminal.) Note the effect of moving the rheostat slide. Record the extreme values of the current. What is the purpose of the friction tape on the rheostat?

Table 5. *Connection of Voltmeters in Circuits.* Draw a wiring diagram to show that the voltmeter is connected in *parallel* with the device whose voltage is to be measured. Note on the diagram the polarity of the cell and the voltmeter terminals. Observe the difference between the way this rheostat is connected to the cell and that of the rheostat at Table 4. Note the effect on the voltmeter of moving the slider of the rheostat. Explain why the voltmeter reading does not change. Note the type of cell used here (Edison storage cell). Record its voltage as indicated by the voltmeter reading.

Table 6. *Ammeters and Voltmeters in a Single Circuit.* The observations at Table 6 are repetitions of the observations at Tables 4 and 5, except that both meters are here connected in one circuit. Draw the wiring diagram; note the polarity of the cell and the meter terminals.

In this setup a battery of two Edison storage cells in series is used. Note the type of battery used; measure and record the voltage of the battery. How does this reading compare with the voltage of the single cell observed at Table 5? Comment on these two values.

Table 7. *Poor Contacts in Electric Circuits.* Make a wiring diagram of the series arrangement of a battery, SPST switch, rheostat, resistance box, and ammeter. Note the effect of removing a plug from the resistance box. (Remove the plug for a large resistance.) Insert the plug very loosely and compare the current thus obtained with the value when the plug is firmly seated. Jiggle the slider of the rheostat and note the effect on the current. (A rheostat with poor contact at the slider is purposely provided.) Loosen, without removing, the wire at one of the terminals in the circuit and note the effect on the current when this wire is jiggled.

Table 8. *Multirange Meters.* Examine a multirange volt-ammeter, noting the fact that it may be used either as a voltmeter, or as an ammeter. Sketch this meter, showing the various ranges available (not a conventional diagram of the meter). Study the two circuits in which a volt-ammeter is used (*a*) as a voltmeter and (*b*) as an ammeter. Combine conventional wiring diagrams of these circuits with actual sketches of the meters to show the ranges and connections used.

Note the type of cell used (dry cell). Record its voltage.

Comment on the advantage of a multirange meter in comparison with separate meters. Examine an alternating-current (a-c) meter and note the difference between its scales and those of a d-c meter. Record the type and ranges of each of the other available volt-ammeters.

Table 9. *Double-pole Double-throw Switch.* Examine the circuit and make a conventional wiring diagram of it. Note the fact that the battery may be connected either to a 15-volt voltmeter or a 150-volt meter. Record the readings of the two voltmeters, first noting carefully the value of the smallest scale division for each range used and then making a careful estimate of fractional divisions for the final readings. (Include the proper number of final zeros if the pointer seems to fall exactly over a scale line.) From the observed data comment on the number of significant figures that may properly be used for the two readings. Which range should be selected when a multirange instrument is to be used?

Note the type of battery (lead-acid) at this table. From its voltage, as read by the voltmeters, what should be the voltage of *one* of the three cells that are connected in series? Compare the voltages of the lead-acid cell, the Edison storage cell, and the dry cell, as observed in Steps 5, 6, 8, and 9.

Table 10. *Reversing Switch.* Trace the current in the circuit of the cell connected through a reversing switch and a resistance box to a zero-center galvanometer. One or more plugs have been removed from the resistance box. This resistance should not be changed. Reverse the switch and note the result. Make two wiring diagrams and show on them by a series of small arrows the direction of the current at various places.

Review Questions: 1. State Ohm's law, both in words and in symbols. Name the units of each concept used. 2. How are ammeters and voltmeters inserted in circuits? Illustrate by wiring diagrams. 3. Show by sketches how a slide-wire rheostat may be used either as a fixed or as a variable resistor. 4. Sketch a plug-type resistance box to illustrate its essential features. How is the total resistance obtained when a number of plugs are removed? Sketch a dial-decade box. 5. Describe the effects of poor contacts in electric circuits. How are plugs properly inserted in and removed from resistance boxes? 6. Distinguish between a cell and a battery; between primary and storage cells. Mention some common types of cells. 7. Describe the essential features and design of the d'Arsonval galvanometer. 8. Differentiate between various types of galvanometers, with respect to the parts producing the stationary and moving fields. Describe two methods of suspending galvanometer coils. What are some methods of control in galvanometers? What is damping and how is it effected in galvanometers? Describe several methods of reading galvanometers. 9. What is the advantage in the use of a multirange meter?

Questions and Problems: 1. What is the present picture concerning the nature of the electric particles that flow in a metallic circuit? in electrolytic conduction? in gaseous discharges? What is the usual agreement with regard to the direction of the conventional current?

2. What determines which one of the terminals of a rheostat is positive when it is connected in a circuit? Illustrate your answer by a diagram.

3. Although the cost of electric energy from dry cells is very high, such sources are widely used. Why is this true?

4. Does a storage cell store up electricity? Why is it properly called a storage cell?

5. Does an electric heater "use up" current? What happens to the current in such a device as time goes on? What is used up in the heater?

6. The voltage applied to a rheostat is doubled, then trebled. What happens to the resistance of the rheostat?

7. State some of the ways in which the design of a galvanometer might be changed in order to increase its sensitiveness.

8. Compare the relative precision with which the following resistors are calibrated with respect to their stated resistances: (*a*) slide-wire rheostat; (*b*) carbon rheostat; (*c*) resistance box.

9. Comment on the relative current-carrying abilities of the resistors mentioned in question 8.

10. A battery with a terminal voltage of 6.00 volts is connected in series with a rheostat and an ammeter that reads 5.00 amp. Neglecting the resistance of the ammeter, what is the resistance of the rheostat?

11. An incandescent lamp is designed for a current of 0.60 amp. If a potential difference of 110 volts is necessary to maintain that current, what is the resistance of the lamp?

12. In a certain rheostat there is a current of 0.45 amp when the difference of potential between the terminals is 60 volts. What is the resistance of the rheostat?

13. A certain wire used in electric heaters has a resistance of 1.75 ohms/ft. How much wire is needed to make a heating element for a toaster that takes 8.25 amp from a 115-volt line?

CHAPTER 26

VOLTAGE, CURRENT, AND RESISTANCE

The relationship between the voltage, current, and resistance in a given metallic conductor is given by the most common and important single principle in the field of current electricity. This relationship, known as *Ohm's law*, is stated as follows: If the temperature and other physical conditions of a metallic conductor are unchanged, the ratio of the potential difference to the current is a constant. This constant ratio is the resistance of the conductor. This word statement may be put into the symbolic form

$$V/I = R \qquad \text{(a constant)} \qquad (1)$$

where V is the voltage across the conductor, I the current in it, and R is its resistance. The usual units of these quantities are current in *amperes*, potential difference in *volts*, and resistance in *ohms*.

The potential difference across a device may be measured by connecting a high resistance voltmeter V in parallel with it. In Fig. 26.1 is shown the conventional wiring diagram in which a voltmeter is used for the measurement of the terminal voltage of a battery.

The current in a device is measured by inserting an ammeter in series in the circuit. All the electricity thus has to pass through the ammeter. Because most

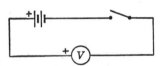

Fig. 26.1 Measurement of voltage of battery by voltmeter

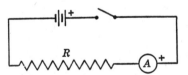

Fig. 26.2 Ammeter measuring current in resistor

ammeters have a very small resistance, they do not greatly affect the current. In Fig. 26.2 an ammeter A is shown in series with a resistor R in a simple circuit.

Ohm's law may properly be applied either to the entire circuit or to any portion of the circuit in which there is no emf. Whenever this law is applied to the entire circuit, the voltage to be used is the net emf E and the resistance is the total resistance R_t of the circuit

$$E/I = R_t \qquad (2)$$

For example, the approximate emf of a storage battery may be measured as in Fig. 26.1. The battery may then be connected to a resistance box R in series with an ammeter, as in Fig. 26.2. From the emf of the battery and the observed current, the resistance R might be calculated from Eq. (2) and compared with the known value of the resistance of the box.

Ohm's law is valid for any portion of a circuit containing resistors only, provided that the individual currents, potentials, and resistances are used strictly for the part of the circuit in question. In this case the symbols of Eq. (1) may be used where the voltage V represents the potential difference across that part of the circuit under consideration, I is the current in that part, and R is the resistance of that part (only).

It is important to remember that Eq. (1) can never be applied to a part of a circuit that contains any device which generates an emf, such as a battery, generator, or motor.

Ohm's law in the form $V/I = R$ may be demonstrated by noting the voltages across a known fixed resistor when the current in the resistor is varied by an outside control. The graph of the measured voltages plotted against observed currents is a straight line through the origin; the slope of this line gives a resistance equal to that of the known resistor.

A second way of studying Ohm's law, in the form $I = V/R$, consists in observing the currents in known resistors across which a constant voltage is maintained. The graph of current against the reciprocal of resistance is a straight line through the origin; the slope of this line is a voltage that equals the known impressed potential difference.

The third demonstration of Ohm's law, in the form $V = IR$, involves the measurement of the voltages across various portions of a known resistor in which there is maintained a known constant current. The curve of voltage against resistance is a straight line through the origin; the slope of this line is the known current.

In the variation of one factor with another it is of course essential to maintain all other quantities constant during the investigation. Otherwise the results are not easily interpreted.

The resistance of a metallic conductor varies directly with the length, inversely with the area of cross section, and depends upon the material of the conductor and its temperature. These items are discussed in detail in Chap. 30.

Experiment 26.1

MEASUREMENT OF RESISTANCE BY OHM'S LAW

Object: To study the application of Ohm's law to resistance measurements and to investigate the factors upon which the resistance of a metallic conductor depends.

Method: 1. The current in a resistance wire is varied by changing the applied voltage. The currents are measured by an ammeter and the voltages by a voltmeter. A voltage-current curve is plotted and the constant slope yields a value of the resistance which can be compared with the known value.

2. The potential drops across various lengths of the known resistance wire are measured while the current is kept constant. The slope of the voltage-resistance curve yields a value of the current, which is compared with the observed value.

3. The resistances of several wires of various sizes and materials are measured by the ammeter-voltmeter method. From these data the variation of resistance with length, area, and material is noted.

Apparatus: Storage battery, 2 to 3 volts (or two dry cells); ammeter with 0.5 and 5-amp ranges; 3-volt voltmeter; SPST switch, fused for 1 amp; 10-ohm rheostat; mounted 2-m slide-wire, with contactor (Fig. 26.3); resistance board (Fig. 26.4).

The resistance board has four wires, each 1 m long. The first wire is No. 30 nickel-silver, diameter 0.0255 cm; the second wire is made of the same material but is No. 24, diameter 0.0510 cm. The third wire is No. 30, iron; the fourth wire is No. 30, copper.

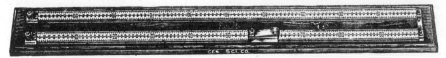

FIG. 26.3 Slide-wire resistor. For convenience the wire is doubled back along a second meterstick

FIG. 26.4 Resistance board

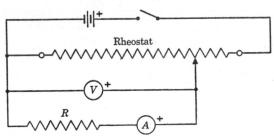

FIG. 26.5 Circuit for studying the variation of potential difference with current (resistance constant)

Procedure: In all parts of this experiment keep the switch closed only long enough to observe the necessary data. In the various steps one terminal of the battery should be left unconnected and the switch left open until the instructor has checked the connections.

1. *Variation of Voltage with Current, Resistance Constant.* Connect the apparatus as in Fig. 26.5; use for R the entire 2-m resistance wire. Arrange the 10-ohm rheostat as a potential divider so that by moving the rheostat slider the voltage across the resistance wire may be varied from zero to the maximum voltage of the battery.

Take a series of five readings of current and voltage for currents of approximately 0.1, 0.2, 0.3, 0.4, and 0.5 amp. Read each meter to the maximum possible number of significant figures. Tabulate the currents and voltages and record the resistance of the 2-m wire.

2. *Variation of Potential along a Resistance Wire, Current Constant.* Connect the apparatus as in Fig. 26.6. Be sure that the terminal A is the zero end of the wire. After the instructor has approved the connections note and record the

voltages across AC for lengths of wire of 40, 80, 120, 160, and 200 cm. Record the current in the wire.

3. *Variation of Resistance with Length, Area, and Material.* Connect the apparatus as in Fig. 26.5 Use for R the first wire of the resistance board (No. 30, nickel-silver). Set the rheostat slider at zero resistance to begin with in each of the following tests. After the instructor has checked the connections adjust the rheostat slider to increase the voltage applied to the resistance wire until the current approaches the maximum of 0.5 amp permitted by the range of the ammeter. Read and record the voltage and current for this maximum setting.

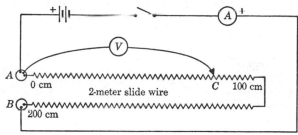

FIG. 26.6 Circuit for studying the variation of potential difference with resistance (current constant)

Return the slider to zero and replace the first wire by the second wire. Gradually increase the applied voltage by moving the slider until the maximum of 0.5 amp is reached. Read and record the voltage and current.

Repeat this process for the third and fourth wires.

Computations and Analysis: 1. Use the data of Step 1 to plot a curve to show the variation of voltage (ordinates) with current (abscissas). Carefully interpret the significance of the shape and intercepts of this curve. Show why the slope of the curve is the resistance of the 2-m wire. Use the origin and a point near the upper part of the curve to calculate the slope. Record the percentage difference between the resistance as calculated from the slope and the value marked on the wire.

2. Use the voltages and current observed in Step 2 to calculate the respective resistances of the various lengths of wire. Plot a curve of resistance against length. Interpret the shape and intercepts of this curve. Calculate its slope. What is the significance of the slope?

3. Use the data of Step 3 to calculate the resistance of each of the wires on the resistance board.

a. Since the first and second wires have the same length and are the same material, how should their resistances vary with diameter? Calculate the ratio R_1/R_2 and compare it with the ratio D_2^2/D_1^2. Record the percentage difference between these ratios.

b. The first, third, and fourth wires have the same length and area, but are of different materials. Calculate their relative resistances. Nickel-silver has a resistance 18 times that of copper, while iron has 5.7 times the resistance of copper (all dimensions being equal). Do you find the observed resistances to be in approximately these ratios?

Experiment 26.2

OHM'S LAW

Object: To study the application of Ohm's law to simple circuits; in particular, to observe the variation of (1) current as a function of potential difference in a given metallic conductor, (2) current as a function of resistance when the voltage is kept constant, and (3) voltage as a function of resistance when the current is maintained constant.

Method: (1) The voltage across a fixed resistor is varied and the currents are observed for a series of values of the voltage. A curve of voltage against current is plotted and interpreted in terms of Ohm's law. (2) The voltage across a resistance box is held constant as the resistance is varied and the currents are measured. A curve of observed currents plotted against the reciprocals of the known resistances is interpreted as an illustration of Ohm's law. (3) The potential drops along various portions of a resistance wire of known resistance per unit length are measured while the current is kept constant. From the slope of the voltage-resistance curve the current is calculated and compared with the measured current in the wire.

Apparatus: Storage battery, emf about 3 volts; heavy-duty resistance box, 0.1 to 111 ohms; 0.5-amp ammeter; 3-volt voltmeter; SPST switch, fused for 1 amp; 10-ohm rheostat; mounted 2-m resistance wire, with contactor (Fig. 26.3).

Procedure: In all parts of this experiment, keep the switch closed only long enough to observe the necessary data. In the various steps one terminal of the battery should be left unconnected and the switch left open until the instructor has checked the connections.

1. *Variation of Current with Voltage, Resistance Constant.* Connect the apparatus as in Fig. 26.5. Use the resistance box for R. Arrange the 10-ohm rheostat as a potential divider so that by moving the rheostat slider the voltage across the resistance box may be varied from zero to the maximum terminal voltage of the battery. Adjust the resistance in the box to 5.00 ohms. After the instructor has approved the connections, take a series of five readings of current and voltage for currents of approximately 0.1, 0.2, 0.3, 0.4, and 0.5 amp. Read each instrument to the maximum possible number of significant figures. Tabulate the respective currents and voltages and record the resistance in the box.

2. *Variation of Current with Resistance, Voltage Constant.* Use the same arrangement as in Step 1. Begin with a resistance of 30 ohms. *Never reduce this resistance to zero.* After the instructor has checked the wiring, vary the resistance to give currents of approximately 0.1, 0.2, 0.3, 0.4, and 0.5 amp. (Do not vary the resistance in equal steps.) Keep the voltage across the resistance box at a constant value of about 2 volts by adjustments of the rheostat slider. Record the voltage, current, resistance, and the reciprocal of the resistance. Record the resistance to three significant figures. (Why is this proper?)

3. *Variation of Potential Difference with Resistance, Current Constant.* Connect the apparatus as in Fig. 26.6. Be sure that the terminal A is the zero end of the wire. After the instructor has approved the connections, note and record the voltages across AC for lengths of wire of 40, 80, 120, 160, and 200 cm. Record the resistance of the 2-m wire and the current in the wire.

Computations and Analysis: 1. Use the data of Step 1 to plot a curve to show the variation of the voltage (ordinates) with current (abscissas), the resistance being constant.   Carefully interpret the significance of the shape and the intercepts of this curve.   Show why the slope of the curve represents the resistance in the resistance box.   Use the origin and a point near the upper part of the curve to calculate the slope.   Note the percentage difference between the resistance as calculated from the curve and the marked resistance of the box.

2. Calculate the reciprocals of the resistances used in Step 2.   (The unit *mho* is used as a unit for reciprocal ohms.)   Use these data to plot a curve showing the variation of the current (ordinates) with the reciprocal of resistance (abscissas). Carefully interpret the meaning of the slope and intercepts of the curve.   Calculate the slope of this curve, using the origin and another point near the extreme end of the curve.   Show why this slope should equal the voltage across the resistance box. • Note the percentage difference between this voltage and that observed from the voltmeter.

3. Use the data of Step 3 to plot a curve showing the variation of the observed voltage (ordinates) with resistance (abscissas).   Calculate the respective resistances from the given resistance of the 2-m wire and the corresponding lengths *AC*.   Carefully interpret the shape and intercepts of this curve.   Since the current was kept constant, what would one expect the slope of this curve to represent?   Compute the slope and compare this value with the observed current. Note the percentage difference between the two values of the current.

Review Questions: 1. State Ohm's law.   Give the law in symbolic form.   What are the common units of the respective factors?   2. Show by the aid of wiring diagrams how current and voltage are measured in a circuit.   3. Describe the application of Ohm's law both to an entire circuit and to a part of a circuit.   4. Show how Ohm's law may be demonstrated in three ways by varying two factors while the third is kept constant.   Describe the graphs obtainable from these three series of observations.   State the significance of the slope of each graph.

Questions and Problems: 1. Is voltage identical with *pressure*, as used in fluid physics?   Is the term *electric pressure* a desirable synonym for potential difference?

2. Why is it sometimes fatal to touch the wires in an electric power line and not nearly so dangerous to come into contact with an electrostatic machine that is developing a much higher voltage?   Can one be "shocked" by touching a wire carrying a very large current in an automobile starting system?   Why?   Why are birds not injured by sitting upon a high-potential power line?

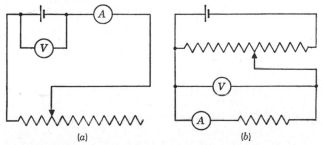

(a)   (b)

FIG. 26.7   Circuits that might be used to demonstrate Ohm's law

3. State some of the advantages and disadvantages of each of the two circuits shown in Fig. 26.7 when they are to be used by an instructor wishing to demonstrate Ohm's law to his class.

4. A cell of emf 3.0 volts and internal resistance 0.10 ohm is connected through an ammeter of 0.05-ohm resistance to a 5.0-ohm rheostat by means of wires having a total resistance of 0.85 ohm.   In the calculation of the current, what percentage error would be made by neglecting all resistances except that of the rheostat?

5. A poorly calibrated voltmeter when connected across 150 cm of a 2-m wire of resistance 16.0 ohms reads 3.92 volts; the wire is connected in series with a 5.0-ohm resistor to a battery of emf 8.00 volts and internal resistance 3.0 ohms.   What is the percentage error in the reading of the voltmeter?

6. A dry cell when short-circuited will furnish about 30 amp for a brief time.   If its emf is 1.5 volts, what must be the internal resistance?   An ordinary household electric lamp takes about 1 amp.   Would it be safe to connect such a lamp directly to the dry cell?   Why?

7. The resistance of the rheostat in Fig. 26.5 is 10.4 ohms and that of the resistance box is 4.50 ohms.   If the slider is at two-thirds of the length of the rheostat and the terminal voltage of the battery is 2.85 volts, what are the readings of the ammeter and voltmeter?

8. In Fig. 26.6 the terminal voltage of the battery is 2.78 volts, the resistance of the ammeter is 0.15 ohm, and the 2-m slide-wire has a resistance of 4.75 ohms.   What is the reading of the voltmeter when C is at 900 mm?   The voltmeter has a resistance of 300 ohms.   What change in the ammeter reading is observed when the voltmeter is attached?

9. A uniform wire 2.00 m long and having a resistance of 11.0 ohms is connected in series with a battery having a terminal voltage of 6.00 volts and a rheostat that has a resistance of 1.00 ohm.   What is the reading of a voltmeter that is placed across 60.0 cm of the wire?

10. A battery having a voltage of 3.00 volts is connected through a rheostat to a uniform wire 100 cm long.   The wire has a resistance of 2.00 ohms.   What must the resistance of the rheostat be in order that the voltage per millimeter of the wire shall be exactly 1 mv?

CHAPTER 27

ELECTRIC-CIRCUIT RELATIONSHIPS

The various relationships between currents, voltages, and resistances of complete circuits or parts of circuits, no matter how complex the circuit, may frequently be satisfactorily analyzed by a proper application of Ohm's law (Chap. 26). In many cases the rules known as Kirchhoff's laws are useful in problems dealing with involved branched circuits. It is helpful to break down complicated circuits including resistors and voltage sources in series and in parallel into equivalent single resistances and voltages.

Resistors in Series. Conductors are said to be connected in *series* when they are joined as shown in Fig. 27.1 so that electricity flows uniformly from one resistor into the next. The basic facts concerning current, voltage, and resistance relationships in series circuits may be summarized in the following statements:

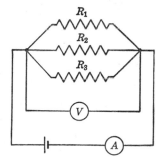

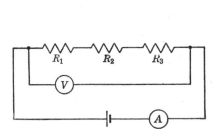

FIG. 27.1   Resistors in series

FIG. 27.2   Resistors in parallel

1. The current in all of the resistors in a series circuit is the same

$$I = I_1 = I_2 = I_3 \tag{1}$$

2. The voltage across a group of series resistors is the sum of the voltage across the individual resistors

$$V = V_1 + V_2 + V_3 \tag{2}$$

3. The total resistance of a group of series resistors is the sum of the individual resistances

$$R = R_1 + R_2 + R_3 \tag{3}$$

Resistors in Parallel. When conductors are arranged as in Fig. 27.2, so that the total current divides among the resistors with the respective ends of the resistors connected to common points, they are said to be in *parallel*. For such cases the following statements apply:

1. The currents in the various resistors are different and are inversely propor-

tional to the respective resistances. The total current is the sum of the individual currents

$$I = I_1 + I_2 + I_3 \qquad (4)$$

2. The voltage across each resistor is the same and is identical with the voltage across the whole group considered as a unit

$$V = V_1 = V_2 = V_3 \qquad (5)$$

3. The reciprocal of the total resistance of the group is the sum of the reciprocals of the individual resistances

$$\frac{1}{R} = \frac{1}{R_1} + \frac{1}{R_2} + \frac{1}{R_3} \qquad (6)$$

Connecting additional resistors in series *increases* the total resistance and connecting additional resistors in parallel *decreases* the total resistance. The joint resistance of a parallel group is always smaller than the resistance of the least resistor in the group.

Experiment 27.1

RESISTORS AND CELLS IN SERIES AND IN PARALLEL

Object: To study the voltage, current, and resistance relationships for circuits containing resistors or cells in series and in parallel.

Method: Two cells of known emf's are connected first in series and then in parallel. Measurements are made of the various currents and voltages. These values are then compared with those to be expected from the established relationships in such circuits. A similar procedure is followed for two known resistors connected first in series and then in parallel.

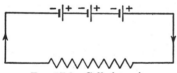

FIG. 27.3 Cells in series

Theory: *Cells in Series.* A group of cells connected in various ways is called a *battery*. When the cells are connected as in Fig. 27.3 with the positive terminal of one cell connected to the negative terminal of the next, the cells are in *series*. For series arrangements the voltage, current, and resistance relations are as follows:

1. The emf of the battery is the sum of the emf's of the various cells

$$E = E_1 + E_2 + E_3 \qquad (7)$$

2. The current in each cell is the same and is identical with the current in the entire battery

$$I = I_1 = I_2 = I_3 \qquad (8)$$

3. The internal resistance of the battery is the sum of the internal resistances of the individual cells

$$r = r_1 + r_2 + r_3 \qquad (9)$$

Cells in Parallel. Cells are connected in *parallel* when all the positive poles are connected to a common junction and all the negative poles are connected to another common junction as in Fig. 27.4. For cells in parallel voltage, current, and resistance relationships are as follows:

1. The emf of the battery is the same as the emf of one cell (when each cell has the same emf)

$$E = E_1 = E_2 = E_3 \tag{10}$$

2. The current in the external circuit is divided between the cells in the inverse ratio of their resistances. The total current is the sum of the currents in the individual cells

$$I = I_1 + I_2 + I_3 \tag{11}$$

3. The reciprocal of the total internal resistance of the battery is the sum of the reciprocals of the resistances of the individual cells

$$\frac{1}{r} = \frac{1}{r_1} + \frac{1}{r_2} + \frac{1}{r_3} \tag{12}$$

Optimum Method of Connecting Cells. A series connection is used when it is desired to have a high voltage in order to maintain a current in a conductor of

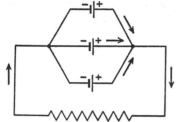

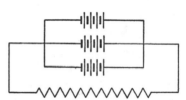

FIG. 27.4   Cells in parallel   FIG. 27.5   Cells in a series-parallel combination

high resistance. Cells may be connected in series regardless of whether the individual cells have identical emf's. In the parallel case it is desirable that the cells have the same emf's. Otherwise currents will circulate in the local branches formed by the cells. Such local currents consume the cells without contributing to the current in the external circuit. Cells are connected in parallel when it is desired to maintain a large current in a low-resistance circuit.

A series-parallel combination of cells is often desirable. Such an arrangement is shown in Fig. 27.5. This grouping has the advantage of giving a fairly high emf, low internal resistance, and ability to furnish a large total current. In general, if a maximum current is to be maintained in a given external resistor by a given number of cells, the cells should be arranged in such a manner that their total internal resistance will most nearly equal the external resistance. This requirement usually necessitates a series-parallel grouping.

Apparatus: Two Edison storage cells (or dry cells); 3-volt voltmeter; 0.5-amp ammeter; two resistance boxes (one with 0.1-ohm coils); SPST switch, fused for 1 amp; two four-way connectors; set of voltmeter leads, with clip connectors.

Procedure: Instead of tabulating the data in this experiment it is probably more convenient and graphic to record the observed values on the wiring diagrams of the various arrangements. Leave the switch closed in all cases only long enough to note the desired data.

I. *Cells in Series and in Parallel.* 1. Connect a voltmeter to the terminals of a fresh cell through a switch, being careful to join the positive pole of the cell to the

marked positive on the voltmeter. Note the reading. Repeat
...d fresh cell.

...ect the two cells in series and attach the battery to the voltmeter.
...he voltage and interpret the result.

...onnect the two cells in parallel and attach the battery to the voltmeter.
Record the voltage and interpret the result.

4. Connect a cell in series with a switch, a resistor of 6 ohms, and an ammeter of
range 0.5 amp. Note the reading of the ammeter. Put the two cells in series
with a 6-ohm resistor and the ammeter. Note the reading and interpret these
data.

5. Put the two cells in parallel and join this battery in series with the 6-ohm
resistor and the ammeter. Note the reading
of the ammeter and discuss the arrangement
as compared with that in Step 4.

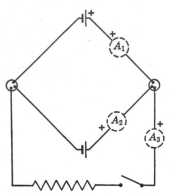

6. With the same arrangement as in Step 5,
insert the ammeter successively in each of the
circuits leading away from the cells; leave the
rest of the circuit unchanged, Fig. 27.6. (It is
very easy to make a wrong connection here and
damage the ammeter. Leave one terminal of
each cell unconnected until the instructor has
checked the wiring.) Note the readings of the
ammeter at A_1, A_2, and A_3 and interpret these
data.

FIG. 27.6 Currents in cells connected in parallel

II. *Resistors in Series.* 7. Connect in series
the two cells, switch, ammeter, and two resistance boxes. Leave one of the battery terminals

unconnected until the instructor checks the wiring. Adjust the resistors to
4.00 and 6.00 ohms, when the 0.5-amp meter is used. Be sure all contacts are
tight and that all plugs in the boxes are firmly seated. Note and record the
current. Open the circuit and place the ammeter *in series* successively at several
other places. Record and comment on the values of the observed currents.

8. Measure the voltage across each resistor and then the voltage across the
entire group; use the voltmeter clip-on leads. Comment on these data insofar as
they show the relationship between the total voltage and the sum of the separate
voltages. Show how these data demonstrate the fact that the voltage is proportional to the resistance when the current is constant. Compute the total resistance from the observed current and total voltage. Note the percentage difference
between this resistance and the sum of the individual resistances.

III. *Resistors in Parallel.* 9. Connect two resistance boxes in parallel. Use
the four-way connectors at the common junctions. Set the resistances at 8.00
and 16.00 ohms. Connect this group to the two-cell battery through the switch
and ammeter A_t, as in Fig. 27.7. Leave one of the battery terminals disconnected
until the instructor checks the wiring. (Never have either of the resistors set at
zero.) Measure the total current by the ammeter A_t. Break the connection
from the resistor R_1 to the four-way connector B and insert the ammeter *in series*
with this resistor, as shown at A_1. Have the instructor check this arrangement
before closing the switch. Note this current I_1. Similarly measure the current

I_2. Compare the total current I_t with the sum of I_1 and I_2. Show from the observed data that the two currents vary inversely with the corresponding resistances; that is, show the equality of the ratio I_1/I_2 to the ratio R_2/R_1. Note the percentage difference between these ratios.

10. Measure the voltage across the parallel group AB. Then measure the individual voltages across R_1 and R_2. How do these three voltages compare?

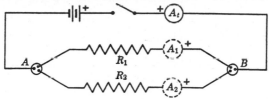

FIG. 27.7   Currents in resistors connected in parallel

11. Determine the joint resistance of the parallel group from the observed total current and voltage. Compare this value with that computed from the reciprocal law, Eq. (6). Note the percentage difference between the two values.

Experiment 27.2

RESISTORS IN SERIES AND IN PARALLEL

Object: To study the resistance, voltage, and current relationships in circuits containing resistors in series and in parallel.

Method: An ammeter and a voltmeter are used to make measurements of the currents and voltages in various series and parallel combinations of known resistors. These values are then compared with those to be expected from the established relationships in such circuits.

Apparatus: Two-cell Edison battery (or two dry cells); 0.5-amp ammeter; 3-volt voltmeter; three resistance boxes, at least one with 0.1-ohm coils; SPST switch, fused for 1 amp; two four-way connectors; set of voltmeter clip-on connectors.

Procedure: Throughout the experiment use only the 0.5-amp range of the ammeter. Keep the switch closed only long enough to observe the necessary data. Instead of tabulating the data it is more convenient and graphic to record the observed values on the wiring diagrams of the various arrangements.

I. *Resistors in Series.*   1. Connect in series the battery, switch, ammeter, and three resistance boxes, as in Fig. 27.1. Leave one of the battery terminals disconnected until the instructor checks the wiring. Suitable values for the resistances are 2.00, 4.00, and 6.00 ohms. Be sure that all contacts are tight and that the plugs in the resistance boxes are firmly seated.

2. Measure the current at various places, as indicated by the ammeters shown at A_1, A_2, A_3, and A_4 in Fig. 27.8. Discuss the significance of the observed values.

3. Using the voltmeter clip-on leads, measure the voltage across each resistor, and then the voltage across the entire group of resistors, AD, as indicated in Fig. 27.9. Note the percentage difference between the reading of V_4 and the sum of V_1, V_2, and V_3. Show from the data that the voltage is directly proportional to the resistance, when the current is constant.

4. Remove two of the resistance boxes from the circuit; retain the box with the tenth-ohm coils. Before closing the circuit decide what the resistance of this box should be in order that it could replace the three resistors R_1, R_2, and R_3. Then adjust the resistance until the current is as nearly as possible equal to that previously observed. Note the percentage difference between this resistance and the sum $R_1 + R_2 + R_3$. Try to account for some of this difference.

II. *Resistors in Parallel.* 5. Connect three resistance boxes in parallel as in Fig. 27.7 except that a third box is added. Use the four-way connectors at A and B. Leave one terminal of the battery unconnected until the instructor

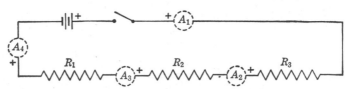

FIG. 27.8   Current in resistors connected in series

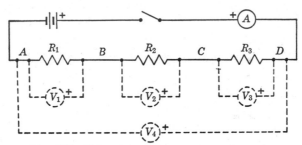

FIG. 27.9   Voltages in resistors connected in series

checks the wiring. Set the resistances at 10.00, 20.00, and 30.00 ohms. Connect the parallel group to the battery through the switch and ammeter A_t. Never have any of the resistors set at zero.

6. Measure the total current indicated by the ammeter A_t. Break the connection from the resistor R_1 to the four-way connector B and insert the ammeter *in series* with this resistor, as shown at A_1. Have the instructor check this arrangement before closing the switch. Note this current I_1. Similarly measure the currents I_2 and I_3 in the other resistors. Compare the total current I_t with the sum of I_1, I_2, and I_3. Show from the observed data that the three currents vary inversely with the corresponding resistances; for example, compare the equality of the ratios I_1/I_2 and R_2/R_1, also I_2/I_3 and R_3/R_2. Note the percentage difference between each pair of ratios.

7. Measure the voltage across the parallel group AB, and then measure the voltages across each of the individual resistors. How do these values compare?

8. Remove two of the resistance boxes. Adjust the resistance of the remaining box (with the 0.1-ohm coils) until a current is obtained as nearly as possible equal to the total current originally observed at A_t. Note how this observed resistance compares with that calculated by the reciprocal law from the values of R_1, R_2, and R_3. Record the percentage difference.

III. *Resistors in Series-parallel Combinations.* 9. Connect three resistance boxes as shown in Fig. 27.10. Set R_1 at 2.50 ohms, R_2 at 4.00 ohms, and R_3 at 8.00 ohms (or other values as directed by instructor). Measure voltage across group AC. (Do not measure any other voltages or currents.) Using voltage across AC and resistances R_1, R_2, and R_3, calculate currents in each resistor. Ask your instructor about your calculations before you finally measure each current with the ammeter. Compare your calculated and measured values of the currents.

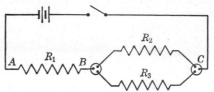

FIG. 27.10   Resistors in series-parallel combination

Experiment 27.3

KIRCHHOFF'S LAWS

Object: To study the application of Kirchhoff's laws to a d-c network by comparing the observed and calculated values of the currents in the circuits.

Method: A network is used in which several batteries are connected to resistance boxes. The currents in the various known resistors are calculated from these resistances and the known voltages of the batteries by the use of Kirchhoff's laws. These currents are checked by measurements with an ammeter.

Theory: The empirical principles known as *Kirchhoff's laws* as applied to d-c circuits may be stated as follows:

1. At any point in a circuit, the sum of the currents directed toward the point is equal to the sum of the currents directed away from the point; or, in other words, the *algebraic sum* of all the currents at a junction is equal to zero. Symbolically

$$\Sigma I = 0 \tag{13}$$

at any point.

2. In any closed circuit, the *algebraic sum* of all the voltages around the loop is equal to zero. In other words, the algebraic summation of all the emf's and the IR drops is equal to zero. In symbols

$$\Sigma E + \Sigma IR = 0 \tag{14}$$

around any closed loop.

With respect to the *algebraic sign* of a current, it is customary to regard currents directed toward a point as positive and currents directed away from a point as negative. With regard to the sign of potential differences in resistors the following convention is useful: If an imaginary journey is taken around a closed mesh or circuit in a given network, positive potential differences in resistors are those met when going opposite to the direction of the current, that is, from a point of lower to one of higher potential. Negative potential differences are those met when going in the direction of the current, that is, from a point of higher to one of lower potential. For an emf the sign of the voltage is considered positive when the journey is from − to + and negative when going from + to −. Internal drops of potential are handled just like those in ordinary resistors. These choices of sign are purely arbitrary and the opposite choices might just as well be made.

although, after a conventional choice has once been adopted, it must be strictly adhered to throughout an entire problem.

The first law is obvious, for it simply states that the total current directed toward a junction is equal to the total current directed away from the junction. This law is illustrated in Fig. 27.11. Since two currents are directed toward the

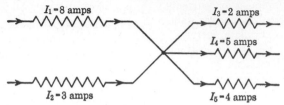

$I_1 = 8$ amps $I_3 = 2$ amps

$I_4 = 5$ amps

$I_2 = 3$ amps $I_5 = 4$ amps

FIG. 27.11   Currents at a junction

junction and three currents are directed away from it, the following equations may be written:

$$+ I_1 + I_2 - I_3 - I_4 - I_5 = 0$$
$$+ 8 \text{ amp} + 3 \text{ amp} - 2 \text{ amp} - 5 \text{ amp} - 4 \text{ amp} = 0$$

As an illustration of Kirchhoff's second law, consider the circuit of Fig. 27.12. The voltage between the points D and F might be calculated from three paths: (1) through R_1, (2) through R_2, and (3) through the battery AC. Thus

$$V_{DF} = R_1 I_1 = \quad 40 \text{ ohms} \times 5.00 \text{ amp} = 200 \text{ volts}$$
$$V_{GH} = R_2 I_2 = 100 \text{ ohms} \times 2.00 \text{ amp} = 200 \text{ volts}$$
$$E_{AC} = 214 \text{ volts} - (2 \times 7) \text{ volts} = 200 \text{ volts}$$

It is clear that the potential difference between D and F is the same regardless of which path is taken. Now apply Kirchhoff's second law to the path $ACHGA$.

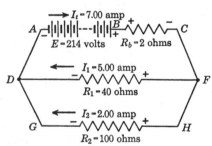

$I_t = 7.00$ amp

A $E = 214$ volts $R_b = 2$ ohms B C

$I_1 = 5.00$ amp

D $R_1 = 40$ ohms F

$I_2 = 2.00$ amp

G $R_2 = 100$ ohms H

FIG. 27.12   Voltages around various meshes in a network

Imagine a journey to be taken around the circuit. Beginning at A and going toward B, the potential changes from a point of lower potential $(-)$ to a point of higher potential $(+)$. Hence the first voltage should be written $+214$ volts. In passing through the IR drop from B to C, the potential changes from $+$ to $-$, since the journey is *with* the current, and hence the voltage is -7×2 volts. In the journey from H to G, the direction is with the current; thus this travel is from a point of higher potential $(+)$ to a point of lower potential $(-)$. Therefore, the IR drop should be written -2×100 volts. Since G is back at the potential of the starting point, the algebraic sum of all the voltages met is put equal to zero, or $+214$ volts $- 14$ volts $- 200$ volts $= 0$.

Suggestions for the Application of Kirchhoff's Laws to Problems: 1. Draw carefully a wiring diagram to a sufficiently large scale. Letter the diagram, using consistent symbols for each unknown current, resistance, and voltage. Plainly indicate each known factor.

2. Indicate the direction of each current by a small arrow. It frequently happens that the direction of some of the currents may be unknown. This lack need cause no difficulty, for the currents may temporarily be assumed in either direction. If the wrong direction has been assumed, this current will have a minus sign when the equations are finally solved. (Be careful to retain any such minus sign through to the end of the solution of the problem.)

3. Place a + sign at the positive terminal of each battery and at the end of each resistor where the electricity is assumed to enter. Place a − sign at the negative terminal of each battery and at the end of each resistor where the electricity is assumed to leave. For this purpose the internal resistance of each battery may be represented by a resistor immediately adjacent to one terminal of the battery.

4. The number of *independent* current equations that may be written is always *one less* than the number of current junctions in the circuit.

5. The number of *independent* voltage equations that may be written is always the same as (never greater than) the number of imaginary meshes or "panels" or "pieces of a jigsaw puzzle" into which the wiring diagram may be considered as being divided. To visualize more clearly the meaning of "panel" choose a point within the wiring diagram (not on a circuit line). Then a panel is composed of the circuit lines on the boundary of the area that includes this point. For example, note the two panels *ACFD* and *DFHG* in Fig. 27.12. The equation written for the circuit *ACHGA* would be a combination of the other two equations and is not independent. The use of the path *ACHGA* is permissible, but if it is used one of the others must be omitted. Each of the three equations may be obtained by combining the other two.

Apparatus: Four heavy-duty resistance boxes; storage or dry-cell battery of two cells; single-cell storage battery or dry cell; 3-volt voltmeter; 0.5-amp ammeter; voltmeter leads with spring clips; two four-way connectors; two SPST switches.

Procedure: 1. Arrange the apparatus as in Fig. 27.13. Leave the switches open until the instructor has checked the circuit.

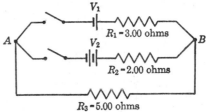

FIG. 27.13 Apparatus arranged to study Kirchhoff's laws

2. With the switches closed, measure with the voltmeter the voltages V_1 and V_2 across the batteries. Leave the switches open except when readings are being made.

3. Considering V_1, V_2, R_1, R_2, and R_3 as known, apply Kirchhoff's laws to the circuits and write the correct number of *independent* current and voltage equations necessary to solve for the (unknown) current in each branch, that is, write one current equation and the fewest number of voltage equations that will include every emf and every resistance at least *once* in the set of equations. Solve the equations for these currents. (Complete the solutions before proceeding with the rest of the experimental work.)

4. Having calculated the current in each branch by an application of Kirchhoff's laws, measure the respective currents experimentally. Break the circuit in turn in each branch and insert the ammeter *in series* to observe the value of the

current.   Give particular attention to polarity.   Have the instructor check the connections before the switches are closed.   Note each current, *with both switches closed.*

5. Calculate the percentage error between the observed values of the currents and those computed by Kirchhoff's laws.

6. Repeat the series of observations and calculations, using the arrangement shown in Fig. 27.14.   Typical suitable values of the approximate resistances are given in the accompanying table.   Select one of these sets or another set of resistances furnished by the instructor.

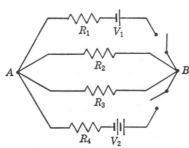

FIG.   27.14   Network   for   studying
Kirchhoff's laws

RESISTANCE VALUES FOR ARRANGEMENT
OF FIG. 27.14

R_1	R_2	R_3	R_4
ohms	ohms	ohms	ohms
0.5	2.5	4.0	3.0
1.5	3.0	5.0	6.0
0.7	2.0	3.0	5.0
2.0	5.0	4.0	3.0
3.0	30.0	10.0	2.0
5.0	10.0	30.0	3.0
5.0	20.0	8.0	1.5
10.0	20.0	15.0	0.5

Review Questions: 1. Make conventional wiring diagrams to show three resistors in series; in parallel.   2. State in words and symbolically the essential voltage, current, and resistance relationships in series circuits; in parallel circuits.   3. Repeat question 1 for three cells in series.   4. Repeat question 2 for three cells in parallel. 5. When is it usually desirable to connect cells in series? in parallel?   When is a series-parallel arrangement used?   6. With a given number of cells available, how should they be connected in order to produce a maximum current in a given external resistor?   7. State Kirchhoff's laws.   8. Explain clearly the conventions with regard to the algebraic signs of currents and voltages in solving involved circuits by using Kirchhoff's laws.   9. How many independent voltage equations may be written for any particular problem?   How many independent current equations?   10. Describe the steps used in the solution of a problem by Kirchhoff's laws.

Questions and Problems: 1. What is the effect caused by the ammeter resistance when an ammeter is inserted into a circuit to measure the current?   What does the answer to this question suggest about a desirable design for ammeters?

2. When a voltmeter is connected across the terminals of a resistor, what usually happens to the potential difference across the resistor?   What does the answer to this question suggest concerning a desirable design for voltmeters?

3. A group of n resistors each of resistance R_1 are connected in series.   What is the total resistance of the group?   What is the total resistance when the resistors are connected in parallel?   A series-parallel group of resistors has n equal resistors in series in each of N rows in parallel.   What is the total resistance?

4. A group of n identical cells in series, each of emf E_1 and internal resistance r_1, is connected in parallel with N similar rows.   The battery is connected to an external resistor R.   What is the symbolic equation for the current delivered to the external circuit?   What is the current in each cell?

5. Do Kirchhoff's "laws" represent any new physical principles, in the same sense that Ohm's law does? Why are these laws so helpful? Are they only approximations or are they rigorously true?

6. When three identical cells are connected in series, one cell is accidentally connected with reversed polarity. What is the effect on the total emf? on the total internal resistance?

7. In Fig. 27.10, R_1 is 3.00 ohms, R_2 is 5.25 ohms, and R_3 is 7.50 ohms. The voltage of the battery is 2.85 volts. What is the current in R_1? the voltage across R_2? the current in R_3?

8. What current would there be in the battery circuit in Fig. 27.13 if there were an open switch in the external circuit of R_3? Assume V_1 to be 1.40 volts and V_2 to be 2.72 volts.

9. In Fig. 27.14, $V_1 = V_2 = 2.85$ volts. The resistors are as follows: $R_1 = 0.65$ ohm, $R_2 = 12.0$ ohms, $R_3 = 25.0$ ohms, and $R_4 = 0.25$ ohm. What is the current in each resistor?

10. Two cells of emf 1.50 and 2.00 volts and internal resistances 0.50 and 0.80 ohm, respectively, are connected in parallel. Determine what current they will deliver to an external resistor of 2.00 ohms. (Use Kirchhoff's laws.)

11. A circuit is made up of the following groups of apparatus, connected in parallel: (a) a 2.00-volt cell of internal resistance 0.40 ohm with the positive terminal at the right; (b) a 6.00-ohm rheostat; (c) a battery of three cells in series, each of emf 1.50 volts and internal resistance 0.30 ohm with the positive terminals at the right; (d) a resistance of 12.0 ohms. Calculate the current in each part of the circuit. (Use Kirchhoff's laws.)

12. Two generators A and B, 400 ft apart, are connected in parallel. Generator A develops 110 volts, and B 120 volts. One of the wires connecting the generators has a resistance of 0.0200 ohm per 100 ft, the other 0.0300 ohm per 100 ft. A bank of four lamps connected in parallel and 100 ft from A is joined between the wires connecting the generators. Each lamp in this group has a current of 10.0 amp. A similar bank of five lamps taking 8.00 amp each is joined 200 ft away from B. Determine the potential difference across each lamp bank. (Use Kirchhoff's laws.)

CHAPTER 28

GALVANOMETERS

The common moving-coil galvanometer consists of a light coil suspended in a magnetic field. When a current is maintained in the coil a torque is set up that tends to turn the coil until its plane is perpendicular to the magnetic field. If the field is uniform, the torque is not constant as the coil turns but is proportional to the cosine of the angle θ between the plane of the coil and the field. Because of this variation the field of a galvanometer is usually made radial so that, no matter what position the coil takes, the torque is proportional to the current and the same for each position. Such an arrangement is shown in Fig. 28.1. The radial field is obtained by making the pole pieces curved and placing between them a soft-iron cylinder. For the radial field the plane of the coil is always parallel to the field and the torque L is given by

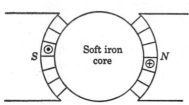

FIG. 28.1 Radial field of galvanometer

$$L = c\mu HNAI \tag{1}$$

where H is the strength of the field, N is the number of turns in the coil, A is the area of the coil, μ is the permeability, and c is a constant. If the instrument is to be used for measurements based on this torque, there must be supplied an opposing torque that is proportional to the angle θ of turn of the coil. That is

$$L = K_0\theta \tag{2}$$

where K_0 is a constant.

Under these conditions the coil will turn until the torque due to the current is equal to the opposing torque

$$c\mu HNAI = K_0\theta \tag{3}$$

or

$$I = \frac{K_0}{c\mu HNA}\theta = b\theta \tag{4}$$

From Eq. (4) we see that the angle turned by the coil is proportional to the current. The constant $b = I/\theta$ is a measure of the sensitivity of the galvanometer. The sensitivity depends upon the strength of the magnetic field, the area enclosed by the coil, and the torsion constant of the spring or fiber.

In a common form of galvanometer the coil is supported by a fine flat metallic suspension that serves not only as a support but also to make electrical connection and to supply the resisting torque. The position of the coil is indicated by means of a mirror M attached to the coil and a telescope and scale mounted a distance D in front of the mirror (Fig. 28.2). For each position of the mirror there is a corresponding reading on the scale. When the mirror turns through an angle

θ, the beam of light is turned through 2θ and the deflection $s = 2D\theta$. The current is thus proportional to the deflection on the scale.

The sensitivity of a galvanometer is described in several ways, each of which is expressed in terms of a standard deflection. The *standard deflection* is usually taken as one millimeter on a scale one meter distant from the mirror.

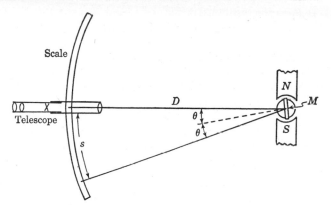

FIG. 28.2   Schematic diagram of galvanometer

The *current sensitivity* k of a galvanometer is defined as the current per millimeter deflection on a scale at the standard distance. It is numerically equal to the current required to produce a standard deflection, $k = I_G/s_1$.

The *voltage sensitivity* is the difference of potential between the terminals per millimeter deflection on a scale at the standard distance. The voltage sensitivity is simply related to the current sensitivity by the product kR_G, where R_G is the resistance of the galvanometer.

The *megohm sensitivity* is the resistance in megohms (10^6 ohms) that must be connected in series with the galvanometer to have a standard deflection for a potential difference of one volt. It follows that the megohm sensitivity = 1 volt/k (in microamperes per millimeter).

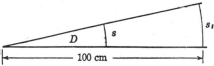

FIG. 28.3   Deflections on scales at different distances

The scale of a galvanometer is frequently not 1 m from the mirror, but at some other distance D. If there is a deflection s on this scale, we must find the corresponding deflection s_1 on a scale at the standard distance (see Fig. 28.3). Since the angle is the same

$$s/D = s_1/100 \tag{5}$$

or

$$s_1 = s\,\frac{100}{D} \tag{6}$$

Hence

$$k = \frac{I_G}{s_1} = \frac{I_G}{s} \times \frac{D}{100} \tag{7}$$

Thus the observed current per unit deflection on the scale used must be multiplied by the ratio of the actual distance D (in centimeters) to the standard distance, 100 cm.

A common method used for measurement of the resistance of a meter is the "half-deflection" method. In this procedure a fixed potential is applied to the meter and the deflection is noted. Resistance in series is added until the deflection is reduced to half the original value. Since the deflection is proportional to the current it follows that the current is reduced to half its original value, $I_2 = I_1/2$. If R_G is the resistance of the meter and R is the added resistance

$$V = I_1 R_G = I_2(R_G + R) \tag{8}$$
$$I_1 = 2I_2$$
$$2I_2 R_G = I_2(R_G + R) \tag{9}$$
$$2R_G = R_G + R$$
$$R_G = R \tag{10}$$

Thus the added resistance is equal to the resistance of the meter.

A small known voltage may be obtained by use of a potential divider circuit,

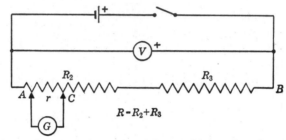

FIG. 28.4 Circuit used for determining sensitivity of a galvanometer

Fig. 28.4. The drop in potential over a large resistance $R = R_2 + R_3$ is measured by means of a voltmeter. A device, here the galvanometer, is connected across a small part r of the total resistance R. Since the current is common in R and r, the galvanometer voltage V_G is given by

$$\frac{V_G}{V} = \frac{Ir}{IR} = \frac{r}{R} \tag{11}$$

or

$$V_G = \frac{r}{R} V \tag{12}$$

Experiment 28.1

THE D'ARSONVAL GALVANOMETER

Object: To determine the current, voltage, and megohm sensitivities of a d'Arsonval galvanometer.

Method: Small known potentials are applied to the terminals of a d'Arsonval galvanometer and the corresponding deflections are observed. A curve is plotted showing the relation between deflection and applied voltage. From the slope of the curve the voltage per unit deflection is determined. The resistance of the galvanometer and the distance from the mirror to the scale are measured. From these three quantities the voltage, current, and megohm sensitivities are computed.

Apparatus: Wall galvanometer with telescope and scale; resistance box, 0.1 to 111 ohms, equipped with two traveling plugs; resistance box, 1 to 1110 ohms;

resistance box, 10 to 11,100 ohms; dry cell; 1.6-volt voltmeter; SPST switch; meterstick.

Procedure: 1. Adjust telescope and scale as described in Appendix, page 328.

2. Connect the apparatus as shown in Fig. 28.4. Leave one of the battery terminals unconnected until the instructor has checked the wiring. Use the box with tenth-ohm coils for R_2 and the 11,100-ohm box for R_3. Make the total resistance $R = R_2 + R_3 = 2000$ ohms. Adjust R_2 to 11 ohms by removing all the tenth-ohm and unit-ohm plugs. Set R_3 at 1990 ohms. Make sure that all plugs remaining in both boxes are tight.

Connect the galvanometer terminals to the two traveling plugs; insert one traveling plug A in the 0.1-ohm position and C in the 0.3-ohm position. The galvanometer is then connected across a resistance r of 0.200 ohm. The insertion of plugs into box changes total resistance slightly; at most by less than a half of 1 per cent. No correction need be made for these minor variations.

Cautiously tap the SPST switch in the cell circuit. If the apparatus is properly connected, a small deflection of the galvanometer is obtained. Then adjust r to a value that will produce nearly full-scale deflection and note whether the coil swings freely throughout its whole range.

3. Set the value of r to produce a deflection of 3 or 4 cm. Record the values of the voltmeter reading and the total resistance R. Record in tabular form the value of r and the deflection s.

4. Take a series of eight readings of r and s. Adjust the successive values of r so that the deflections increase uniformly to a full-scale value. Record the values of r and s in the table.

5. Use same arrangement as in Step 4 to measure R_G, the resistance of the galvanometer. Remove one of the galvanometer terminals and insert 1110-ohm resistance box in series with the galvanometer. Using at first zero resistance in this box, adjust r to produce approximately full-scale deflection. Then remove plugs from the box in series with the galvanometer until deflection is reduced to just half its initial value. Record this resistance and explain clearly why it is equal to R_G.

6. Measure and record distance D from scale to mirror (center of suspension tube).

Computations and Analysis: 1. Calculate voltage V_G (in microvolts) across the galvanometer for each observation, by the use of Eq. (12). Then compute galvanometer currents I_G (in microamperes).

2. Plot a curve of deflection s, in millimeters, against I_G, in microamperes. Discuss the significance of the shape of this curve. Calculate the reciprocal of the slope and state its significance.

3. Compute the current sensitivity of the galvanometer. This may be determined from the slope of the curve obtained in Part 2 and the use of Eq. (7).

$$k = \frac{I_G}{s_1} = \frac{I_G}{s} \times \frac{D}{100} = \frac{1}{\text{slope}} \times \frac{D}{100}$$

4. Calculate the voltage sensitivity, in microvolts, by the use of the current sensitivity and the galvanometer resistance.

5. Compute the megohm sensitivity by taking the reciprocal of the current sensitivity, in microamperes.

Review Questions: 1. What is meant by the standard deflection of a galvanometer? 2. Upon what factors does the deflection of a galvanometer depend? 3. Define current, voltage, and megohm sensitivities. Give appropriate units for each. How are they related to each other? 4. How is the necessary small known voltage for the galvanometer obtained? 5. Why is it permissible to neglect the small changes in resistance that occur when the positions of the traveling plugs are changed?

Questions and Problems: 1. What might be done to a galvanometer to change its sensitivity?

2. Are the deflections of a d'Arsonval galvanometer strictly proportional to the current when a straight scale is used? Why?

3. Would a high battery resistance cause an error in the determination of the galvanometer sensitivity in Exp. 28.1? Why?

4. Would thermoelectric effects in the circuit cause an error in this experiment? Explain.

5. In a typical experiment performed to measure the current sensitivity of a galvanometer, the following data were obtained: 3.00-volt battery; galvanometer tapped across 0.300 ohm of a 1000-ohm box connected across the battery; galvanometer resistance, 125 ohms; deflection on a scale 48.0 cm from mirror, 15.0 cm. Calculate the current, voltage, and megohm sensitivities for this instrument.

6. A galvanometer with a resistance of 80.0 ohms has a current sensitivity of 1.00×10^{-7} amp/mm. Find the voltage and megohm sensitivities.

7. A certain d'Arsonval galvanometer has a resistance of 150 ohms and a sensitivity of 33.3 megohms. Calculate its current and voltage sensitivities. What current would be necessary to produce a deflection of 3.00 cm on a scale 60.0 cm from the mirror?

8. Does the use of an external damping resistor change the current sensitivity of a galvanometer? the voltage sensitivity? Why is the use of such a resistor a convenience.

9. A tap key is often placed in parallel with a galvanometer and manipulated to bring the coil quickly to rest when the current is interrupted. Describe the technique used, and show why this happens. Is such a circuit underdamped, overdamped, or critically damped?

10. A galvanometer is to be used (as in a Wheatstone bridge experiment) to detect a small difference of potential. Should an instrument of high current sensitivity be chosen or would one of high voltage sensitivity be better? Explain.

11. A galvanometer has a current sensitivity of 0.105 μa/mm. What resistance must be connected in series with a cell of 1.25 volts emf and the galvanometer in order for the instrument to give a deflection of 150 mm on a scale 250 cm from the mirror?

CHAPTER 29

AMMETERS AND VOLTMETERS

Most d-c ammeters and voltmeters utilize the basic principles employed in the d'Arsonval galvanometer. Such instruments are of the permanent-magnet, movable-coil type with readings indicated by a light pointer attached to the

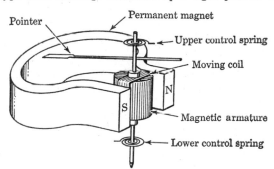

Pointer

Permanent magnet

Upper control spring

Moving coil

N

S

Magnetic armature

Lower control spring

FIG. 29.1 Essential features of d'Arsonval meter

movable coil (Fig. 29.1). In well-designed instruments the deflections of this type of meter are directly proportional to the current

$$I = k\theta \qquad (1)$$

The scale of such a meter has uniformly spaced divisions. The torque turning the coil is a result of the force between the field of the permanent magnet and the field produced by the current in the coil. From the twisting of the spiral springs that conduct the electricity into and out of the coil, a countertorque is established that is also proportional to the deflection; hence the coil comes into equilibrium with a deflection proportional to the current.

There is no essential difference in the construction of the magnet, movable coil, and indicating system of a portable galvanometer, an ammeter, or a voltmeter. In the usual portable galvanometer only the movable coil is in the current circuit.

A voltmeter differs from a galvanometer in that a high-resistance *multiplier* R_m is inserted in *series* with the movable coil, Fig. 29.2. The potential difference across this resistor does not differ greatly from the voltage to be measured, since the voltage across the movable coil is comparatively low. The resistance of a

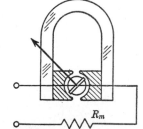

R_m

FIG. 29.2 Voltmeter multiplier

voltmeter multiplier may be calculated in terms of the resistance R_V of the original meter and the number of times N it is desired to increase the range of the instrument by the use of the multiplier. The conventional wiring diagram of a voltmeter multiplier circuit is shown in Fig. 29.3.   Assume that the current I produces a full-scale deflection of the basic meter, which may be a galvanometer or a low-range voltmeter.   The voltage across the meter is $R_V I$ and the voltage across

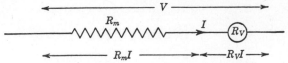

FIG. 29.3   Circuit of a voltmeter multiplier

the multiplier is $R_m I$.   The total voltage V across both multiplier and meter is $N(R_V I)$ or

$$N(R_V I) = R_V I + R_m I$$

Dividing through by I gives

$$R_m = NR_V - R_V = (N - 1)R_V \qquad (2)$$

Example: The resistance of model 45 Weston voltmeter is 100 ohms per volt of range.   If it were desired to convert a 2-volt voltmeter of this type into a 20-volt instrument, it would be necessary to insert in series a multiplier of resistance

$$R_m = \left(\frac{20}{2} - 1\right) \times 200 = 1800 \,\text{ohms}$$

A multirange voltmeter is made by inserting a number of multipliers in series with the basic meter, as shown in Fig. 29.4.   When the terminals A and B are

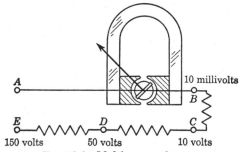

FIG. 29.4   Multirange voltmeter

used the instrument is a millivoltmeter of 10-mv range.   When the connections are made to A and C, the range of the meter is 10 volts; A and D give a 50-volt range; and A and E provide a range of 150 volts.   The range of an ordinary voltmeter may be increased by inserting in series with it a resistance box containing the desired resistance.

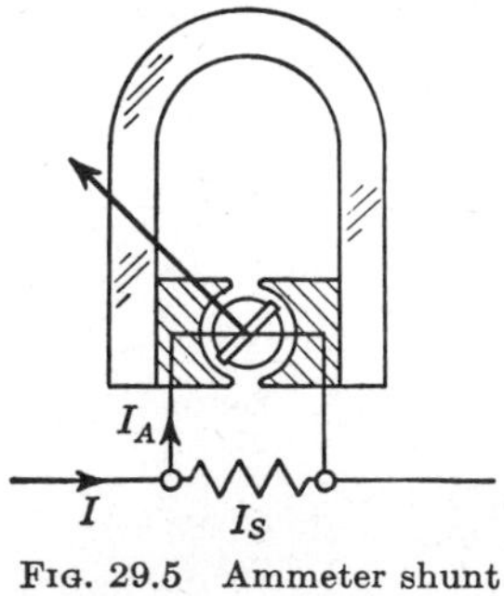

In an ammeter a low-resistance *shunt* is placed in parallel with the moving coil of the basic instrument, as shown in Fig. 29.5.   The total current is divided between the shunt and the coil; the shunt current is much the greater.   The resistance R_S of the shunt necessary to increase the range of a meter by a factor of N may easily be computed.   In Fig. 29.5 the total current I to be measured is

FIG. 29.5   Ammeter shunt

NI_A. This current divides, with the current I_A in the ammeter and I_S in the shunt,

$$I_S = I - I_A = NI_A - I_A = (N - 1)I_A$$

From Ohm's law and the fact that the voltage across parallel resistors are the same it follows that

$$R_S I_S = R_A I_A$$

whence

$$R_S = R_A \frac{I_A}{I_S} = R_A \frac{I_A}{(N - 1)I_A}$$

or

$$R_S = \frac{R_A}{N - 1} \tag{3}$$

Example: The voltage across a model 45 Weston ammeter is 50 mv for full-scale deflection. Hence a 2-amp meter has a resistance $R_A = V/I = 0.050$ volt/2.0 amp = 0.025 ohm. In order to increase the range of this meter to 20 amp a shunt would have a resistance given by

$$R_S = \frac{R_A}{N - 1} = \frac{0.025 \text{ ohm}}{(20 \text{ amp}/2.0 \text{ amp}) - 1} = 0.0028 \text{ ohm}$$

A multirange ammeter may be constructed by the use of several self-contained shunts, as shown in Fig. 29.6. For example, if the terminals A and B are used, the meter is calibrated for 10 ma; when A and C are connected, the range is 500 ma; and A and D are designed to furnish a range of 5 amp.

Both ammeters and voltmeters are frequently designed to be provided with external shunts or multipliers. The term *multiplying ability* of an ammeter shunt or a voltmeter multiplier is sometimes used to designate the number of times N the range of the meter is increased by the use of the shunt or multiplier.

Many combination instruments, called *volt-ammeters*, are made in such a way that either a shunt may be placed in parallel with the movable coil or a multiplier may be inserted in series with the coil. Such a meter may be used either as a voltmeter or as an ammeter. These meters are often provided with several ranges each of current and voltage.

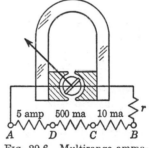

FIG. 29.6 Multirange ammeter

Measurement of Meter Resistance. A convenient method of determining the resistance of a *voltmeter* is the *half-deflection method*, as follows: The voltmeter is connected across a battery and the reading taken. A resistance box is then introduced into the circuit and adjusted until the voltmeter reading is one-half its former value. The current in the second case must be just half the current in the first case. Since the total voltage in the two cases is the same, the resistance of the second circuit (voltmeter and added resistance) is just twice the resistance of the first circuit (voltmeter alone). Therefore the voltmeter resistance must be just equal to the added resistance. (See page 196.)

A rapid method of obtaining the resistance of an *ammeter* is to maintain a current in the meter and to measure the PD across it with a low-range voltmeter. The current in the instrument is given by its reading. The resistance is computed by using Ohm's law.

Alternating-current Meters. The d'Arsonval type of meter cannot be used for a-c measurements since every reversal of the current reverses the direction of the torque on the movable coil. Hence a-c meters are constructed by utilizing various other design principles. There are two common types: the electrodynamometer and the repulsion-vane meters.

In *electrodynamometer* instruments the permanent magnet of the d'Arsonval meter is replaced by a fixed coil, arranged in series with the movable coil, Fig. 29.7. The movable coil deflects because of the torque due to the interaction of the magnetic fields of the fixed and movable coils. Since the two fields reverse simultaneously, the torque is always in the same direction. The design of these instruments is such that the scale is not uniform; the divisions are crowded together near the zero end and expanded near the upper end of the scale.

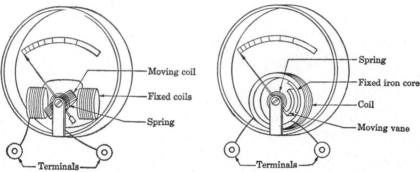

FIG. 29.7   Electrodynamometer meter   FIG. 29.8   Repulsion-vane meter

In *repulsion-vane* instruments a movable soft-iron vane, pivoted on jewelled bearings and mounted within a fixed coil, is repelled by a fixed iron core fastened to the inside of the coil, Fig. 29.8. Current in the coil induces like magnetic poles in the fixed iron core and in the movable iron vane. This action results in a repulsion tending to rotate the movable vane. The deflection depends upon the current in the coil, but the torque is not a linear function of the current and the scale is not uniform.

Both electrodynamometer and repulsion-vane meters may be manufactured for use either as ammeters or voltmeters, depending upon the inclusion of shunts or multipliers, as in d-c meters. These a-c meters may be used also for d-c measurements but they are greatly inferior to the d'Arsonval type because the currents required to energize the fixed coils are much greater then those needed in meters provided with strong permanent magnets. Hence the resistance of these a-c voltmeters is much less than the corresponding d-c meters and the resistance of a-c ammeters is much higher than the resistance of d-c ammeters of similar ranges. In a-c meters it is therefore more important to correct for the variations introduced by the insertion of the meters than in the case of d-c meters.

Experiment 29.1

CONSTRUCTION OF A VOLTMETER AND AN AMMETER
FROM A GALVANOMETER

Object: To study the basic features and physical principles involved in the design and use of d-c voltmeters and ammeters.

Method: A portable galvanometer is used as a basic instrument. A voltmeter is constructed by inserting a high-resistance multiplier in series with the galvanometer. An ammeter is made by connecting a low-resistance shunt in parallel with the galvanometer. The constructed voltmeter is tested by placing it in parallel with a conventional voltmeter. The constructed ammeter is likewise checked by inserting it in series with a regular ammeter.

Apparatus: Portable galvanometer, three-push-button type; storage cell (or dry cell); SPST switch, fused for 1 amp; resistance box, 10 to 11, 110 ohms; resistance box, 0.1 to 111 ohms; 3-volt voltmeter; 0.5-amp ammeter; resistance wire for ammeter shunt.

The portable galvanometer, Fig. 29.9, is provided with three push buttons at the base of the instrument. When the first button is depressed, a 10,000-ohm

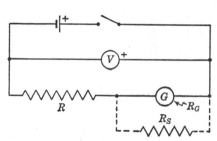

FIG. 29.9 Portable galvanometer, three-button type

FIG. 29.10 Circuit for measuring galvanometer resistance

resistor is inserted in series with the meter. Closing the second key places a 2000-ohm resistor in series. When Key 3 is pressed, the meter alone is in the circuit.

Procedure: 1. Arrange the apparatus for the measurement of galvanometer resistance by connecting in series the cell, SPST switch, and 11,110-ohm resistance box (with all plugs removed), and the galvanometer, Fig. 29.10. Leave one terminal of the cell unconnected and the switch open until the instructor checks the wiring. Since galvanometers are delicate instruments, they must be handled very carefully and no current that is too large for the range of the instrument should be used. Always close the keys in turn, beginning with Key 1. When the circuit is properly arranged and checked by the instructor, adjust the resistance box R until the galvanometer deflection is full scale when Key 3 is closed. Record this deflection. Now place in parallel with the galvanometer the resistance box which has the 0.1-ohm coils, as shown by the dashed lines in Fig. 29.10. Adjust this shunt resistance R_S until the galvanometer reads one half of full-scale; the remainder of the circuit should stay unchanged. Show why this value of R_S is just equal to the resistance R_G of the galvanometer.

2. The *sensitivity* of a galvanometer equipped with a fixed pointer is defined as the current per unit scale division on the meter. Calculate this sensitivity from the data observed in Step 1 and the additional measurement of the voltage of the

cell. Measure this voltage with the voltmeter. The current in the galvanometer circuit for full-scale deflection may be calculated from

$$I_G = \frac{V}{R + R_G}$$

Divide this full-scale current by the number of galvanometer divisions (on one side of the zero) to obtain the galvanometer sensitivity. This sensitivity is conveniently expressed in microamperes per division (1 μa = 10^{-6} amp).

3. Construct and test an ammeter of range 0.500 amp (or other value directed by the instructor) from the galvanometer, as follows: Calculate by the use of Eq. (3) the shunt resistance R_S that must be placed in parallel with the galvanometer (Fig. 29.5). Fabricate this shunt by using wire (supplied by the instructor) of known resistance per unit length. In order to check the calibration of the constructed ammeter thus made, insert a conventional ammeter in series with the main circuit of Fig. 29.10 and adjust R from a very high value to lower ones until nearly full-scale deflection is obtained. Compare the reading of the conventional ammeter with that of the constructed ammeter. (Have the instructor check the wiring before connection is made to both of the cell terminals.)

4. From the galvanometer construct and test a voltmeter of range 2.50 volts, as follows: Calculate by the use of Eq. (2) the resistance R_m of the multiplier that must be inserted in series with the galvanometer (Fig. 29.2). Check the constructed voltmeter thus made by connecting it and the conventional voltmeter across a cell, as in Fig. 29.10 (omitting the shunt). Compare the readings of the constructed voltmeter and the conventional voltmeter. (Have the instructor check the wiring before connection is made to both cell terminals.)

Experiment 29.2

AMMETER SHUNTS AND VOLTMETER MULTIPLIERS

Object: To study the physical principles involved in the design and operation of some common types of ammeters and voltmeters; to construct an ammeter shunt and a voltmeter multiplier that will increase the ranges of the meters.

Method: The voltages and currents required to operate typical ammeters and voltmeters are measured. Their respective resistances are determined from these observations. A voltmeter multiplier is made by the insertion of the calculated resistance in series with the meter. An ammeter shunt is constructed by the use of a calculated resistance obtained from wire of length determined from its resistivity and area of cross section.

Apparatus: Resistance box, 0.1 to 111 ohms; resistance box, 10 to 11,100 ohms; 0.5-amp ammeter; two-range voltmeter, 1.5 and 3 volts; two-cell Edison storage battery (or two dry cells); SPST switch, fused for 1 amp; 10-ohm rheostat; meterstick; No. 18 copper wire; obtain from the instructor a 50-mv millivoltmeter and various other d-c and a-c ammeters and voltmeters.

Procedure: 1. Measure the voltmeter resistance as follows: Connect the 3-volt range of the voltmeter to the battery and note the reading. Insert both resistance boxes in series with the voltmeter and adjust the resistance until the voltmeter reading is one-half its former value. Show why this added resistance is equal to the voltmeter resistance (half-deflection method).

Connect the ammeter and the voltmeter in series with the battery and record the reading of each meter. (This arrangement is not conventional, but the results are instructive.) How is the reading of the voltmeter (as previously observed) affected by the insertion of the ammeter? Explain. Comment on the value of the current required to operate the voltmeter. Is it desirable for a voltmeter to have a high or a low resistance? Explain clearly the reasoning for your answer.

2. Construct a voltmeter multiplier as follows: Using the observed value of the voltmeter resistance, calculate the resistance of the multiplier that must be used to increase the range of the voltmeter tenfold. Check this value by noting first the reading of the voltmeter connected directly to the battery and then the reading when the multiplier is inserted in series.

3. Measure the ammeter resistance as follows: Connect the battery, the SPST switch, the rheostat (adjusted for maximum resistance), and the 0.5-amp ammeter in series. Leave one terminal of the battery disconnected until the instructor checks the wiring. Connect the 1.5-volt range of the voltmeter in parallel with the ammeter. Observe the reading of both instruments after the rheostat has been adjusted to give approximately full-scale deflection on the ammeter. Following this preliminary observation, obtain a 50-mv millivoltmeter from the instructor and insert it in place of the voltmeter originally used. Record the readings of the millivoltmeter and the ammeter and from them calculate the resistance of the ammeter. Comment on the value of this resistance. Is it desirable for an ammeter to have a high or a low resistance? Clearly explain the reasoning.

4. Construct an ammeter shunt as follows: Calculate the resistance of the shunt that should be used to increase by tenfold the range of the ammeter. Fabricate this shunt with copper bell wire. Determine the diameter of the wire by the use of a micrometer caliper. Record the resistivity of this wire as given in Table 6 in the Appendix. Calculate the length of wire necessary for the shunt; allow for a total contact resistance of 0.0010 ohm at the binding posts where the shunt is attached. Test the shunt as follows: Adjust the rheostat to give a full-scale current, attach the shunt, and note the ammeter reading. Determine the percentage difference between the reading with the shunt attached and the expected value.

5. Compare a-c and d-c voltmeters as follows: Obtain an a-c voltmeter, preferably of 3-volt range, from the instructor and measure its resistance by the half-deflection method. Compare the resistance of this meter with that of the d-c meter of the same range. Which instrument is preferable for d-c circuits? State reasons for your answer. Explain why a-c and d-c instruments of the same range have different resistances.

6. Compare a-c and d-c ammeters as follows: Obtain an a-c ammeter, preferably of 0.5-amp range, from the instructor. Measure its resistance by using the technique of Step 3, being careful to insert a suitable resistor to control the current. Compare the resistance of the a-c and the d-c meters of the same range. Explain why one instrument is far superior to the other for measuring currents in d-c circuits.

7. Repeat Steps 5 and 6 for such other instruments as are available. Compare the resistance characteristics of the different models of d-c voltmeters; repeat for d-c ammeters.

Review Questions: 1. What are the essential parts of the common type of d-c ammeter? of d-c voltmeter? How do each of these meters differ from a galvanometer? 2. Explain why the deflection of a good d-c meter is proportional to the current in the moving coil. 3. What is a voltmeter multiplier? How is it connected? How is its resistance related to the voltmeter resistance in order to give a multiplying ability of N-fold? 4. Describe the multiplier used for a multirange voltmeter. 5. What is an ammeter shunt? How is its resistance related to the ammeter resistance to produce an N-fold multiplying ability? Describe the arrangement of shunts used in a multirange ammeter. 6. Describe a volt-ammeter. What is its advantage over separate instruments? 7. Explain how the resistance of a voltmeter might be measured by the half-deflection method. 8. Describe a method of measuring the resistance of an ammeter by the use of Ohm's law. 9. Describe the features and operation of an electrodynamometer meter. How does its scale compare with that of a d'Arsonval meter? Explain. 10. Describe the features and operation of a repulsion-vane meter. 11. How do the resistances of a-c meters compare with d-c meters of the same range? Show why the difference exists.

Questions and Problems: 1. Ammeters and voltmeters are sometimes protected from overloads by the insertion of the proper size of fuse. Would the same fuse be suitable for a 3-volt and a 300-volt voltmeter? Would a fuse properly designed for a 5-amp ammeter be suitable for a 50-amp meter? Explain.

2. Some types of fuses used to protect electric meters have resistances of several ohms. Is this objectionable (a) in voltmeter circuits; (b) in ammeter circuits? Why?

3. Is it desirable for an ammeter to have a high resistance or a low one? Should a voltmeter have a high resistance or a low one? Why?

4. How will the readings of two ammeters compare if they are placed in series in a circuit? Will the answer to this question be different if the resistances of the two ammeters are not the same?

5. Compare the readings of two voltmeters of unequal resistance connected in series and attached to a battery. Compare their readings when they are placed in parallel.

6. Compare the readings of two ammeters of unequal resistance when placed in parallel with each other and the combination inserted in series in a circuit.

7. A portable galvanometer has a zero-center scale and 20 divisions on each side of zero. A pointer attached to the coil deflects 15 divisions for a current of 38 μa. What is the current sensitivity of this instrument?

8. A portable galvanometer is given a full-scale deflection by a current of 1.00 ma. If the resistance of the meter is 7.0 ohms, what series resistance must be used with it to measure voltages up to 50 volts?

9. A certain 3.00-volt voltmeter requires a current of 10.0 ma to produce full-scale deflection. How may it be converted into an instrument with a range of 150 volts?

10. A voltmeter reads full scale when connected to terminals whose difference of potential is maintained at 5.00 volts. When a resistance of 500 ohms is placed in series with the meter, it indicates half of full-scale deflection. What is the resistance of the meter?

11. An ammeter with a range of 5.00 amp has a voltage drop across it at full-scale deflection of 50.0 mv. How could it be converted into a 20.0-amp meter?

CHAPTER 30

MEASUREMENT OF RESISTANCE BY WHEATSTONE'S BRIDGE

The most widely used device for the measurement of resistance is the *Wheatstone bridge*. It consists essentially of a network of four resistors, three of which are known and the fourth is to be measured. The conventional diagram of a Wheatstone bridge is shown in Fig. 30.1. The values of the resistances are adjusted until the current in the galvanometer is made zero, as indicated by zero deflection. The bridge is then said to be *balanced*. For this condition the points A and B are at the same potential and $V_{MA} = V_{MB}$. Also $V_{AN} = V_{BN}$. In terms of currents and resistances

$$I_1 R_1 = I_3 R_3$$
and
$$I_2 R_2 = I_4 R_4$$

With the current in the galvanometer zero, $I_1 = I_2$ and $I_3 = I_4$. When we divide the second equation by the first, we obtain

$$\frac{R_2}{R_1} = \frac{R_4}{R_3} \quad \text{or} \quad R_2 = R_1 \frac{R_4}{R_3} \quad (1)$$

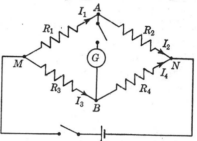

Fig. 30.1 Conventional diagram of Wheatstone bridge

Slide-wire Form Wheatstone Bridge. The principle of the Wheatstone bridge is simply portrayed in the slide-wire form, often used in student laboratories (Fig. 30.2). The branch MBN is a uniform resistance wire of manganin, usually 1 m

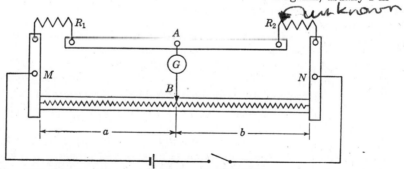

Fig. 30.2 Diagrammatic representation of slide-wire Wheatstone bridge

long, so that the ratio of R_4 to R_3 is simply the ratio of the two lengths BN to MB. The point A is fixed, and the galvanometer contact is moved along MN until the proper point of balance B is found. A good-quality resistance box may be used for R_1. The unknown resistor is placed at R_2.

Certain errors are inherent in the design of the slide-wire bridge but they are usually small if proper precautions are taken.   The uniformity of the slide-wire is assumed; it may be calibrated, but the process is laborious.

An inspection of Fig. 30.2 shows that R_2 includes all the resistance between A and N, including contact resistances and the resistance of the lead wires.   These resistances are negligible when R_2 is large, but in case R_2 is small, they become exceedingly important.   The same is true of R_1, R_3, and R_4.   Great care should be taken to have good contacts and to use large, heavy connecting wires to reduce this error to a minimum.   Then the working equation becomes

$$R_2 = R_1 \frac{b}{a} \tag{2}$$

where b is the length of the slide-wire from N to B and a is the length from M to B.

Because of contact resistances at the ends of the slide-wire, best results are obtained if the sliding contact B is near the middle of the bridge wire when the bridge is balanced.   In order to obtain this condition, B is first arbitrarily set at the 50-cm mark and R_1, the known but variable resistance, is adjusted until the bridge is approximately balanced; then B is slightly moved for an exact balance.

The battery key should be closed *before* the galvanometer key and opened *afterward*.   Because of a property called "inductance," the currents in the two branches of the circuit may grow from zero to their maximum values at different rates, so that the galvanometer key should not be closed until this steady state is reached.   Ordinarily, the steady state is reached in a fraction of a second.

Thermal and contact emf's are sometimes a source of error.   When the galvanometer circuit is closed, the battery circuit remaining open, a small galvanometer deflection may be observed.   This deflection may be caused by thermal emf's in different parts of the galvanometer circuit.   If the wires, switches, and binding posts are made with a single metal, such as copper, these difficulties will be minimized.   One remedy for this and certain other errors is to interchange the known and unknown resistors and to average the observed values of the unknown resistance.

Since the value of the known resistance R_1 includes the contact resistances of the plugs, it is essential to have the plugs clean and well seated.

Factors upon Which the Resistance of a Conductor Depends.   The resistance of a wire depends (1) on the material—iron, for example, has a resistance about seven times that of copper; (2) on the length—a wire 20 ft long offers four times as much resistance as a wire 5 ft long; (3) on the diameter of the wire—the larger the cross-sectional area, the less the resistance; a wire 2 mm in diameter has one fourth the resistance of a wire 1 mm in diameter; (4) on the temperature—heating a copper wire from 0 to 100°C increases its resistance about 40 per cent.

These facts may be stated in symbols as follows

$$R = \rho \frac{L}{A} \tag{3}$$

where R is the resistance of the wire of length L and cross-sectional area A.   The factor ρ is a dimensional constant for a given material at constant temperature. It is called the resistivity of the material, and its value for a given temperature depends on the kind of material used and the units in which R, L, and A are measured. If the factors in Eq. (3) are rearranged into

$$\rho = R\frac{A}{L} \tag{4}$$

it is seen that ρ is numerically equal to R when L and A are unity. Hence, the *resistivity* of a substance is defined as numerically the resistance offered by unit length and unit cross section of the material.

In the metric system the length is usually expressed in centimeters, the area in square centimeters, and the resistance in ohms. Hence the resistivity is given in ohm-centimeters.

The unit of resistivity in the British engineering system differs from that just given, in that different units of length and area are employed. The unit of area is the *circular mil*, the area of a circle 1 mil (0.001 in.) in diameter. The unit of length is the foot. In this system of units the resistivity of a substance is numerically equal to the resistance of a sample of that substance 1 ft long and 1 circular mil in area and it is expressed in ohm-circular mils per foot. The area of a circle in circular mils is equal to the square of its diameter in mils, $A = d^2$. The abbreviation CM is often used for circular mils. This abbreviation should not be confused with the abbreviation cm used for centimeters. It should be emphasized that the value of ρ in metric and British engineering units is vastly different; thus for copper, ρ has a value of 10.5 ohm-circular mils/ft and a value of 1.74×10^{-6} ohm-cm. The metric value is frequently expressed in microhm-centimeters (one microhm is one-millionth of an ohm). Hence, the value for copper is 1.74 microhm-cm. A table of the resistivities of certain substances is given in the Appendix.

Experiment 30.1

RESISTIVITY BY SLIDE-WIRE WHEATSTONE BRIDGE

Object: To study the design and use of a slide-wire Wheatstone bridge; to determine with it the variation of the resistance of a conductor with length, diameter, and material; also to measure the resistivities of several conductors.

Method: A slide-wire form of Wheatstone bridge is used to measure the resistance of a copper wire of known length and diameter. A similar wire having

FIG. 30.3   Slide-wire Wheatstone bridge

twice the length but the same diameter is measured, and the variation of resistance with length is thus determined. The resistance of a third wire having the same length as the first but one-half the diameter is measured, and the variation of resistance with diameter noted. Another wire of the same length and diameter as the first but of nickel-silver is used, and its resistivity is computed. This resistivity is compared with that of copper.

Apparatus: Slide-wire Wheatstone bridge (Fig. 30.3); dry cell; portable galvanometer; resistance box (with 0.1-ohm coils); SPST switch, with 10,000-ohm protective resistor; tap key; inductive resistor (transformer); set of resistance spools for "unknowns"; four short, heavy wires.

The resistance spools to be measured (Fig. 30.4) have the characteristics given in the accompanying table.

Spool number	Material	Length m	Wire size (B & S) number	Diameter	
				in.	cm
1	Copper	10	22	0.0254	0.0644
2	Copper	10	28	0.0126	0.0321
3	Copper	20	22		
4	Copper	20	28		
5	Nickel-silver (german silver)	10	22		

Procedure: 1. Arrange the apparatus as shown in Fig. 30.2. Have the instructor check the connections before closing any keys. Place the tap key in

series with the cell and the SPST switch with 10,000-ohm resistor in series with the galvanometer. When this switch is left open for the preliminary trials, the 10,000-ohm resistor is in series with the galvanometer. For greater sensitivity after an approximate balance has been obtained, the SPST switch should be closed, thus short-circuiting the protective resistor. The galvanometer coil should be made

Fig. 30.4   Resistance spools

free to swing by sliding the clamp in a direction opposite to that indicated by the arrow on the instrument. Adjust the pointer to zero by rotating the zero-adjusting device. Use the first spool for the unknown resistor R_2. Connect it to the bridge by very short, heavy wires; be sure that the contacts are good. Connect the resistance box R_1 to the bridge by similar wires. Set the sliding contactor B at the 50-cm mark on the slide-wire. For a balance at this point R_2 equals R_1. For the first spool a value of $R_1 = 0.5$ ohm is appropriate. For this and all future values of R_1 be certain that all plugs in the resistance box are firmly seated.

2. First close the tap key in the cell circuit. Then give the galvanometer contactor key B a light tap. This action will prevent a too-violent deflection of the galvanometer. The key in the cell circuit should always be closed first and the galvanometer key closed later. Move the contactor B to the right or left until the galvanometer shows practically no deflection. Raise the contactor each time to prevent excessive wear of the center portion of the slide-wire. After a rough balance is made, close the SPST switch, short-circuiting the protective resistor to increase the sensitivity. It is a good technique to have the galvanometer leads so arranged that a galvanometer deflection to the right requires a movement of B to the left to produce a balance. Since the resistance of copper changes somewhat

with temperature, it is desirable to keep the tap key in the cell circuit open except when a test for balance is being made.

The data may be tabulated as in the accompanying table.

Spool number	Known resistance R_1	Length of slide-wire		Unknown resistance		Resistivity
		a	b	R_2	R_2 average	
	ohms	cm	cm	ohms	ohms	ohm-cm
1						
1 reversed						
2						
2 reversed						
etc.						

3. In a similar manner measure the resistance of the other four wires. Each time, when a new resistance is to be measured, the protective resistor should be included in the galvanometer circuit until an approximate balance has been obtained. The best technique is always to begin by setting B at 50 cm and then adjusting R_1 to a value which is estimated to be equal to R_2. For example, for spool 1 and spool 2 the length and material are constant but the diameter of the wire on spool 2 is just one-half as large as that on spool 1. This fact should suggest to the student the appropriate value of the resistance of wire 2 in comparison with wire 1. Set R_1 at the estimated value and note the direction of the galvanometer deflection when testing for a balance. If a further increase of R_1 results in a larger deflection, it is apparent that the value of R_1 is too great and should be reduced. Continue to adjust R_1 until the galvanometer is nearly balanced with the contactor at the 50-cm mark. Make the final adjustment by moving the contactor on the slide-wire.

4. Interchange the positions of R_1 and R_2 and repeat the series of measurements. Note the fact that the working equation, Eq. (2), now becomes

$$R_2 = R_1 \frac{a}{b}$$

5. (Optional) Observe the effect of closing the key in the cell circuit *after* the galvanometer key is closed. Select as the unknown resistor a coil having a considerable inductive effect, such as one of the windings on a transformer. Balance the bridge originally in the conventional manner, closing first the key in the cell circuit. Observe the usual precautions for measuring an unknown resistance, that is, insert the protective resistor in the galvanometer circuit and tap the contactor B lightly. When the bridge is finally balanced, there is no deflection of the galvanometer on closing and opening the keys in the regular manner. If the galvanometer key is first closed, however, and then the key in the cell circuit is closed, there will be a temporary deflection of the galvanometer, even though the bridge is balanced for steady currents. Observe the effect obtained when the key in the cell circuit is opened before that in the galvanometer circuit. Comment on these observations.

Computations and Analysis: 1. From the data of Steps 2, 3, and 4 calculate the values of R_2; average the two measured values of R_2 for the final value.

2. Compute the ratio of the resistances of coils 1 and 3 and compare this value with the ratio of the lengths of the wires in these coils. Determine the percentage difference between the values of these two ratios. Is $R \propto L$? Repeat for coils 2 and 4.

3. Calculate the ratio of the resistance of coils 1 and 2 and the inverse ratio of the square of the diameters of the wires in these coils. Determine the percentage difference between the values of these ratios. Is $R \propto 1/A$? Repeat for coils 3 and 4.

4. Compute the resistivity of the wire in spools 1 and 2, using metric units. Calculate the resistivity of the wire in spools 3 and 4, using British units. Determine also the resistivity of the wire in spool 5. These values may agree only approximately with the values given in the Appendix because specimens of copper and nickel-silver often vary considerably in composition.

Experiment 30.2

TEMPERATURE COEFFICIENT OF RESISTANCE BY BOX FORM OF WHEATSTONE'S BRIDGE

Object: To study the design and operation of the box form of Wheatstone bridge and to measure the temperature coefficient of resistance of copper.

Method: The resistance of a copper wire is measured by a box form of Wheatstone bridge at temperatures varying from room temperature to 100°C. A resistance-temperature curve is plotted and its extrapolation to 0°C gives the resistance of the wire at that temperature. From this value and the slope of the curve the temperature coefficient of resistance is determined.

Theory: The electric resistance of all substances is found to change more or less with changes of temperature. Three types of change are observed: (1) The resistance may increase with increasing temperature, which is true for all pure metals and most alloys. (2) The resistance may decrease with increase of temperature, which is true for carbon, glass, and many electrolytes. (3) The resistance may be independent of temperature, which is approximately true for some special alloys, such as manganin (Cu 0.84; Ni 0.12; Mn 0.04).

Experiments have shown that the change of resistance with temperature of conductors can be represented, for a small range of temperatures, by the equation

$$R_t = R_0 + R_0\alpha t = R_0(1 + \alpha t) \qquad (5)$$

where R_t is the resistance at temperature t, R_0 is the resistance at 0°, and α is a quantity characteristic of the substance and known as the temperature coefficient of resistance. The defining equation for α is obtained by solving Eq. (5), giving

$$\alpha = \frac{R_t - R_0}{R_0 t} \qquad (6)$$

The *temperature coefficient of resistance* is defined as the change in resistance per unit resistance per degree rise in temperature, based upon the resistance at 0°.

Although Eq. (5) is only approximate, it may be used over medium ranges of temperature for all but very precise work.

Since $(R_t - R_0)$ and R_0 have the same units, their units will cancel in the frac tion of Eq. (6). Hence, the unit of α depends only upon the unit of t. For instance, for copper $\alpha = 0.004$ per °C, or $\frac{5}{9} \times 0.004$ per °F.

The value of α for many pure metals is roughly the same, approximately $\frac{1}{273}$ or about 0.004 per °C. This value is the same as the coefficient of expansion of an ideal gas. At the absolute zero of temperature (-273°C) such a conductor would therefore offer zero resistance to a current; that is, a current once started would continue indefinitely without the expenditure of any energy. Low-temperature research has shown that near 0°K a current persists in certain cases without an applied potential and there is no generation of heat. A table of values of the temperature coefficients of resistance of certain materials is given in the Appendix.

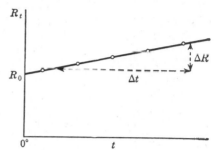

Fig. 30.5   Variation of resistance with temperature

From Eq. (5) it follows that, if the resistance of a pure metal is plotted as a function of temperature over a limited range, a curve similar to that shown in Fig. 30.5 will be obtained. This curve does not pass through the origin, that is, at 0°C the resistance is not zero. Hence we cannot say that $R \propto t$. The slope of the curve $\Delta R/\Delta t$ is a constant. Since

$$\alpha = \frac{\Delta R/\Delta t}{R_0} = \frac{\text{slope}}{R_0} \tag{7}$$

it is clear that the value of α depends upon the base temperature chosen for R_0. In computations involving temperature variation of resistance, the value of R_0 must first be obtained for use in Eq. (5).

If a curve of resistance against absolute temperature is plotted, an approximately straight line is obtained which, if extrapolated to 0°K, passes through the

Fig. 30.6   Dial-decade Wheatstone bridge

Fig. 30.7   Universal galvanometer shunt

origin. Such a curve shows that the resistance of pure metals is approximately proportional to the temperature *on the absolute scale.*

Apparatus: Dial-decade form of box-type Wheatstone bridge (Fig. 30.6); wall galvanometer; telescope and scale; universal galvanometer shunt (Fig. 30.7); dry cell; 100°C thermometer; Bunsen burner; double connector; ring stand with clamp; test coil of wire in a calorimeter vessel.

The specimen to be used for the measurement of temperature coefficient of resistance is a copper wire. It is wound on an insulated cylinder, within a thin brass tube filled with oil, Fig. 30.8. This tube is immersed in a calorimeter vessel containing water to which heat is applied.

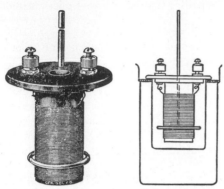

FIG. 30.8 Temperature coefficient of resistance coil

The dial-decade box form of Wheatstone bridge is popular because of its ease of manipulation and accuracy of measurement. The known resistance R_1 consists of four decade groups; one dial varies the resistance in steps of 1 ohm from 1 to 9 ohms, the second dial in steps of 10 ohms from 10 to 90 ohms, the third dial in 100-ohm steps from 100 to 900 ohms, and the fourth dial in 1000-ohm steps from 1000 to 9000 ohms. Thus a total resistance range from 1 to 9999 ohms is available for R_1.

The resistances R_4 and R_3 are known as "ratio coils." In this box the arrangements are such that the ratio R_4/R_3 may be varied from 1000 to 0.001 in powers of 10 by manipulating only one dial. The value of R_2 will therefore equal R_1 multiplied by some convenient decimal ratio.

Procedure: 1. Connect the apparatus as shown in the diagrammatic sketch of Fig. 30.9. Connect the cell and unknown resistor (temperature coefficient

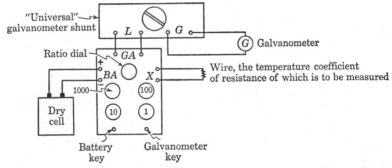

FIG. 30.9 Arrangement of apparatus for measuring temperature coefficient of resistance

of resistance coil) directly to the terminals on the Wheatstone bridge box marked BA and X, respectively. The galvanometer terminals on the box (marked GA) should be connected to the line (L) posts on the universal galvanometer shunt. The posts marked G on the shunt box are to be connected directly to the wall galvanometer. The shunt dial should first be set at 0.001 for the preliminary adjustments, which means that only $\frac{1}{1000}$ of the sensitivity of the galvanometer is being used. When the bridge has been partly balanced, the shunt dial may be successively turned to higher factors until the full sensitivity (marked 1) is being utilized.

2. The resistance of the unknown coil is approximately 1.5 ohms.   Hence the final adjustments are made by utilizing the 0.001 ratio of the ratio dial on the bridge and all four dials on the resistance decades.

It is unnecessary to wait for the galvanometer to come to rest, except during the final adjustment; just note which way the galvanometer starts to deflect.   Tap the galvanometer key (marked GA) with a quick motion, so that a too-violent fling of the galvanometer is avoided.

3. Measure the resistance of the coil at room temperature and record the value. Stir the oil in which the coil is immersed and be sure that the temperature is constant and that the balance on the bridge remains at a constant value before taking readings.

4. Heat the water about 10°C and determine the resistance as before, using all the previous precautions.   It is very convenient to keep the bridge approximately balanced while the temperature is being raised, so that a reading can be taken very quickly when desired.   Remove the flame well before the temperature reaches the desired value and allow the system to come into thermal equilibrium before making final readings.

As the water begins to heat, the form for a temporary graph of resistance against temperature should be made.   Both scales should be so chosen that the data to be plotted will give a curve extending over practically all the graph paper in both directions.   The resistance axis should not begin at zero but should begin at a value somewhat less than the resistance observed at room temperature, in order that the curve may be extrapolated to R_0.   Estimate the value of R_{100} in order to select the best scale for the resistance axis.

5. Continue heating the water in approximately 10°C intervals, following the technique of Step 4.   Plot the observed resistances on the temporary graph as the data are recorded.   If any point does not lie on or very near the straight line joining the other points, that observation should be rechecked.

6. In Step 5, the system was allowed to come into thermal equilibrium before taking each reading.   The following is an alternative procedure.   Knowing the initial resistance and the approximate final resistance to be expected at, say, 90°C, divide the resistance interval into 10 equal parts.   Set the bridge at the next value above room temperature and slowly heat the vessel, meanwhile testing occasionally for a balance.   Keep the oil thoroughly stirred all the time.   When the galvanometer shows a balance, quickly read the thermometer; this reading gives the temperature for that particular value of the resistance.   Then continue in the same manner until the maximum temperature is reached.

7. Remove the connecting wires from the heater, connect them together with a double connector, and measure the resistance of these lead wires; remember to use all needed precautions to prevent an excessive throw of the galvanometer, since this resistance is only a few thousandths of an ohm.

8. Disconnect the apparatus and leave everything in good shape.   Pour out the hot water, and fill the vessel again with cold water.   Mop up all spilled water and oil.

Computations and Analysis: 1.   Subtract the resistance of the lead wires observed in Step 7 from each measured resistance of the coil.

2. Draw a final curve of the corrected coil resistance plotted against temperature.   Extrapolate the curve to the resistance axis at zero temperature.   What is

the significance of this intercept? Calculate the slope of the curve. What is its significance?

3. Compute by Eq. (7) the value of α from the results of Part 2 of the Analysis. Note the percentage deviation between this experimental value and the standard value for copper.

4. Plot a curve to show the variation of the resistance with the temperature expressed on the Kelvin scale. Begin both coordinate scales with zero for this curve. Extrapolate the observed portion of the curve by a dashed line until it intersects the axis. Give a full interpretation of the significance of this curve.

Review Questions: 1. Make a conventional wiring diagram of a Wheatstone bridge. State the working equation for the bridge. 2. Sketch a slide-wire form of Wheatstone bridge. How is this bridge balanced? State the working equation for the slide-wire bridge. 3. Mention the precautions to be observed in the use of the Wheatstone bridge. 4. In what order must the keys be closed in using a Wheatstone bridge? Why is this necessary? 5. What are the factors upon which the resistance of a metallic conductor depends and how does the resistance vary with each factor? 6. Define resistivity and give the defining equation. What are the usual metric and British units? What is a mil? a circular mil? 7. Describe how the resistance of various classes of materials changes with temperature. State the equation that gives this variation. 8. Define temperature coefficient of resistance; give the defining equation, and state the usual British and metric units. 9. What is the significance of the numerical value of the temperature coefficient of resistance of pure metals? 10. Describe the differences between the dial-decade Wheatstone bridge and the slide-wire type. 11. Explain the procedure used to measure the temperature coefficient of resistance. What precautions are desirable? 12. Sketch a graph of resistance against centigrade temperature for copper. Explain the significance of the intercepts and slope of this curve. How can the value of α be obtained from such a curve?

Questions and Problems: 1. Which of the following statements are true and which are false? (a) When a Wheatstone bridge is balanced, there is no current in the resistor being measured. (b) A Wheatstone bridge cannot measure accurately resistances as low as a microhm. (c) A common metric unit of resistivity is the resistance of a cubic centimeter of the conductor. (d) The resistance of a wire of circular cross section is inversely proportional to the square of its diameter. (e) The resistivity of a conductor is directly proportional to its temperature in degrees centigrade.

2. State whether the following factors introduce (1) zero, (2) negligible, or (3) appreciable errors in the Wheatstone-bridge measurement of resistance: (a) using a battery of appreciable internal resistance; (b) contact resistance in the circuit of the measured resistance; (c) temperature fluctuation during measurements; (d) contact resistance in the galvanometer circuit; (e) nonuniformity of slide-wire; (f) contact resistance at junction of slide-wire and brass block.

3. Assuming an uncertainty of ± 0.02 cm in locating the position of the contact on the slide-wire in Fig. 30.2, what percentage error is made in determining (a) MB, (b) BN, and (c) MB/BN, when the contact B is at 50 cm? at 5 cm? What fact does this problem emphasize?

4. Suppose that an error of 1% had been made in determining both the length and the diameter of one of the wires whose resistivity was measured in this experiment. Discuss the errors that would have been introduced into the value of the resistivity.

5. With a standard resistance of 1.60 ohms connected at the zero end of the meterstick in a simple Wheatstone bridge, the balance point is found to be at 53.1 cm. The unknown resistance is a 5.00-m wire, 0.200 mm in diameter. What is the resistivity of this wire?

6. The diameter of No. 20 wire is about three times that of No. 30 wire. A certain No. 30 wire 100 cm long has a resistance of 6.35 ohms. What is the resistance of 400 cm of No. 20 wire made of this material?

7. A certain wire has a resistivity of 16.8 ohm-circular mils/ft. What is its resistivity expressed in metric units?

8. The resistance of a wire is 1.562 ohms at 25°C and 1.835 ohms at 80°C. Determine its temperature coefficient of resistance; its resistance at 100°K.

9. What is the resistance at 80.0°C of a copper wire 10.0 m long and 0.0500 cm in diameter, if the resistivity of copper is 1.70 microhm-cm at 0°C and the temperature coefficient of resistance is 0.00405 per °C?

10. Explain whether each of the following factors introduces an error that is (a) zero, (b) negligible, or (c) appreciable in the measurement of the temperature coefficient of resistance of copper: (1) contact resistance in battery circuit, (2) lag of temperature of oil behind temperature of water in calorimeter vessel, (3) high thermal capacity of thermometer, and (4) nonuniformity of diameter of wire of test coil.

11. Which of the following statements are true and which are false? The resistance of pure copper (a) increases linearly with increase of temperature; (b) is directly proportional to the temperature if expressed in degrees centigrade; (c) is approximately zero at 0°C; (d) is directly proportional to temperature measured on the absolute scale; (e) changes with temperature in approximately the same proportion as silver; (f) changes with temperature in about the same proportion as manganin.

12. Make rough sketches of the current against voltage curves for (a) a tungsten-filament electric lamp; (b) a carbon-filament lamp. Give reasons for the shapes of the curves.

13. A conductor at 32°C has a resistance of 12 ohms. Its temperature coefficient is 0.0038 per °C. Find its resistance at 0°C and 194°F.

14. The resistance of a conductor at 40°C is 11.68 ohms; at 60°C it is 12.52 ohms. What is its resistance at 0°C? at 100° absolute? its temperature coefficient?

CHAPTER 31

RESISTANCE BY VOLTMETER-AMMETER METHOD

The resistance of a conductor may easily be measured with moderate precision by the use of an ammeter and a voltmeter to determine the simultaneous values of the current and the voltage. From these observations the resistance is obtained by the use of Ohm's law, $R = V/I$. There are two alternative methods for connecting the meters, as shown in Figs. 31.1 and 31.2.

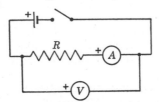

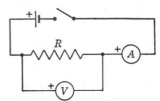

Fig. 31.1 Resistance and power by voltmeter-ammeter, Method 1

Fig. 31.2 Resistance and power by voltmeter-ammeter, Method 2

When the connection shown in Fig. 31.1 is used, the ammeter properly records the current in the resistor. But here the voltmeter indicates the PD over both the resistor and the ammeter. When R is large in comparison with the ammeter resistance R_a, the percentage error caused by neglecting the voltage IR_a across the ammeter is not serious. The arrangement of Fig. 31.1 should be used when R is large in comparison with R_a, and R_v is not necessarily large in comparison with R.

When the voltmeter is connected as in Fig. 31.2, it correctly indicates the PD over the resistor. But the ammeter does not read the exact current in the resistor, since it shows the sum of that current and the current in the voltmeter. The percentage error caused by neglecting the voltmeter current is small when R is negligible in comparison with the voltmeter resistance R_v. The connection of Fig. 31.2 is therefore used when R is small in comparison with R_v and is not large in comparison with R_a.

In some cases one might satisfactorily measure the current in the resistor with the voltmeter disconnected and then use the arrangement of Fig. 31.2 to measure the voltage. This technique would not give good results if the applied voltage fluctuated between the respective readings or if the internal resistance of the source were such that the TPD (terminal potential difference) would vary because of the change in line current as a result of the added voltmeter current. This technique would be especially poor if there were a number of resistors in the circuit and the voltage across only a part of the circuit was to be measured. This voltage may change appreciably when the voltmeter is inserted, because of the added drop of potential in the other resistors.

It is necessary to have some idea of the relative values of the resistances of the ammeter, voltmeter, and the unknown resistor before deciding which of the two alternative connections to use. The resistances of meters are often given on the instrument cases. The resistance of the widely used Weston model 45 instruments may be determined by noting the fact that these voltmeters have resistances of 100 ohms per volt of range and this model of ammeters require a PD of 50.0 mv for full-scale deflection. For example, a 3-volt voltmeter has a resistance of 300 ohms; a 5-amp ammeter has a resistance of $0.050/5 = 0.010$ ohm.

In each of the sets of measurements made as in Fig. 31.1 (method 1) and in Fig. 31.2 (method 2) a correction must be applied to the observed values if the best determination of the unknown resistance is to be obtained. These correction terms are derived as follows:

Method 1. The observed voltage V_1 includes both the voltage I_1R across the resistor and the voltage I_1R_a across the ammeter

$$V_1 = I_1R + I_1R_a$$

Hence the value of R is given by

$$R = \frac{V_1}{I_1} - R_a = R_1 - R_a = R_1\left(1 - \frac{R_a}{R_1}\right) \tag{1}$$

where $R_1 = V_1/I_1$ and represents the *apparent* resistance of the unknown as determined from the uncorrected readings of the voltmeter and the ammeter. From Eq. (1) it follows that the true resistance is less than the apparent resistance by the following:

$$\text{Percentage error made in method 1} = \frac{R_a}{R_1} \times 100\% \tag{2}$$

Method 2. The true current I in the resistor is $I_2 - I_v$, where I_2 is the observed current and I_v is the current in the voltmeter. Designating the apparent resistance, as calculated from the observed readings V_2 and I_2, by R_2 (where $R_2 = V_2/I_2$), we have

$$R = \frac{V_2}{I_2 - I_v} = \frac{V_2}{I_2 - V_2/R_v} = \frac{V_2}{I_2(1 - V_2/I_2R_v)}$$

If the term $(1 - V_2/I_2R_v)^{-1}$ is expanded by the binomial theorem, the approximate equation for R may be written

$$R = R_2(1 + R_2/R_v) \tag{3}$$

The conclusion follows from Eq. (3) that the true resistance R is greater than the uncorrected observed resistance R_2 by the percentage:

$$\text{Percentage error made in method 2} = \frac{R_2}{R_v} \times 100\% \tag{4}$$

The curves in Fig. 31.3 have been plotted for a voltmeter of range 3.00 volts and resistance 300 ohms and an ammeter of range 0.500 amp and resistance 0.100 ohm. These curves indicate the percentage by which the observed resistances R_1 and R_2 must be corrected to obtain the true resistance. When method 1 is used, the proper correction must be *subtracted from* the observed value, and for method 2 the correction must be *added to* the observed value.

The two methods give equally accurate results when the correction factors are equal in magnitude; that is, when

$$\frac{R_a}{R_1} = \frac{R_2}{R_v}$$

In Fig. 31.3 the corrections are equal for a resistance of slightly less than 6 ohms.

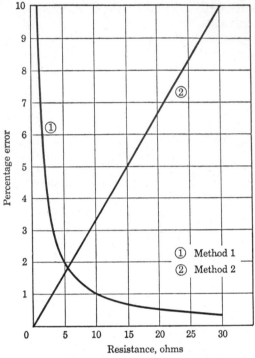

FIG. 31.3   Percentage errors in the two alternative connections when the ammeter-voltmeter method is used with $R_v = 300$ ohms and $R_a = 0.100$ ohm

For this condition of equal errors neither of the corrections is large and R is approximately equal to both R_1 and R_2.   Therefore

$$R = \sqrt{R_1 R_2} = \sqrt{R_a R_v} \tag{5}$$

If the unknown resistance is larger than the right-hand member of Eq. (5), the arrangement of Fig. 31.1 produces the smaller error.   If R is smaller than this member, the arrangement of Fig. 31.2 is better.

The errors discussed here are by no means the only ones that enter into the experimental measurements of resistance by the voltmeter-ammeter method. Errors caused by incorrect calibration of instruments and contact resistances may be far greater than the errors discussed.

When electric lamps are operated at different voltages the filaments have widely different temperatures.   Since resistance is a function of temperature, such observations enable a study to be made of the variation of resistance with temperature for lamps of different materials.

Experiment 31.1

RESISTANCE BY VOLTMETER-AMMETER METHOD

Object: To study the application of the voltmeter-ammeter method to the measurement of resistance.

Method: The resistances of several electric lamps are measured from observations with a voltmeter and an ammeter.   The two alternative methods of connecting the ammeter and voltmeter are used and a critical study is made of these methods to determine the optimum one to be employed in various cases.   The resistances of tungsten- and carbon-filament lamps are measured for various applied voltages and the variation of resistance with voltage is plotted.

Apparatus: Voltmeter, with 3- and 150-volt ranges; ammeter, with 0.5- and 5-amp ranges; two-cell Edison storage battery (or two dry cells); 115-volt power-line source; DPST switch, fused for 2 amp; 1200-ohm rheostat; "unknown" resistor (2 to 3 ohms); several sizes of tungsten-filament gas-filled lamp bulbs; carbon-filament lamp; mounted lamp socket, with binding posts.

Procedure: I. *Measurement of Resistance.*   1. Connect the apparatus as in Fig. 31.2.   Use a battery of about 3 volts as a source and the unknown resistor for R.   Insert the variable rheostat and the 0.5-amp ammeter in series with R. Leave one terminal of the battery disconnected until the instructor checks the connections.

2. Take simultaneous readings of the current and the voltage across the unknown resistor, with the rheostat adjusted to give a nearly full-scale deflection on the ammeter.   Compute the approximate value of R from the observed current and voltage.

3. Move the voltmeter terminal to the connection shown in Fig. 31.1.   Record the current and voltage and calculate the resistance.   Is the difference between the values of R found in Steps 2 and 3 caused merely by errors of observation?

4. Calculate the resistance of the voltmeter and the ammeter, using the data for these instruments given above (or furnished by the instructor).   Determine the percentage error caused by the instruments for both methods 1 and 2, by the use of Eqs. (2) and (4).   Which is the better method?   Apply Eqs. (1) and (3) to determine the corrected value of R.   Request the instructor to furnish the value of the unknown resistance and to check your conclusions before proceeding with the experiment.   Calculate the percentage difference between your observed value of R and the standard value.

II. *Effect of Temperature upon Resistance.*   5. Connect the wires from an outlet plug to the DPST switch, but *do not insert the plug into the power socket until the instructor checks the wiring.*   The probability of damage to the equipment or of shocks to the experimenter are much greater with the 115-volt source than when one is using low-voltage cells.   Insert the 1200-ohm rheostat (with the slider adjusted for maximum resistance) in series with the lamp socket, ammeter, and switch.   Connect the 150-volt voltmeter in parallel with the lamp, as in Fig. 31.2. Use a 50-watt, tungsten-filament, gas-filled lamp for this test.   After the instructor has checked the wiring close the switch and adjust the rheostat until the voltage across the lamp is the rated value.   Record the voltage and current.   Compute the resistance of the lamp at its rated voltage.

6. Rearrange the voltmeter connection to that of Fig. 31.1.   Repeat the readings at the rated voltage.   Compute the resistance and compare it with the value

determined in Step 5.   Calculate the percentage errors introduced by the meter currents and voltages, by the use of Eqs. (2) and (4).   Which method yields better results for this lamp?

7. Use the more accurate setup and take a series of readings of voltage and current at intervals of about 20 volts.   Observe and record the following: type of lamp, lamp voltage, and current.

8. Repeat the observations of Step 7, using a carbon-filament lamp.   (Always have the maximum rheostat resistance in the circuit before beginning the series of observations.)

Computations and Analysis: 1. From the data of Steps 7 and 8 compute the resistances of the lamps at the various voltages.

2. Plot a curve for each lamp showing the variation of the resistance with voltage.   Give reasons for the shape of each curve.

Review Questions: 1. Draw the conventional wiring diagrams for the two alternative methods of measuring resistance by the voltmeter-ammeter method.   2. Describe the errors made by each method because of the instrument currents and voltages.   3. Derive the equations for the correction terms that must be applied to the observed resistances in order to obtain the best values for the unknown resistances.   4. State the expressions for the percentage errors made in each method. 5. Discuss the use of the equation $R = \sqrt{R_a R_v}$ for the determination of the optimum method to be used.

Questions and Problems: **1.** In view of the observations made in this experiment what is the probability of burning out a tungsten-filament lamp as compared with a similar size of carbon-filament lamp if they are both subjected to the same overvoltage?

2. An unknown resistance of about 98 ohms is to be measured by the ammeter-voltmeter method.   A Weston model 45 voltmeter of 150-volt range and a similar model of ammeter of range 0.750 amp is to be used.   What percentage error is made by each method and which is the better one to use?

3. An ammeter of resistance R_a is connected through a rheostat to a cell of negligible internal resistance.   When the rheostat is adjusted to a value R_1, the ammeter reads 0.75 of full-scale value.   If the rheostat is changed to a value R_2, the ammeter reading is 0.25 of full-scale value.   Compute the value of R_a in terms of R_1 and R_2.

4. From the data observed for the resistances of tungsten- and carbon-filament lamps when at white-hot temperatures and at nearly room temperatures and the use of the temperature coefficients of resistances found in tables, compute the temperatures of the carbon and the tungsten filaments when operated at their rated voltages.

5. In a simple series circuit a battery is connected through an ammeter to an electric lamp.   A voltmeter is connected in parallel with the lamp.   The ammeter indicates 0.75 amp and the voltmeter 50 volts.   What is the resistance of the lamp, neglecting the fact that the voltmeter carries a small part of the current?   If the current in the voltmeter is 0.0010 amp for each volt indicated by it, what is the actual current in the lamp and the corrected value of its resistance?

6. In question 5 the current is increased to 1.00 amp.   What will now be the reading of the voltmeter?

7. What is the resistance of the voltmeter of question 5?

8. A 3-volt voltmeter connected to a battery reads 2.84 volts.   A 275-ohm rheostat is inserted in series with the voltmeter, reducing its reading to 1.42 volts.   What percentage errors are made when this voltmeter is used with a 0.125-ohm ammeter to measure an unknown resistance of approximately 5 ohms by each of the methods of connection used in this experiment?

CHAPTER 32

EMF AND TERMINAL POTENTIAL DIFFERENCE

A cell or a generator develops an emf that is the potential difference between its terminals when the source has no current in it. When there is a current in the source, the terminal potential difference (TPD) is not equal to the emf; it is smaller than the emf when the source is delivering current to an external circuit and larger when there is a reverse current in the cell.

In Fig. 32.1 a battery of emf E and internal resistance r is shown connected to a voltmeter. With the switch open the voltmeter reads a value nearly equal to E, since it may be assumed for the present that the voltmeter current is negligible.

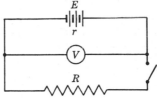

FIG. 32.1 Terminal PD and emf

When the switch is closed, the battery maintains a current I and the voltage V across its terminals is less than the emf by the amount of the internal drop of potential Ir across the internal resistance. The TPD is given by

$$V = E - Ir \tag{1}$$

This important equation may be put into the alternate form

$$E = V + Ir \tag{2}$$

which states that the voltage generated by the source is equal to the sum of the voltage V across the external circuit and the internal drop of potential Ir. Note that $V = IR$ and hence the PD at the cell terminals is the same as the voltage drop across the external circuit.

Equation (1) may also be written in the form

$$I = \frac{E}{R + r} \tag{3}$$

which is merely the Ohm's-law statement that the current in a circuit is the net emf divided by the total resistance of the circuit.

From Eq. (1) it is easy to see why a voltmeter cannot read the *exact* emf of a cell, since some current is always required to operate the voltmeter. A high-resistance voltmeter requires so little current that the Ir term in Eq. (1) is negligible for the usual cell of low internal resistance.

When the TPD of a Daniell cell is plotted against the current delivered to an external resistor, a curve of the type shown in Fig. 32.2 is usually obtained. The Y intercept of the curve shows $V = E$ at $I = 0$. The value of I is zero when the cell is on open circuit, that is, when R is infinite. The X intercept shows $V = 0$

when $Ir = E$. The value of V is zero when the cell is short-circuited, that is, when $R = 0$. The slope of this curve is the negative of the internal resistance of

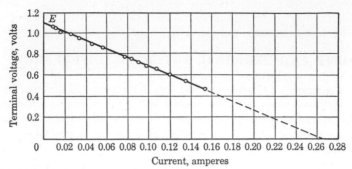

FIG. 32.2 Variation of TPD with current delivered by a high-resistance cell

the cell. This may be seen by solving Eq. (1) for r, giving

$$r = \frac{E - V}{I} \qquad (4)$$

Experiment 32.1

TERMINAL POTENTIAL DIFFERENCE BY VOLTMETER AND RESISTANCE BOX

Object: To study the variation of the TPD of a cell as a function of current delivered and to measure its internal resistance.

Method: The voltage of a cell is measured with a voltmeter for various values of current delivered to a resistance box. The current is calculated from the TPD and the known external resistance. The internal resistance is obtained from the negative of the slope of the curve of TPD against current. Observations are made first for a cell of high internal resistance and second for a cell of low internal resistance; the variations of terminal voltage with current are then compared.

Apparatus: Single storage cell (or dry cell); single storage cell (or dry cell), with artificially added "internal resistance"; 1.5-volt voltmeter; SPST switch, fused for 1 amp; resistance box, 0.1 to 111 ohms; 20-ohm rheostat; 5-amp ammeter.

Procedure: 1. Select for the cell to be tested the one with the artificially added high internal resistance. Arrange the apparatus as in Fig. 32.1. Note the reading of the voltmeter with the switch open, that is, with R infinite. What relationship does this reading have to the emf of the cell?

2. Adjust the resistance box to make R about 20 ohms. Close the switch and note what happens to the voltmeter reading. Because the switch and fuse have appreciable resistances that are not easily measured they should next be removed from the circuit. Then adjust R until the voltmeter reading is about 1.25 volts. Record the voltmeter reading and the resistance. Take a series of simultaneous readings of V and R for values of the voltmeter readings of approximately 1.00, 0.75, 0.50, and 0.25 volt. Finally make $R = 0$ and record the corresponding voltmeter reading. Note the difference between the voltmeter readings when R

is made zero, (1) by tightly inserting all the plugs in the resistance box and (2) by removing the box and substituting for it a short piece of copper wire.

3. Select a storage cell (or a fresh dry cell) of low internal resistance. Connect the apparatus as in Fig. 32.3. Leave one terminal of the cell disconnected until the instructor checks the wiring. Use the 5-amp range of the ammeter, and a rheostat for the resistor. Set the slider of the rheostat for maximum resistance for the preliminary trial. Observe, without recording, the readings of the voltmeter as the current is varied from zero (open circuit) to approximately 5 amp. Record the maximum and minimum values of the TPD.

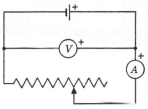

FIG. 32.3 Measurement of TPD and internal resistance of cell by voltmeter and ammeter

Computations and Analysis: 1. Calculate the current for each of the observations of Steps 1 and 2. These currents are the sums of the currents in the voltmeter and the resistance box and hence are given by

$$I = I_V + I_R = \frac{V}{R_V} + \frac{V}{R} \tag{5}$$

The resistance of the model 45 Weston voltmeter is 100 ohms per volt of range; therefore a 1.5-volt voltmeter has a resistance of 150 ohms.

2. Plot a curve of TPD against current. Calculate the slope of this curve. Compare the negative of this slope with the internal resistance of the cell as marked on the cell. Carefully interpret the significance of the shape of the curve, the slope, and the values of the intercepts.

3. Plot a similar curve for the low-resistance cell, using the data of Step 3. Compute the internal resistance of this cell from the negative of the slope of the curve.

4. Comment on the difference between the variation of the TPD with current for the high-resistance cell used in Step 2 as compared with the low-resistance cell used in Step 3. Explain the reason for this difference.

Experiment 32.2

SLIDE-WIRE POTENTIOMETER

Object: To study the potentiometer principle; in particular, to calibrate a slide-wire potentiometer and to use this potentiometer to measure the emf of a cell and the terminal voltages of this cell when it supplies energy to a circuit.

Method: A uniform resistance wire is connected into a circuit containing a battery and a rheostat. The resistance of the circuit is adjusted to produce a unit drop of potential per unit length of wire. This calibration is accomplished by balancing the potential drop in the wire produced by the battery against the emf of a standard cell connected across a span of wire numerically equal to the emf of the standard cell. The calibrated wire is then used as a direct-reading potentiometer to measure the emf of the test cell and also its terminal voltages when the test cell supplies energy to a circuit. The internal resistance of the test cell is determined from a graph of these readings.

Theory: The essential principle of the potentiometer is the balancing of one voltage against another in parallel with it. In the diagram of Fig. 32.4a a branched circuit is shown in which there is a current in the lower branch, resulting from the PD along the slide-wire AC to which it is connected. In Fig. 32.4b a cell has been introduced into the lower branch. Depending upon the voltage

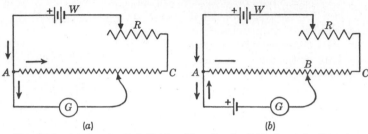

FIG. 32.4   (a) A potential divider; (b) a simple slide-wire potentiometer

of this cell the current in the lower branch may now be in either direction, as indicated by the arrows. As a very special case, the current in the lower branch may be zero when the emf of the cell just equals the PD between A and B and the positive terminal of the cell is connected to the same end of the slide-wire as the positive terminal of the working battery W. In the actual potentiometer the unknown emf, which is to be measured, is balanced against a potential drop along a calibrated slide-wire.

The arrangement in Fig. 32.4b shows the essentials of an elementary, but practicable, form of direct-reading slide-wire potentiometer. A storage battery serves as a working source of emf to maintain a constant drop of potential along the slide-wire resistor AC. A standard cell of known emf is connected to the slide-wire; the position of the sliding contact B is carefully chosen to make the reading on the scale at B equal to the emf of the cell. For example, an emf of 1.018 volts might be represented by 1018 scale divisions. By means of the rheostat R the current in the slide-wire is then adjusted so that the PD between A and B is just equal to the emf of the standard cell; this condition is indicated by zero deflection of the galvanometer. The PD per scale division is then 1 mv and the potentiometer is said to be *direct reading*. A cell of unknown voltage is then substituted for the standard cell and the sliding contact B is adjusted until the potentiometer is again balanced. The emf of the unknown cell may then be read directly on the scale.

A great advantage of the potentiometer is that, at the moment of balance, there is no current in the cell under test. Hence the lead wires carry no current, nor do errors due to line drop or contact resistances occur. But even more important is the fact that the *true emf* of the cell is obtained and not just the *TPD*, which differs from the true emf by the drop of potential over the internal resistance of the cell in accordance with Eq. (1). At the moment of balance $I = 0$ and hence $V = E$, regardless of the value of the internal resistance of the cell. For instance, an old dry cell may have such a high internal resistance that its apparent emf as read by a voltmeter may be very small, whereas the potentiometer will give an emf as high as 1.4 volts in many cases.

Standard cells are sources of especially constant emf. These cells are not designed to supply energy, in fact, every precaution must be made to ensure that only very minute momentary currents are ever present because the cells are damaged by improper use.   Currents greater than 0.0001 amp must not be permitted in the standard-cell circuit.   For this reason a very high resistance of 10,000 ohms or more is usually added to the galvanometer circuit to protect the cell during the initial steps while a near balance position is being located.   After this near balance position is found, the high resistance may be greatly reduced or even eliminated for a more accurate final location of the zero-current position.

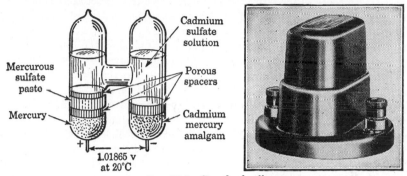

FIG. 32.5   Standard cell

The form of standard cell in general use in student laboratories is the unsaturated Weston cell, Fig. 32.5.   The emf of this cell is about 1.0186 volts.   It has a negligible change of emf with temperature.

Apparatus: Two-meter slide-wire (Fig. 26.3); 6-volt storage battery; standardized dry cell; fresh dry cell; old dry cell; storage cell or dry cell, with artificial "internal" resistance; voltmeter, with 1.5- and 3-volt ranges; three rheostats of resistances 10, 23, and 180 ohms; SPST switch, fused for 0.5 amp; ammeter, 0.5-amp range; DPDT switch; SPST switch, with 10,000-ohm protective resistor; portable galvanometer; Weston standard cell; two four-way connectors.

Procedure: Arrange the apparatus as in Fig. 32.6.   Leave one terminal of the working battery disconnected until the instructor checks the wiring.   For R,

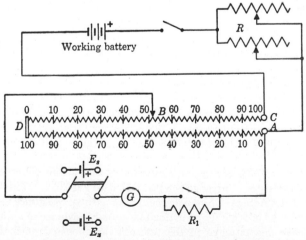

FIG. 32.6   Two-meter, slide-wire potentiometer

use the 23-ohm and the 180-ohm rheostats in parallel. For R_1, use the switch with the 10,000-ohm protective resistor. Make terminal A the zero end of the slide-wire. In the preliminary test, use the standardized dry cell for E_s. A fresh dry cell may be used for the first unknown E_x. Be sure the polarity of E_s and E_x and the working battery are as shown in the wiring diagram. See that all connections are very tight in the circuit containing the working battery and the slide-wire.

2. Make a check of the setup by connecting a voltmeter across the ends AC of the slide-wire. With the DPDT switch open, adjust the rheostats R until the voltmeter reads just 2.00 volts. This condition indicates that the slide-wire is calibrated for 1 mv/mm. Adjust R by placing the slider of the 180-ohm rheostat at about the midpoint of the rheostat. Make the initial setting with the 23-ohm rheostat and the final adjustments with the 180-ohm rheostat. This arrangement provides a convenient "micrometer" adjustment. When these adjustments are complete and *have been checked by the instructor*, remove the fused switch in the working-battery circuit and keep the potentiometer circuit closed for the remainder of the experiment. This closed circuit eliminates the variable resistance in the fuse contacts and gives a steady voltage across the slide-wire.

3. Remove the voltmeter. Set the contactor key B at a place on the slide-wire that will make the length AB in meters numerically equal to the emf E_s of the standardized dry cell. For example, if $E_s = 1.562$ volts, set AB at 1.562 m or 1562 mm. Close the DPDT switch on the E_s side. With the protective resistor switch open, that is, with R_1 in the galvanometer circuit, lightly tap the contactor B and observe the direction of the galvanometer deflection. Adjust the rheostats R until the deflection becomes zero. Short-circuit the protective resistor for the final settings. After the bridge is balanced, the slide-wire is precisely calibrated for a potential difference of 1 mv/mm.

4. Throw the DPDT switch over to the unknown cell E_x. Without changing R, adjust the contactor on the slide-wire to B', the position where the emf E_x is just equal to the PD from A to B'. Make a recheck of the setting for E_s at intervals to be sure that the current in the working battery circuit has not changed.

5. When you have become familiar with the technique of operating the potentiometer, replace the standardized dry cell by the Weston standard cell. Place the contactor B at the proper place (about 1018 mm) determined by the emf of the standard cell and readjust R until a balance is obtained. Recheck the measurement of the emf of the fresh dry cell.

6. Measure the emf of an old dry cell with the potentiometer. In order to illustrate the difference between the measurement of the emf of a cell by a potentiometer and that by a voltmeter, connect the old dry cell to the voltmeter and record the reading. Account for the difference in the values observed from the voltmeter and the potentiometer. Try a similar test with the fresh dry cell. Never connect the Weston standard cell or the standardized dry cell to the voltmeter. Why?

7. Measure the TPD of a cell as a function of the current delivered by the cell to an external resistor. Arrange the test cell as in Fig. 32.7. The test cell to be used consists of a storage cell (or a dry cell) with a coil of resistance wire connected in series with it to provide for an artificial internal resistance of appreciable magnitude. Leave one terminal of the test cell disconnected until the instructor

checks the wiring. Use the potentiometer to measure the emf of the test cell on open circuit, that is, with zero current. This voltage is the emf E of the cell. Use the potentiometer to make five or six readings of the terminal voltage V of the cell for currents of approximately 0.1, 0.2, 0.3, 0.4, and 0.5 amp.

Computations and Analysis: 1. From the data of Step 7 plot a curve of TPD against current. Explain clearly the significance of each of the intercepts of this curve. What is the significance of the shape and the slope of the curve? Calculate the internal resistance from the observed slope of the curve and note the percentage difference between this value and that given on the cell.

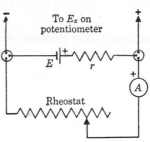

FIG. 32.7 Measurement of internal resistance of cell by slide-wire potentiometer

2. From the data of Step 6 calculate the internal resistance r of both the old and the fresh dry cells. The value of the current I to be substituted in Eq. (4) may be obtained from V/R_V, where V is the reading of the voltmeter and R_V is the voltmeter resistance. The model 45 Weston voltmeters have a resistance of 100 ohms per volt of range. Hence a 1.5-volt meter has a resistance of 150 ohms. Comment on the value of r for the old dry cell as compared with the internal resistance of a fresh cell of the same type.

Review Questions: 1. Explain what is meant by the emf of a source; by TPD. 2. State the equation relating emf and TPD of a cell. Express this equation in several alternate forms. 3. Show why a voltmeter does not indicate the exact emf of a cell. Under what circumstances is this error negligible? 4 Sketch a typical curve to show how the TPD of a cell varies with current. What is the significance of the slope of this curve? of the intercepts? 5. Describe how the internal resistance of a cell may be measured by the use of a voltmeter and a resistance box. 6. Describe by the aid of diagrams the essential potentiometer principle. 7. Show how a slide-wire potentiometer may be calibrated to be direct reading. 8. Explain why a potentiometer is better than a voltmeter for the measurement of voltages free from errors caused by internal resistances or contact resistances. 9. Describe a standard cell. What is its chief virtue? What precautions must be observed in its use?

Questions and Problems: 1. What differences would be observed in the curves of TPD plotted against current for a cell of low internal resistance as compared with a cell of high internal resistance?

2. For the curve of Fig. 32.2 show that $R = r$ when $V = \frac{1}{2}E$.

3. A good measure of the internal resistance of a dry cell may be obtained by short-circuiting the cell through a suitable ammeter. Why is this not feasible in the case of the measurement of the resistance of a commercial generator? Describe how Eq. (4) may be used for this latter case.

4. A dry cell has an emf of 1.518 volts and an internal resistance of 0.125 ohm. A voltmeter of resistance 300.0 ohms is connected to the cell. What reading does the voltmeter indicate? What is the percentage difference between this reading and the emf of the cell? What would this percentage be if the internal resistance were 1.25 ohms?

5. What main feature should a voltmeter have in order that it may accurately read the emf of a cell? Show why this feature cannot be carried out indefinitely in the design of a voltmeter.

6. In what respect is a voltmeter more desirable than a potentiometer for the measurement of voltages?

7. How is the precision of Exp. 32.2 affected by contact resistances at the end A of the slide-wire (Fig. 32.6)? at B? at D? by internal resistance in the working battery?

8. The slide-wire of a simple potentiometer is 2.00 m long and has a resistance of 5.00 ohms. The emf of the working battery is 6.00 volts; its internal resistance is 0.20 ohm. What resistance must be added to the working-battery circuit in order for the wire to be direct-reading at 1 mv/mm?

9. A potentiometer is used to measure the voltage of a cell. On open circuit the cell voltage is found to be 2.218 volts. When a resistor of 12.0 ohms resistance is connected to the cell, the potentiometer reading is 2.154 volts. What is the internal resistance of the cell?

10. A battery is connected to the ends of a rheostat so that the maximum resistance of the rheostat is included. A voltmeter is connected to one of these end terminals and the other side of the voltmeter is connected to the slider of the rheostat. Make a conventional wiring diagram of this apparatus. Explain what happens to the reading of the voltmeter as the slider is moved. This arrangement is properly called a *potential divider* (see Fig. 26.5). Show why this term is appropriate. In radio circuits this arrangement is usually called a "potentiometer." Explain why this designation is not comparable with the use of this term in Exp. 32.2.

11. Explain how a slide-wire potentiometer might be used for the measurement of voltages without having the wire calibrated to be direct reading.

12. In the circuit of a potentiometer that is *not* designed to be direct reading, Fig. 32.4b, AB and BC are adjusted to 64.0 cm and 36.0 cm, respectively, in order to produce zero deflection of the galvanometer when a standard cell of emf 1.0182 volts is in the circuit. When the standard cell is replaced by a dry cell, the point of balance is found at the 95.0-cm point. What is the voltage of the dry cell?

CHAPTER 33

THERMAL EMF

In 1822 J. T. Seebeck announced the discovery of a novel method for the direct transfer of heat into electric energy, a phenomenon now known as the *Seebeck*, or *thermoelectric, effect*. In Fig. 33.1 the ends of a Constantan wire are shown joined to two pieces of copper wire. The copper wires are connected to a galvanometer. When the two junctions are kept at different temperatures, an emf is generated that can maintain an electric current in the circuit. Such a device is called a *thermocouple*. The energy associated with the current is derived from the heat required to keep one junction at a higher temperature than the other.

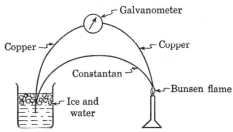

FIG. 33.1   A simple thermocouple circuit

It is now known that a thermal emf is caused by the tendency of two dissimilar metals to transfer electrons across the boundary of contact at unequal rates, thus producing a difference of potential between the two metals. When the junctions are at the same temperature, the two emf's are equal in magnitude and opposite in direction, but when the temperatures are different there is a net emf.

The magnitude of a thermal emf is a function of the metals used and of the difference in temperature between the two junctions. The curve of emf against temperature is never exactly a straight line but for a well-chosen pair of metals the curve shows a close approximation to a linear relationship for a moderate range of temperatures. One commonly used combination of copper and an alloy known as Constantan (0.60 Cu, 0.40 Ni) produces emf's of about 43 μv/°C in the range of 0 to 100°C. Platinum combined with a 90% platinum–10% rhodium alloy is used for high temperatures (up to 1400°C). A bismuth-antimony thermocouple yields the greatest emf for a given temperature difference, but the metallurgical properties of these metals limit their use.

The chief use of thermocouples is in the measurement of temperatures where mercury-in-glass thermometers are too massive to respond quickly to temperature changes, or where it is impossible or inconvenient to introduce them. Any thermoelement may be calibrated with a millivoltmeter or galvanometer to read temperatures directly. Such a device is called a *thermoelectric pyrometer*. The hot junction may be arranged in any location, such as in a furnace, smokestack, or pot of molten metal or glass, with the cold junction kept in an ice-water bath or other fixed-temperature device. The temperatures are read at any convenient station.

The thermal emf developed by thermocouples is relatively small; the values are usually only a few microvolts per degree difference in temperature between the two junctions. An indicating instrument, calibrated for use with a given thermocouple, should be calibrated whenever it becomes necessary to replace one thermocouple with another, even though the new one is made of the same materials.

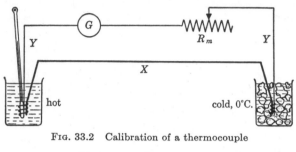

FIG. 33.2   Calibration of a thermocouple

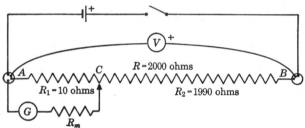

FIG. 33.3   Calibration of a galvanometer as a voltmeter

The calibration of a thermocouple does not differ greatly from the methods used for the calibration of any thermometer. It is essentially a process of relating the readings of the indicating instrument in the thermocouple circuit (usually a potentiometer or a millivoltmeter) to a series of known temperatures to which the junctions of the couple are exposed. These calibration temperatures are frequently the melting or boiling points of selected materials having sharp and well-known transition temperatures.

A thermocouple may also be calibrated by a direct comparison of the instrument readings with the readings of a thermometer placed in contact with one of the junctions; the other junction is immersed in ice water, Fig. 33.2. The wire X is the Constantan wire and the wires YY are copper. The galvanometer may then be calibrated as a voltmeter, in order that the emf generated by the thermocouple may be determined.

The galvanometer may be calibrated as a voltmeter by the use of the circuit shown in Fig. 33.3. The resistor R_1 should be a resistance box provided with traveling plugs so that R_1 may be varied while the total resistance R remains fixed. The multiplying resistor R_m (as in Fig. 33.2) must be kept fixed in the galvanometer circuit. The calibration consists in applying a series of suitable small and measurable voltages to the galvanometer and observing the respective deflections. The voltage is applied by tapping the galvanometer across a small, known portion R_1 of the resistance R. From Ohm's law it follows that the voltage V_1 across the galvanometer circuit is given by

$$V_1 = \frac{R_1}{R} V \tag{1}$$

where V is the voltage across R, as indicated by the voltmeter. [In Eq. (1) it is assumed that the galvanometer current is negligible in comparison with the current in R.] A curve showing the galvanometer deflections plotted against V_1 gives the desired calibration curve to convert the galvanometer into a voltmeter.

Experiment 33.1

CALIBRATION OF A THERMOCOUPLE

Object: To study the phenomenon of thermoelectricity and to calibrate a thermocouple so that either temperature or emf may be read directly from the calibration curves.

Method: A thermocouple has one of its junctions maintained at a constant temperature in an ice-water bath. The other junction, placed in a container of water, is subjected to various known temperatures. A study is made of the effect of temperature change on the emf generated in the thermocouple, and calibration curves are plotted with these data.

Apparatus: Copper-Constantan thermocouple; wall galvanometer; resistance box, 0.1 to 10,000 ohms, with traveling plugs (or two resistance boxes 0.1 to 111 ohms and 10 to 11,110 ohms); dial or plug-type resistance box, 1 to 999 ohms; water boiler; ice bath; 100°C thermometer; Bunsen burner; dry cell; SPST switch; 1.6 or 3-volt voltmeter; two four-way connectors; fine copper wire.

Procedure: I. *Calibration of the Thermocouple.* 1. Arrange the apparatus as in Fig. 33.2. Place one junction of the thermocouple in ice water. Adjust the multiplier R_m to several hundred ohms before placing the other thermocouple junction in a bath of boiling water. Fasten this junction of the thermocouple to the thermometer with fine copper wire, to ensure the same temperature for the thermometer and the junction. If a metal boiler is used, the junction should be kept from touching the metal when readings are taken. Adjust the multiplier R_m until a practically full-scale galvanometer deflection is obtained with the hot junction at the temperature of the boiling water. Do not change this value of R_m throughout the remainder of the experiment.

2. Cool the hot water about 15° and record the galvanometer reading and the temperature. Continue successive observations as you lower the temperature until room temperature is reached. Add small amounts of ice to the warm water to hasten cooling between observations. See that the bath is well stirred and that equilibrium temperature is obtained before readings are recorded. (Does the thermometer reading reach equilibrium as rapidly as the thermocouple, as indicated by a constant galvanometer deflection?)

II. *Calibration of the Galvanometer as a Voltmeter.* 3. Connect the apparatus as in Fig. 33.3. Leave one terminal of the cell disconnected until the instructor checks the wiring. Adjust the total resistance R to 2000 ohms. A single box may be used for R if one is available having 0.1-ohm coils and sufficiently high resistance. Otherwise use a 0.1- to 111-ohm box for R_1 and another box adjusted to 1990 ohms for R_2. Remove all the tenth-ohm and other low-resistance plugs from R_1 to give a resistance of 11 ohms. Keep R_m at the value used in Steps 1 and 2. For the first trial insert the traveling plugs across a resistance of a few tenths of an ohm.

4. Take a series of six readings of galvanometer deflection for various values of R_1. Choose the resistances so that the deflections will vary in approximately equal intervals over the entire scale.

Computations and Analysis: 1. Plot a curve of galvanometer deflection against temperature, using the data obtained in Steps 1 and 2. This is the calibration curve for the thermocouple, used in connection with the particular galvanometer and multiplier.

2. Compute the voltages across the galvanometer circuit for the data of Step 4, by the use of Eq. (1). Express these voltages in microvolts.

3. Plot a curve of galvanometer deflection against voltage.

4. Use the data of the two curves to plot a third curve to show the relation of the thermal emf and temperature difference between the junctions of the thermocouple. What is the value and the significance of the slope of this curve?

Review Questions: 1. What is the thermoelectric effect? State the energy transformations that occur in thermoelectricity. What is the modern explanation for this effect? 2. Upon what factors does a thermal emf depend? What is the order of magnitude of such voltages? 3. What is the chief use of thermocouples? Why are thermoelectric pyrometers so convenient? 4. Describe a method of calibrating a thermocouple for the direct reading of temperature. 5. Describe a method of calibrating a galvanometer for use as a voltmeter to measure the emf of a thermocouple.

Questions and Problems: 1. Show how a galvanometer scale could be changed to read temperature directly, utilizing data like that observed in Steps 1 and 2 of this experiment.

2. If one junction of the thermocouple used in this experiment had been kept at room temperature, about 20°C, instead of 0°C, what difference would have been obtained in the results? What differences would have been observed in the calibration curves?

3. How should the method of calibrating the galvanometer for use as a voltmeter be modified if the resistance of the galvanometer circuit is not considered large in comparison with R_1?

4. A thermocouple has one junction in an ice bath and the other in a beaker of water at room temperature. Will the indicating meter show any different reading when the junction is removed from the water and held (wet) near the beaker? Explain.

5. Explain reasons why thermobatteries have not been used as a commercial source of electric energy.

6. What is the effect on the emf of putting a number of thermocouples in series? in parallel?

7. The emf generated by a thermocouple is independent of the size of wire used. Is this true of the *internal resistance* of such a thermoelectric generator? Discuss briefly.

8. A *thermopile* consists of a group of thermocouples joined in series. A considerable voltage may be thus developed. What about the internal resistance of such a thermopile? its efficiency? its convenience? its life?

9. An iron-Constantan thermocouple gives about 1.80 μv for each centigrade degree difference of temperature between the two junctions. If a thermopile is made of 100 such couples, each having a resistance of 0.100 ohm, what current is furnished to an external resistor of 2.00 ohms when the two faces of the thermopile are maintained at a difference of temperature of 200°C?

10. One junction of an iron-Constantan couple is placed in ice, the other junction in an oil bath at 55°C. What thermal emf is generated? Use data of problem 9.

11. How many thermocouples like the one used in this experiment would be required to give an emf equal to the voltage of a dry cell when the temperature difference is 90°C?

CHAPTER 34

ELECTROLYSIS

Salts, bases, and acids, fused or in solution, which may be decomposed by an electric current are called *electrolytes*. The chemical action associated with the passage of electricity through an electrolyte is called *electrolysis*.

When some substances go into solution, they break up into their constituent *ions*, a process called *dissociation*. No electric current is required to produce this type of *ionization*. An *ion* is an electrically charged particle of atomic or molecular magnitude. In electrolysis the motion of these oppositely charged carriers effects the electrolytic conduction.

According to modern theory, when certain molecules are formed there are electron transfers from one atom to another so that the constituent parts of the molecule are of ionic character. For example, NaCl consists of Na^+ ions and Cl^- ions in the ordinary crystal form of the material. When the solid is melted or goes into solution in water, the crystal forces are broken down and the free ions are liberated. Each ion carries a charge equal to the electronic charge times the valence of the ion. The following dissociations (in water) are familiar examples of this process:

$$NaCl \rightleftarrows Na^+ + Cl^-$$
$$AgNO_3 \rightleftarrows Ag^+ + NO_3^-$$
$$CuSO_4 \rightleftarrows Cu^{++} + SO_4^{--}$$

The double arrows conventionally show that the process is reversibly in equilibrium, that is, recombination is taking place at the same rate as dissociation.

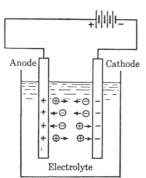

Anode

Cathode

Electrolyte

FIG. 34.1   Migration of ions in electrolysis

There is no tendency for the ions, in the absence of an applied potential difference, to move in any particular direction in preference to other directions. In short, their motion is a random wandering through the solution.

The situation is greatly changed when a potential difference is maintained across the electrolyte. In Fig. 34.1 a battery is shown connected to two electrodes in a solution. The positive ions are attracted by the negative electrode and the negative ions by the positive electrode. For example, if the electrolyte is a solution of $AgNO_3$, the negatively charged NO_3^- ions are attracted to the positive terminal and the positively charged silver ions are attracted to the negative terminal. The current that exists in the cell is the result of the net motion of the ions caused by these attractions. This conduction differs from that in a solid in that both negative and positive ions move through the solution. The electrode at which the conventional current enters the cell is called the *anode;* the electrode by which it leaves is called the *cathode.*

If the anode of the electrolytic cell shown in Fig. 34.1 is made of silver, it will be gradually used up as the electrolysis proceeds. The solution will remain unchanged. The cathode will have metallic silver deposited upon it and will increase in mass.

Michael Faraday in 1833 established by quantitative measurements the following two laws, which are known as Faraday's laws of electrolysis:

Faraday's first law: The mass of a substance separated in electrolysis is proportional to the quantity Q of electricity that passes.

Faraday's second law: For a given charge the mass of a substance deposited is proportional to the chemical equivalent of the ion, that is, to the atomic mass of the ion divided by its valence.

Faraday's laws may be expressed by the following symbolic statements:

$$m \propto Q \qquad (Q = It)$$

$$m \propto c \qquad c = \frac{\text{atomic mass}}{\text{valence}}$$

whence

$$m = kcQ = zQ = zIt \qquad (z = kc) \quad (1)$$

where k is a proportionality constant whose value depends only upon the units involved, m is the mass deposited, and z is a constant

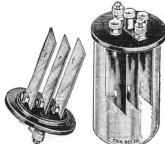

FIG. 34.2   A coulometer

for a given substance (but different for different substances) that is known as the electrochemical equivalent of the substance considered. The *electrochemical equivalent* of a substance is the mass deposited per unit charge; in the usual units it is numerically the number of grams deposited in one second by an unvarying current of one ampere. A table of electrochemical data is given in the Appendix.

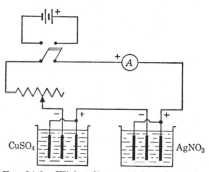

FIG. 34.3   Wiring diagram for study of electrolysis of copper and silver

The *ampere* is legally defined in terms of the electrochemical equivalent of silver; that is, it is the unvarying current that will deposit in one second 0.00111800 gram of silver from an aqueous solution of silver nitrate. If an unknown current is maintained in a solution whose electrochemical equivalent is known, we may determine the value of the current from the measured mass of the material deposited and the time. In fact, the calibration of ammeters is based essentially upon such experiments. The *coulometer* used for this purpose (Fig. 34.2) consists essentially of a pair of plates, connected together to form the anode, and a single *gain* plate, the cathode, placed midway between the positive plates. In practice, this method of calibrating ammeters is rarely used, since it is slow and the experimental technique rather difficult when precise data are to be obtained.

If electrolytic cells are connected in series, as in Fig. 34.3, the current and time (and hence the quantity of electricity) are necessarily the same for each cell. Therefore the masses of material deposited upon the cathodes are proportional to the chemical equivalents of the respective ions. These facts, as well as the first law of electrolysis, may be tested as in the following experiment.

Experiment 34.1

FARADAY'S LAWS OF ELECTROLYSIS

Object: To study the application of Faraday's laws to the electrolysis of silver and copper.

Method: A current is maintained in two coulometers containing solutions of copper sulphate and silver nitrate. The masses of electrolytically deposited copper and silver are obtained by weighing the gain plates (negative electrodes) before and after the current is established. The masses deposited are compared with the values given by Faraday's laws. A different measured quantity of electricity is then used, and the ratio of the observed masses deposited is compared with that obtained from Faraday's laws.

Apparatus: Six-volt storage battery; switch, fused for 1 amp; 0.5-amp ammeter; 5-amp ammeter; stop watch or clock; 22-ohm rheostat; a glass or enameled tray; sandpaper; solutions of $AgNO_3$ and $CuSO_4$; two small jars of alcohol; extra copper and silver cathodes; a copper and a silver coulometer; triple-beam balance.

A coulometer (Fig. 34.2) consists of a glass jar in which are suspended three plates. Two of the plates are anodes, or loss plates, and are connected in parallel. They are arranged on either side of the central cathode, or gain plate. This arrangement facilitates an even coating of the cathode. The top of the coulometer is of molded bakelite, which is acid resisting. Perforations in the plates assist in their being handled by a wire hook to avoid touching with the fingers.

A solution of copper sulphate for electrolysis may be made by dissolving about 25 gm of pure copper sulphate in 100 cm³ of distilled water. About 1 cm³ of concentrated sulphuric acid should be added. For the silver coulometer a neutral silver nitrate solution is made by dissolving about 18 gm of silver nitrate crystals in 100 cm³ of distilled water. The solution should be kept in a dark bottle.

Procedure: 1. As economy of time is essential in this experiment, one student should immediately start weighing the gain plates. These plates should have been left clean by the preceding group but, if they are not clean, follow the instructions given below. The second observer should connect the apparatus and make a preliminary test run.

2. Weigh the copper and silver gain plates on a triple-beam balance. Hang the plates from the hook on the balance arm by means of a copper wire. Be sure to use this same wire in all future weighings.

3. Connect in series the battery, the switch, the copper coulometer, the silver coulometer, the 0.5-amp range ammeter, and the rheostat, as in Fig. 34.3. *Never put the copper plates in the silver nitrate solution.* (Why?) Be sure to make the anode (the pair of plates) the positive terminal of each coulometer and the cathode (the single gain plate) the negative terminal. Clean the under side of the coulometer tops with sandpaper and water until they are free of metallic deposits. The electrolyte should be kept at the proper levels in each coulometer, well below the

plate clamps but covering the major portion of the plates. Too large a current density (current per unit area) results in a flaking off of the silver deposit. For copper about 20 ma/cm² and for silver about 3 ma/cm² are optimum values of the current density. Adjust the rheostat until a current of about 0.5 amp is obtained. See that the rheostat is so arranged that it will be possible later either to increase or decrease the current. Ask the instructor to check the circuit.

4. After the preliminary observation, remove the gain plates and clean them. Lightly scrape off any loosely adhering granules of metal with sandpaper. Rinse in clean water, dip them in the appropriately labeled jar of alcohol, and dry by twirling in the air. Weigh carefully. The plates should be placed in the coulometers not more than a few minutes before the current is started. The copper, particularly, is slightly soluble in the electrolyte.

5. Replace the weighed gain plates and, at an accurately noted time, start the current. One observer should constantly watch the ammeter and adjust the resistance so as to keep the current constant. Maintain the current for about 20 min, noting carefully the time when the switch is opened.

NOTE: If sufficient plates are available, it is well to clean and weigh the extra cathodes while this part of the experiment is being performed.

6. Remove the cathodes, rinse them in clean water, then in alcohol, and dry them by holding them over a radiator or other source of heat. The strictest precautions should be taken in order not to jar off any of the loose granules of metal. Weigh the plates carefully and record these masses.

Remove the silver coulometer and use only the copper coulometer in the remaining tests.

7. Substitute the 5-amp ammeter for the sensitive one. Take a second set of observations in which the product of current and time is sufficiently altered to test the variation of the mass deposited as a function of the quantity of electricity. Use a current of 1.5 or 2 amp.

8. Before leaving the laboratory, carefully clean a copper and a silver plate for the use of the next group. Leave these plates on a clean sheet of paper.

Computations and Analysis: 1. From the data of Steps 2, 6, and 7 compute the gain in mass of the plates.

2. Make a check of Faraday's first law of electrolysis by comparing the ratio of the masses of copper deposited with the ratio of the charges. Determine the percentage difference between the ratios.

3. Test Faraday's second law by computing the ratio of the mass of silver deposited by a given charge to the mass of copper deposited by the same charge and comparing this ratio with the ratio of the known chemical equivalents. Note the percentage difference between the ratios.

4. Using the observed current, time, and gain in mass, calculate the electrochemical equivalent of copper from Eq. (1) for each observation. Repeat for silver. Compute the percentage error between the average of each pair of values and the standard values.

Review Questions: 1. Explain the meaning of the terms: electrolyte, electrolysis, and ion. 2. Explain dissociation and recombination from the point of view of modern theory. 3. State what happens in an electrolytic solution when a PD is maintained between two plates immersed in the electrolyte. 4. State Faraday's laws of electrolysis. 5. Define electrochemical equivalent. In what units is it usually expressed?

6. Define the legal ampere.   7. Show how a coulometer may be used to measure current.

Questions and Problems: 1. The terms *cation* and *anion* are frequently used to designate certain ions.   Identify the sign of the charge on each of these ions.   Show why these terms are appropriate.

2. In addition to the gram per coulomb, what other units might be used for electrochemical equivalent?

3. Describe what would happen if pure platinum plates were used for the electrodes instead of the copper and silver plates used in this experiment.

4. State the various chemical reactions occurring in the electrolysis of sodium chloride.

5. Describe the phenomena occurring when a current is maintained in an acidulated water coulometer.   How do the masses of the materials deposited compare? How do the volumes compare?

6. Is it possible to tell from the rate of evolution of gas at the terminals of a pair of electrodes in an acidulated water cell which is the positive terminal?   Explain.

7. A current of 2.00 amp deposits 2.78 gm of a substance in 2.00 hr.   If the atomic weight of the substance is 55.8, what is its valence?

8. What is the cost per 24-hr day of operating a silver-plating vat across which a PD of 3.00 volts is maintained and 4.00 kg/hr is deposited, if the rate is 2 cents per kilowatt-hour?

9. If the specific gravity of copper is 8.90 and the area of the gain plate is 35.2 cm², what thickness of copper is deposited on each side of the plate in a coulometer in which there is a current of 0.475 amp maintained for half an hour?

10. Assuming as known the value of z for silver, calculate the quantity of electricity necessary to deposit a mass equal to the chemical equivalent.

11. How much charge is required to deposit 1.00 gm of silver?

CHAPTER 35

HEATING EFFECT OF ELECTRIC CURRENT

Potential difference is defined as work per unit charge; in symbols

$$V = W/Q \tag{1}$$

where W is the work done in moving the charge Q through the circuit. The most frequently used units are V in volts, W in joules, and Q in coulombs.

If Eq. (1) is written in the form $W = VQ$, and the substitutions $Q = It$ and $V = IR$ are made, the basic equation for electric energy may be written in the forms

$$W = VQ = VIt = I^2Rt \tag{2}$$

This equation indicates that 1 joule of work must be expended in maintaining for 1 sec a current of 1 amp in a resistor of 1-ohm resistance.

In a circuit containing only resistance, a direct proportion exists between the expenditure of electric energy W and the heat H developed. This fundamental law is represented by the conservation-of-energy equation

$$W = JH \tag{3}$$

where J (after *Joule*) is the proportionality factor called the *mechanical (or electrical) equivalent of heat*. In Eq. (3) J has the value of approximately 4.18 joules/cal.

A useful form of the equation for the heat developed in a resistor is obtained by combining Eqs. (2) and (3), namely,

$$H = \frac{W}{J} = \frac{VIt}{4.18} = \frac{I^2Rt}{4.18} \tag{4}$$

If one wishes to change calories to joules, he *multiplies* the number of calories by 4.18 joules/cal, whereas, if he wishes to change joules to calories, he *divides* the number of joules by 4.18 joules/cal.

The variation of the heating effect of an electric current with various factors as represented by the equation $H = I^2Rt/J$ is usually referred to as *Joule's law*. From this equation it is evident that the heat developed in a resistor varies directly with (1) the square of the current (when R and t are constant), (2) the resistance (if I and t are constant), and (3) the time (when I and R are constant).

Experiment 35.1

JOULE'S LAW

Object: To study the factors upon which the heating effect of an electric current depends, and to measure Joule's electrical equivalent of heat by the use of an electric calorimeter.

Method: A measured potential difference and current are maintained for a given time in an electric calorimeter. The heat evolved is determined by the usual calorimetric methods. The current is varied, all other factors being kept constant, and the variation in the heating effect is observed. Similar observations are made of the variation of the heat with time and with resistance. The value of the electrical equivalent of heat, Joule's constant J, is determined from the ratio of the electric energy expended to the heat produced.

FIG. 35.1 An electric calorimeter

Apparatus: Electric calorimeter; extra coil for calorimeter; trip scales; set of masses; ice, for cooling water; voltmeter, 15- and 150-volt ranges; 5-amp ammeter; 22-ohm rheostat; 3-ohm rheostat; 20- to 30-volt battery (or commercial power source); switch, fused for 5 amp; 50°C thermometer; stop watch, or clock with sweep-seconds hand; distilled water.

The electric calorimeter, Fig. 35.1, consists of a double-walled vessel with water in the inner container. A heating coil, a thermometer, and a stirrer are inserted in the water.

Procedure: I. *Variation of Heating Effect with Time.* 1. Connect the apparatus as in Fig. 35.2. Leave one terminal of the battery unconnected and the switch open until the instructor has checked the wiring. Arrange the apparatus so that one observer can conveniently watch the ammeter and control the rheostat R_1. *Never close the switch unless the heating coil is immersed in water,* as otherwise the resistance element might be burned out.

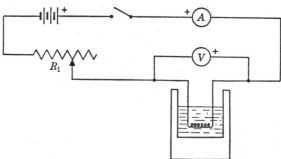

FIG. 35.2 Arrangement of apparatus for studying Joule's law

2. Make a short test run, using tap water in the calorimeter, and a current of about 2 amp. Practice the technique of carefully stirring the water and noting the rise of temperature.

3. Empty the calorimeter and refill it with 250 gm of pure water to which a little ice has been added to reduce the temperature to about 10°C. Stir the water thoroughly until the temperature approaches the equilibrium value. At an accurately noted time, close the switch, quickly adjusting R_1 if necessary in order to bring the current to the desired value (2.00 amp). Continue stirring the water and record the temperature every minute until about ten observations have been made. Record the current and the voltage.

II. *Heating Effect as a Function of Current.*   4. Take observations of temperature rise of same mass of water as in Step 3, beginning with cold water each time. Note the temperature rise in 5 min, recording final temperature after water has risen to its highest temperature.   Repeat observations for a total of five different currents.   Data recorded in Step 3 can be used for one observation.   Do not heat the water above 40°C, as otherwise the effects of radiation will be large. It is well to have the initial temperature about as much below room temperature as the final temperature is above that of the room.   If necessary the water should be cooled with ice to produce this condition.   Record each voltage.

III. *Variation of Heating Effect with Resistance, Current Constant.*   5. It will be necessary to take only one more observation in order to test the variation of heating effect with resistance since any one of the observations previously taken may be used.   Measure the resistance of the heating coil used in Steps 3 and 4, by the use of the voltmeter-ammeter method (Chap. 31).   Select another coil of materially different resistance and measure its value.   Use the second coil and adjust the rheostat to give the same current as selected for one of the observations in Step 4.   Maintain this current for the same time as that of the previous observation.   Note and record the rise in temperature of the same mass of water as that previously used.

Computations and Analysis: 1. Use the data of Step 3 to plot a curve showing the variation of the heating effect with time, all other factors being constant.   In this case the rise in temperature may be plotted since the rise in temperature is a direct measure of the heat developed.   Discuss the significance of the shape and intercepts of the curve.

2. Use the data of Steps 3 and 4 to plot a curve showing the variation of the heat developed as a function of current, all other factors being constant.   Because this curve will not be a straight line it will be difficult to interpret.   Can you change one of the variables so that a linear relation is obtained?   If not, request advice from the instructor.   Plot the straight-line curve and carefully interpret its significance.

3. Use the data of Steps 4 and 5 to check the variation of H with R when all the other factors are constant.   Compare the ratio $\Delta T_1 / \Delta T_2$ with the ratio R_1 / R_2 by noting the percentage difference between them.   Comment on the significance of the approximate identity of these ratios.

4. Use the straight-line curve of ΔT against I^2 which was plotted in Part 2 as the basis for the computation of the average value of J from the observed data. The working equation may be written as

$$J = \frac{W}{H} = \frac{I^2 R_t}{(M_w S_w + M_c S_c)(t_f - t_i)} \tag{5}$$

where the symbols have the usual significance of those used in calorimetry.   The reciprocal of the slope of the curve gives $I^2/(t_f - t_i)$.   Use 6 gm for the water equivalent of the heating coil, stirrer and electrodes.   Determine the percentage difference between the average value of J obtained by the use of the curve and Eq. (5) and the standard value of 4.18 joules/cal.

Review Questions: 1. Define potential difference and show how this definition can be put into the form of an expression for energy.   Write the energy equation in several alternate forms.   State the usual unit of each of the factors in these equations. 2. State Joule's law.   Explain what is meant by the electrical equivalent of heat.

3. State in several alternate forms the equation for the heat developed in a resistor. Give the common unit of each factor in these equations. 4. Describe how the electrical equivalent of heat can be measured with an electric calorimeter.

Questions and Problems: 1. Use the data of one of the steps in Exp. 35.1 to calculate the efficiency of the electric calorimeter as a water heater.

2. Show how Exp. 35.1 constitutes an *absolute* measurement of potential difference and thus ideally could be used to calibrate a voltmeter.

3. In an experiment performed to measure J by the electric-calorimeter method, the following data were obtained: resistance of coil, 55.0 ohms; potential difference across coil, 110 volts; mass of water, 156 gm; mass of calorimeter, 60 gm; specific heat of calorimeter, 0.100 cal/gm °C; time of run, 1.25 min; initial temperature of water, 10.0°C; final temperature of water, 35.0°C. Calculate the value of J from these data and, comparing the result with the standard value, determine the percentage error.

4. An electric water heater of the kind often used for heating small quantities of water operates at 110 volts. If its efficiency is 80% and 1 liter of water can be heated from 20 to 80°C in 10 min, what is the cost of operation at 6 cents per kilowatt-hour?

5. Two wires of radii in the ratio of 1 to 2, but otherwise identical, are connected in parallel and joined to a battery. Compare the quantities of heat produced in the two wires. How would the quantities of heat compare if the wires were connected in series?

6. What would be the power rating for an electric lamp bulb to replace the electric calorimeter used in Step 5? Would this bulb be an equally effective heater, as compared with the resistance coil? Explain.

7. An electric heater is immersed in 500 gm of oil of specific heat 0.214 cal/gm °C. A current of 5.32 amp produces a temperature rise of 3.25°C/min. What is the resistance of the coil and the power used in it?

8. Explain clearly why the precision of calibration of some fuse links is improved by designing them to have a constricted portion in the center.

9. Prove the fact that the rate of heating a conductor with a constant voltage impressed on it varies directly with the cross section of the conductor.

10. A 5-ohm resistor A and a 20-ohm resistor B are connected in series. A 3-ohm resistor C and a 6-ohm resistor D are connected in parallel and this group is connected in series with A and B. Compare the rates of heating in A and B; in C and D; in A and C.

11. Two rheostats A and B are connected in series and joined to a battery. If A has twice the resistance of B, how will the rates of heating in the two rheostats compare? These rheostats are connected in parallel and attached to the same battery. Compare their rates of heating. Compare the rate of heating in A in the series case with that of A in the parallel arrangement.

CHAPTER 36

ELECTROMAGNETIC INDUCTION

Among the most significant discoveries in the field of electricity of the nineteenth century were those having to do with the mutual relationships between electricity and magnetism. The first of these was the observation, by H. C. Oersted in 1820, that a wire in which there is a current has associated with it a magnetic field.

The second significant discovery in this field was made by Michael Faraday in a series of experiments lasting from 1824 to 1831. Faraday reasoned that if a current has an associated magnetic field, then conversely, if a magnetic field is set up around a wire, the wire should acquire a current. However, when Faraday attempted to find such a current, he met with failure as long as the field was static relative to the wire. He finally observed that the expected current is produced *when there is motion of the wire relative to the field or when the field is changing in strength.*

Whenever a conductor moves through a magnetic field in such a manner as to cut across the magnetic flux, an emf is induced in the conductor. The magnitude of the induced emf is directly proportional to the rate of cutting lines of force. Similarly an induced emf is produced whenever there is a change in the flux threading a coil. The emf is proportional to the rate of change of flux and to the number of turns N. In symbols

$$\bar{E} \propto \frac{\Delta\phi}{\Delta t} \quad \text{and} \quad \bar{E} \propto N$$

whence

$$\bar{E} = kN\frac{\Delta\phi}{\Delta t} \tag{1}$$

where $\bar{E}$ is the average emf induced when there is a change of flux $\Delta\phi$ (or flux cut) in time Δt. The value of the constant of proportionality k depends upon the units used. If $\bar{E}$ is in volts, $\Delta\phi$ in maxwells, and Δt in seconds, k has the value 10^{-8}. For this system of units Eq. (1) becomes

$$\bar{E} = \frac{N}{10^8}\frac{\Delta\phi}{\Delta t} \tag{2}$$

If the N pole of a magnet is moved toward a coil, Fig. 36.1a, an emf is induced in the coil and there is a current in the closed circuit in the direction indicated by the arrows. The field of the current is opposite in direction to that of the magnet and thus tends to oppose the increase of the field. When the N pole is moved away from the coil, Fig. 36.1b, the current reverses in direction. In this case the field caused by the current is in the same direction as the field of the magnet and opposes the decrease in field.

Lenz's law states the general condition for the direction of an induced emf:

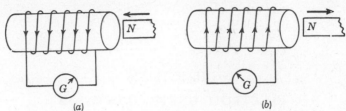

FIG. 36.1   Magnet and coil to show induced currents

Whenever there is an induced emf, it is in such a direction as to tend to oppose the change that caused it.   This law is a form of the law of conservation of energy.

Experiment 36.1

INDUCED EMF'S

Object: To study induced emf's and the factors that affect their production.

Method: A coil of wire is connected to a galvanometer.   Observations are made of the galvanometer deflections produced when a bar magnet is inserted into the coil and when it is withdrawn.   The polarity of the magnet, the direction of the motion, and the direction of the current are correlated with Lenz's law. The permanent magnet is replaced by a second coil of wire connected to a battery through a switch and rheostat.   By observing the deflections produced when the

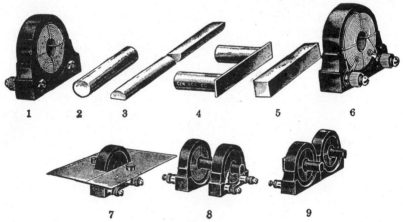

FIG. 36.2   Induction apparatus

switch is opened or closed and when the current is increased or decreased, the relationship between primary and secondary currents is determined.   The effect of the intervening medium is demonstrated by inserting cores of iron and brass and by placing sheets of iron and brass between the coils.

Apparatus: Galvanometer, 1 or 2 μa per division; 9999-ohm resistance box to use as a shunt on the galvanometer; rheostat, 22 ohms; dry cell; tap key; pair of coils (Fig. 36.2); bar magnet, preferably Alnico; soft-iron rod; brass rod; soft-iron sheet, 3 by 3 in.; brass sheet, 3 by 3 in.; long flexible conductor; compass; support rod; right-angle clamp; horseshoe magnet.

Procedure : 1. In order to determine the directions of the induced currents, it is first necessary to determine the direction of the galvanometer deflection corresponding to a known direction of current.

Connect the galvanometer into a potential divider circuit, Fig. 36.3. Use a dry cell as a source and set the slider of the rheostat to tap off one or two wires of the slidewire. Note the direction of the current when the tap key is closed, and note the direction of deflection of the galvanometer. Arrange polarity of cell so that galvanometer deflects toward terminal marked positive on the meter. Use this information

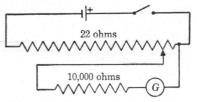

FIG. 36.3   Potential-divider circuit

throughout the experiment to determine the direction of the induced current.

2. Mount the horseshoe magnet in a right-angle clamp on a support rod. Connect the two ends of the long flexible conductor to the terminals of the galvanometer. Use the compass to determine the polarity of the magnet and mark with chalk the N pole. Move the wire quickly into the space between the N and S poles and note the deflection of the galvanometer. What is the direction of the induced current in the wire? Use the right-hand rule to show the direction of the field of the induced current. In front of the wire, is the field of the current in the same direction as the field of the magnet? Is the direction of the field of the current found to be in accord with Lenz's law? Make a sketch of this arrangement, showing clearly the direction of the field of the magnet and the field of the induced current. Show why the resultant field opposes the moving of the wire into the field of the magnet.

Make similar sketches and give similar explanations for each of the succeeding steps of this experiment.

3. Make a single loop in the flexible conductor, thrust it quickly onto the N pole of the magnet and note the deflection of the galvanometer. Is the direction of the current that predicted by Lenz's law? Explain. Observe and explain the behavior of the galvanometer when the loop is held stationary around the magnet. Withdraw the loop and note and explain the deflection of the galvanometer. Repeat these observations using first two, then three loops. Compare the three deflections. What do these observations show regarding the dependence of the induced emf upon the number of turns?

4. Connect one of the coils to the galvanometer. Connect the 9999-ohm resistance box across the terminals of the galvanometer to use as a shunt to vary the sensitivity of the meter. Note the way the wire is wound on the coil. Record the direction of the current each time as clockwise or counterclockwise as viewed from the side on which the binding posts are mounted.

Thrust the N pole of the magnet toward the coil and note the deflection of the galvanometer. Adjust shunt to give a readable deflection. Explain deflection in terms of Lenz's law. Hold magnet stationary near the coil and note galvanometer reading. Explain. Withdraw the N pole and explain the result. Repeat, using the S pole.

Move the N pole of the magnet slowly toward coil. Compare deflection thus obtained with that observed when magnet was thrust in quickly. Explain.

5. Connect the second coil in series with a dry cell, a 22-ohm rheostat, and a tap key. Connect the cell so that the conventional current in the coil is clockwise

as viewed from the side opposite to the binding posts. Set this coil immediately adjacent to the coil connected to the galvanometer and with their axes in line. Set the resistance box at 9999 ohms and the rheostat at zero. Close the switch and note the deflection. Is the induced current in the secondary coil in the same or opposite direction to that in the primary coil? Explain in terms of Lenz's law. With the key closed note the reading of the galvanometer. Explain. Open the switch and note the direction of the induced current. Is the induced current in the same or the opposite direction to that in the primary coil? Explain in terms of Lenz's law. Repeat with the coils separated by a half inch. Explain the effect of the separation.

6. With the coils in the initial position of Step 5 close the switch of the primary circuit. After the pointer has come to rest, quickly move the primary coil away from the secondary. Note and explain the deflection. Quickly replace the primary coil. Note and explain the deflection.

Repeat these readings with the coil moved slowly. Explain the difference in the two effects.

7. With the coils in the initial position of Step 5 close the switch. When the pointer has come to rest, move the slider of the rheostat quickly to the position of maximum resistance. Record and explain the deflection. Move the slider back to the position of zero resistance and explain the deflection.

8. Set the two coils in the position of 8 in Fig. 36.2. Decrease the sensitivity of the galvanometer by reducing the resistance in the resistance box to about 5 ohms. Close the switch and note the deflection. Insert the soft-iron core and again close the switch and note the deflection. What does this indicate as to the effect of the presence of the iron core? Change the distance between the coils with the iron core still joining them. Close the switch and compare the reading with the one previously taken by use of the iron core. Does the distance between the coils materially affect the reading? Explain.

Repeat this step using a brass core in place of the iron.

9. Set the two coils in the position of 9 in Fig. 36.2. Close the switch and note the deflection. Insert the U-shaped soft iron and again close the switch. Explain the difference. What does this experiment show regarding the path of the flux?

10. Set the two coils adjacent as in Step 5. Set the galvanometer for maximum sensitivity. Close the switch and note the deflection. Insert a flat soft-iron plate between the coils, again close the switch, and note the deflection. Carefully explain the difference.

Repeat, using a brass plate in place of the iron plate.

11. Summarize the results of all these experiments. What factors are common to all the experiments? State the conclusions that you can reach concerning the factors upon which the value of the induced emf depends.

Review Questions: 1. What was Oersted's discovery regarding the relation between electricity and magnetism? 2. Describe Faraday's discovery of induced emf's. 3. Under what circumstances may an induced emf be obtained? an induced current? 4. Upon what factors does the magnitude of an induced emf depend? Write the symbolic equation for the average induced emf. 5. What determines the magnitude of an induced current? 6. State Lenz's law. Show that it is a form of the law of conservation of energy.

Questions and Problems: **1.** What is the source of the energy of the induced current when a bar magnet is inserted into a coil of wire?

2. Discuss the emf induced in the coil of Fig. 36.1 as the magnet is passed through the coil and withdrawn from the other side.

3. A circular coil lies on a horizontal table. Discuss the emf and current induced by cutting the vertical component of the earth's field as the coil is slid along the table.

4. A metal clothesline stretched east and west falls to the ground. In what direction is the emf induced by cutting the earth's field? Illustrate by a diagram.

5. Explain how a coil, such as the primary coil used in Exp. 36.1, could be wound noninductively, that is, so that it would have no effect upon the secondary coil.

6. Two secondary coils are identical in every respect except that one is wound with copper wire and the other with nickel-silver wire. Compare the induced emf's and the induced currents when these coils are used with the same primary coil.

7. The magnet in Fig. 36.1 is quickly withdrawn from the coil. Discuss the situation in detail with respect to Lenz's law and Oersted's relationship.

8. A rectangular coil whose plane is perpendicular to a magnetic field is 20 cm long, 12 cm wide, and is composed of 100 turns of wire. Find the average induced emf when the coil is given a quarter revolution in 0.10 sec. Assume a field strength of 10,000 oersteds.

9. If a coil composed of 20 turns of wire and having an area of 100 cm² is rotated at a rate of 120 rpm in a field of 1000 lines/cm², what is the average emf over a half revolution beginning with the plane of the coil perpendicular to the field?

10. Two coils are arranged as in Step 5 and the induced emf observed when a current is established in one coil. Then the coil connected to the galvanometer is placed at an angle of 45° with respect to the first coil and the same current is established in the first coil. Compare qualitatively the induced emf for this case with that originally observed. Repeat for the case when the angle is 90°. Explain.

11. Show that the quantity of electricity Q flowing in the coil connected to the galvanometer circuit of Steps 5 to 10 is given by

$$Q = \frac{N \, \Delta\phi}{10^8 R}$$

where N is the number of turns in the coil and R is the resistance of the circuit in which it is connected.

12. Describe the way in which the observations of Step 8 illustrate the basic principles of the electric transformer. Explain what would happen if the first coil were connected to an a-c power source and the second coil were connected to an a-c meter.

13. The coil connected to the galvanometer in Step 5 has 300 turns. The resistance of the galvanometer-and-coil circuit is 85 ohms. When the switch in the second coil is closed, a flux of 1.25×10^4 maxwells (lines) is linked with the first coil. What charge flows in the galvanometer circuit? What is the average current during the first tenth second while this charge is flowing in the galvanometer circuit?

CHAPTER 37

CAPACITANCE

When an insulated conducting body is charged, it may retain this charge for a considerable time. Any such system that acts as a device to store electricity is called a *capacitor*. (The term *condenser* was formerly used for this concept.) The most common design of capacitor is the parallel-plate type, Fig. 37.1.

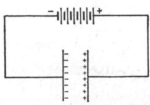

FIG. 37.1 A simple parallel-plate capacitor

When a capacitor is connected to a battery, electricity flows for a brief time from the battery into the plates of the capacitor. However, the PD across the plates quickly becomes equal to the emf of the source and the current drops to zero.

If the battery were then removed, the capacitor would retain its charge for some time. The leaks through the insulation finally reduce the voltage of the capacitor to zero.

The *capacitance* of a device is defined as the ratio of the charge on the capacitor to the potential difference across it. In equation form

$$\text{Capacitance} = \frac{\text{charge}}{\text{potential difference}} \qquad C = Q/V \qquad (1)$$

The capacitance of a device depends upon its geometrical constants, particularly the area of the plates and the distance between them, and also depends upon the nature of the insulator separating the plates.

The unit of capacitance in the practical system is the farad. A *farad* is the capacitance of a capacitor that acquires a PD of 1 volt when it is given a charge of 1 coulomb. The farad is such a large unit that the *microfarad* (μf, 10^{-6} farad) is the unit most frequently used.

In Fig. 37.2 three capacitors are shown connected in parallel and joined to a

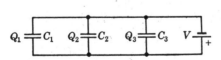

FIG. 37.2 Capacitors in parallel

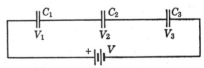

FIG. 37.3 Capacitors in series

cell of terminal voltage V. The voltage relationships for capacitors in parallel are represented by

$$V = V_1 = V_2 = V_3$$

The charge relationships are

$$Q = Q_1 + Q_2 + Q_3$$

250

When these relationships are considered together with Eq. (1), the total capacitance is given by

$$C = C_1 + C_2 + C_3 \tag{2}$$

Capacitors are connected in parallel when a high capacitance is desired.

When capacitors are connected in series, Fig. 37.3, the following relationships apply:

$$V = V_1 + V_2 + V_3$$
and
$$Q = Q_1 = Q_2 = Q_3$$

The total capacitance is therefore given by

$$\frac{1}{C} = \frac{1}{C_1} + \frac{1}{C_2} + \frac{1}{C_3} \tag{3}$$

Capacitors are connected in series when it is desired to divide up the total voltage and a low total capacitance is not objectionable.

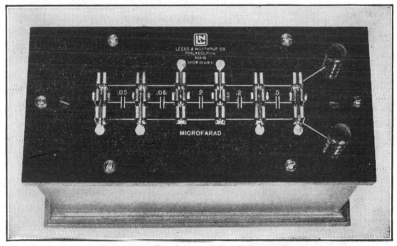

Fig. 37.4   Precision subdivided capacitor

One can observe the correspondence between the relationships for the joint resistance of conductors in series and the joint capacitance of capacitors in parallel; also between resistors in parallel and capacitors in series.

High-grade, precision capacitors are made with thin tin-foil plates, separated by mica insulators. Careful design to minimize leakage and precision calibration result in accurate capacitors, often used for secondary standards. These capacitors are frequently arranged in boxes, Fig. 37.4, similar in form to resistance boxes. The white lines engraved on the hard-rubber top of the box show the electrical connections. By a proper manipulation of the switches individual capacitors may be selected, or the several capacitors may be placed in series, in parallel, or in various series-parallel combinations. In order to use any single capacitor the experimenter closes the switches on either side of the desired capacitor, with one switch closed upward and the other one downward. All other switches are left open. When the experimenter wishes to place capacitors in parallel, he closes the switches in a staggered manner, as shown in the dashed lines of Fig. 37.5a. Series connections are made by closing the switches, respectively, upward at the right and downward at the left of the group of capacitors that are to be put in series, as shown in Fig. 37.5b. A series-parallel arrangement is shown in Fig. 37.6a. The

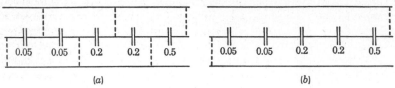

(a)   (b)

FIG. 37.5 Subdivided capacitor box, with capacitors (a) in parallel, (b) in series

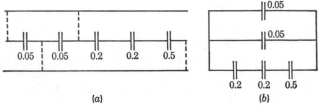

(a)   (b)

FIG. 37.6 Subdivided capacitor box, with capacitors in series-parallel arrangement

conventional wiring diagram for this series-parallel combination is given in Fig. 37.6b.

The capacitance of a pair of parallel plates separated by an insulator of dielectric constant K, Fig. 37.7, is given approximately by

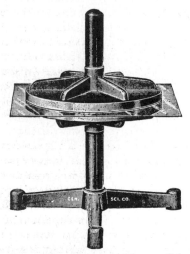

FIG. 37.7 Experimental parallel-plate capacitor

$$C = \frac{KA}{4\pi s \times 9 \times 10^5} = \frac{8.84KA}{10^8 s} \quad (4)$$

where C is in microfarads; A is the area (of one plate), in square centimeters; and s is the thickness of the insulator, in centimeters.

The value of K may be determined by measuring the capacitance C_x of a parallel-plate capacitor with the unknown material as an insulator, then measuring the capacitance C_0 of the same capacitor with the same thickness of air as an insulator, and applying Eq. (4) to give the following proportion

$$\frac{C_x}{C_0} = \frac{K_x}{K_0} \quad \text{or} \quad K_x = K_0 \frac{C_x}{C_0} \quad (5)$$

Because K_0 is nearly unity for air it is seen that the dielectric constant of a substance is numerically nearly equal to the ratio of the capacitance of a capacitor using the given substance as a dielectric to the capacitance of the same capacitor with air as the insulator.

Experiment 37.1

CAPACITANCE BY BRIDGE METHOD

Object: To study the bridge method for the comparison of capacitances; in particular, to measure the capacitance of capacitors connected in series, in parallel, and in series-parallel arrangements; and to measure the dielectric constant for mica.

Method: A Wheatstone-bridge circuit is formed by the use of two capacitors and two resistors. An alternating voltage is applied and the circuit balanced by varying the resistors until there is a minimum sound audible in the telephone receiver used as the detector. Several unknown capacitors are measured. Various series, parallel, and series-parallel combinations of known capacitors are prepared and their joint capacitances measured. The measured values are then compared with the values obtained by calculation. The dielectric constant of mica is measured by the use of an experimental parallel-plate capacitor.

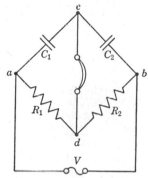

FIG. 37.8 Bridge method of comparing capacitances

Theory: An unknown capacitance may be measured by the use of a known capacitor and two known resistors, using the arrangement shown in Fig. 37.8. A source of alternating or intermittent voltage V is impressed upon a bridged circuit of two branches. One branch consists of two capacitors C_1 and C_2 in series. The other branch has two resistors R_1 and R_2 in series. A telephone receiver,

FIG. 37.9 Precision capacitor, for standard

connected to the points c and d, is used to determine the condition of balance, that is, when there is no hum in the receiver the points c and d are at the same potential. For this condition $V_{ac} = V_{ad}$ and $V_{cb} = V_{db}$. If we apply Eq. (1) and Ohm's law to these equalities, we obtain

$$Q_1/C_1 = R_1 I_1$$
and
$$Q_2/C_2 = R_2 I_2$$

If there is no current in the telephone receiver, $Q_1 = Q_2$ and $I_1 = I_2$. Hence if we divide the first of the foregoing equations by the second, we obtain

$$\frac{C_1}{C_2} = \frac{R_2}{R_1} \quad \text{or} \quad C_1 = C_2 \frac{R_2}{R_1} \tag{6}$$

This working equation is similar to that of the Wheatstone-bridge method of comparing resistances, except that here the capacitances are in *inverse* ratio to the resistances.

Apparatus: Subdivided capacitor; 0.5-μf standard capacitor (Fig. 37.9); two dial-type resistance boxes, 0 to 9999 ohms; audio oscillator (or other source of alternating or intermittent voltage); SPST switch; telephone receiver; four three-way connectors; experimental parallel-plate capacitor; several unknown capacitors; micrometer caliper; paraffined paper; meterstick; sheet of mica.

Procedure: I. *Measurement of Capacitance of Unknown Capacitor.* 1. Arrange the apparatus as in Fig. 37.8. Use for C_2 the precision capacitor of fixed capacitance (0.5 μf). Use a telephone capacitor for the unknown. Ask the instructor to check the wiring before closing the circuit to the voltage source. If an audio oscillator is used, select the output terminals that give a maximum intensity of sound in the receiver for the circuit being utilized.

2. Set R_1 at some convenient value, say 250 ohms. Adjust R_2 until a balance is obtained; for this condition the sound in the receiver has a minimum intensity. The best value of R_2 is obtained by approaching the value for balance from the side of a too-high resistance and also from a too-low value and averaging these observations. Record the final values for R_1, R_2, and C_2. Compute the value of C_1 from these data. In the same manner measure the capacitance of several other unknowns. Tabulate all of the data in ruled columns.

II. *Capacitors in Series and in Parallel.* 3. Adjust the switches on the subdivided capacitor box to put several of the capacitors in this box in parallel. Record this arrangement by a sketch. Measure the joint capacitance with the bridge. Calculate the joint capacitance from the known individual values and the use of Eq. (2). D termine the percentage difference between the observed and calculated values of the joint capacitance.

4. Repeat Step 3 for the 0.5-, 0.2-, and 0.2-μf capacitors connected in series.

5. Make several series-parallel combinations of the capacitors in the subdivided box and repeat the procedure of Step 3 for each case. Be careful to record by means of a simple diagram the arrangement of the capacitors in each test.

III. *Determination of Dielectric Constant.* 6. Use the experimental parallel-plate capacitor for the unknown. For the known capacitor use the standard box and the subdivided box in series. Arrange the switches in the subdivided box so that all the capacitors are in series. Measure the capacitance of the unknown with mica between the plates. Hold the plates firmly together during the final measurement, in order to ensure good contact with the mica. Remove the mica and replace it by three small pieces cut from the mica plate. Measure the capacitance of this air capacitor. Calculate the value of K for mica by the use of Eq. (5).

7. Measure the thickness of the mica pieces with a micrometer caliper. Measure the diameter of the plate with a meterstick and calculate its area. Calculate the capacitance of the air capacitor by the use of Eq. (4). Record the percentage difference between this value and the observed capacitance.

8. (Optional) Measure the capacitance of the experimental parallel-plate capacitor when the insulator is one sheet of paraffined paper. Comment on the observed value of this capacitance, as compared with that found in Step 6, where mica was used for the insulator.

Experiment 37.2

BALLISTIC GALVANOMETER

Object: To measure the charge sensitivity of a ballistic galvanometer and to use the galvanometer to determine unknown capacitances, including capacitors connected in series, parallel, and in series-parallel arrangements.

Method: The ballistic constant of a galvanometer is measured by charging a standard capacitor to a known voltage, discharging it through a ballistic galvanometer, and observing the resulting deflection. The capacitance of an unknown capacitor is determined by charging it to the same voltage and observing the galvanometer deflection when it is discharged, the ratio of the deflection to that for the standard capacitor giving the ratio of the capacitances. Various unknown capacitors, including series, parallel, and series-parallel combinations of known capacitors are charged and discharged in a similar manner. The measured capacitances of these combinations are then compared with the values obtained by calculation from the known separate capacitances.

Theory: A ballistic galvanometer gives deflections that are proportional to the quantity of electricity discharged through the coil, provided that the moment of inertia of the coil is sufficiently high that the coil does not move appreciably during the time the charge is flowing. When a charged capacitor is discharged through the comparatively low resistance of a galvanometer, the time of discharge is ordinarily negligible in comparison with the time required for the galvanometer coil to turn.

Ballistic Constant of Galvanometer. If a charged capacitor C is discharged through a ballistic galvanometer the deflection s of the galvanometer, which is proportional to the charge Q, is also proportional to the capacitance, since $Q = CV$. Therefore

$$Q = CV = k_b s \qquad (7)$$

where k_b, the ballistic constant of the galvanometer, is defined by $k_b = Q/s$, the charge per unit deflection. Hence an experimental determination of k_b can be made by using a capacitor of known capacitance, charging it to a measured voltage, and then discharging it through the ballistic galvanometer and observing the resulting deflection.

Capacitance by Ballistic Galvanometer. The capacitances of capacitors may be compared by measuring the ratio of the deflections produced when the unknown capacitor and a standard capacitor are successively charged from a given source and then discharged through the galvanometer. The working equation is

$$\frac{Q_1}{Q_2} = \frac{C_1 V}{C_2 V} = \frac{k_b s_1}{k_b s_2} \quad \text{or} \quad C_1 = C_2 \frac{s_1}{s_2} \qquad (8)$$

Apparatus: Subdivided capacitor (Fig. 37.4); standard 0.5-μf capacitor (Fig. 37.9); several unknown capacitors; ballistic galvanometer; 7- to 10-volt battery; 10-volt voltmeter; SPST switch, fused for 0.5 amp; 22-ohm rheostat; reversing switch; SPST switch (or charge-discharge key); tap key; four four-way connectors; stop watch (or clock with sweep-seconds hand).

The ballistic galvanometer used in this experiment differs from the ordinary current galvanometer only in the fact that the short-circuiting rectangular loop of wire frequently attached to the coil of the current galvanometer has been removed. This modification enables the coil to oscillate with a minimum of damping, a condition essential in ballistic experiments. The oscillations may be stopped, when the coil approaches zero deflection after a reading, by short-circuiting the coil with a damping key connected in parallel with the coil.

Procedure: 1. Connect the apparatus as in Fig. 37.10. Leave one battery terminal disconnected and all switches open until the instructor approves the wiring. Switch S_2 is used to reverse the voltage across the capacitor; S_3 is a charge-discharge switch or key; S_4 is a short-circuiting key for damping the oscillations of the galvanometer. The potential divider R enables any desired fraction of the total voltage to be applied to the capacitor. Use the standard capacitor for C.

2. Move the slider of the potential divider until the voltmeter shows a PD of about 1 volt. Close first S_2 and then S_3 in the charging (downward) position. After a few seconds discharge the capacitor through the galvanometer by moving S_3 to the discharging (upward) position. Note the maximum deflection of the

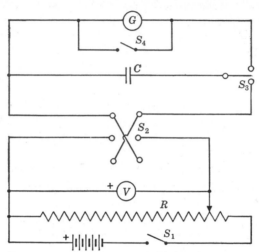

FIG. 37.10   Capacitance by ballistic galvanometer

galvanometer. The galvanometer coil can be brought to the zero position after a reading by a skillful manipulation of the damping key S_4. Close this key just before the deflection approaches zero. If the galvanometer deflection is not 20 cm or more, increase V and repeat the observations. Observe the effect produced when the reversing switch S_2 is closed in the opposite direction and the observations repeated. After the adjustments are completed, record the readings of the voltmeter, the direct and the reversed readings of the galvanometer, and the value of C_0. Adjust the potential divider to a value of V that is one-half the former value. Note the deflection produced when the capacitor is discharged. What relation do you observe between the deflections and the corresponding voltages?

3. Replace the standard capacitor by an unknown capacitor C_x and repeat the manipulation of Step 2. Reduce V to a suitably low value each time a new capacitor is used, but for the final observations it is imperative that the same value of V be used for both the unknown and the standard capacitors. This requirement may necessitate several trials for both C_0 and C_x until suitably large deflections are obtained. Record the data for several unknown capacitances.

4. Insert the subdivided box (or several of the measured capacitors if such a box is not available) for C. Adjust the switches on the box to put several of the capacitors in parallel. Record the arrangement by a simple sketch. Measure the joint capacitance by the use of the technique of Step 2.

5. Adjust the switches to place several of the capacitors (such as 0.5, 0.2, and

0.2 μf) in the subdivided box in series and measure the joint capacitance of the arrangement.

6. Make several series-parallel combinations of the capacitors in the subdivided box and measure the joint capacitance of each arrangement. Be careful to record by means of a simple diagram the arrangement of the capacitors in each case.

7. (Optional) Test the quality of a capacitor by noting the time rate of loss of charge by leakage. This determination is performed by observing the deflections produced when the discharge is started at intervals of 0, 30, 60, 90, and 120 sec after the capacitor has been charged and then disconnected from the voltage source. Select a high-loss capacitor for this test. Comment on the results obtained. It is also very instructive to observe and plot the rate of discharge of a capacitor when a large resistor (several megohms) is connected in series with the capacitor.

Computations and Analysis: 1. Use the data of Step 2 and Eq. (7) to determine the value of the ballistic constant k_b of the galvanometer.

2. Use the value of k_b and the data of Step 3 to determine the capacitances of the unknown capacitors.

3. Calculate the measured capacitance of each of the arrangements used in Steps 4, 5, and 6. Compute the respective joint capacitances from the known individual values by the use of Eqs. (2) and (3). Record the percentage difference between the observed and the computed values of the joint capacitance for each arrangement.

Review Questions: 1. What is a capacitor? 2. Define capacitance. What are some of the commonly used units? 3. State the relationships between the total and individual voltages, charges, and capacitances for capacitors connected in parallel; for capacitors in series. When is each arrangement used? 4. Describe a subdivided capacitor. Explain how to manipulate the switches to produce series, parallel, and series-parallel combinations. 5. Sketch and describe the bridge method of comparing capacitances. Derive the working equation. 6. State the equation for the capacitance of a parallel-plate capacitor. 7. Show how the dielectric constant of a material may be measured by the use of an experimental parallel-plate capacitor. 8. Describe the method of the comparison of capacitances by the ballistic galvanometer.

Questions and Problems: 1. Explain why it is not essential that the applied voltage in Exp. 37.1 be sinusoidal. Could an intermittent d-c source be used? Explain. Why is a sinusoidal voltage desirable?

2. The standard 0.5-μf capacitor used in these experiments is guaranteed by the manufacturer to be accurate to 0.25%. Were the measurements you made equally reliable? Would it be helpful in these experiments to use a standard capacitor that is more accurately calibrated?

3. What is the effect upon the precision of Exp. 37.1 of contact resistances at the terminals where the source voltage is applied? at the junction a in the branch ad (Fig. 37.8)? at the junction a in the branch ac? in the branch cd?

4. A capacitance bridge circuit (Fig. 37.8) is balanced in the normal manner. The connections to the source voltage are then moved to cd and the telephone receiver to ab. Show why the bridge remains in balance.

5. An experimental parallel-plate air capacitor is charged to a certain voltage. It is then immersed in oil. What happens to the capacitance? to the charge? to the PD?

6. Calculate the joint capacitance of each of the subdivided capacitors shown in Figs. 37.5 and 37.6. What is the capacitance of the capacitor of Fig. 37.4?

7. An experimental parallel-plate capacitor has plates 15.0 cm in diameter. Small mica strips 0.872 mm thick are used to make an air capacitor. The capacitance is measured to be 912 $\mu\mu$f when a large piece of mica of the same thickness as the small strips is used as the insulator between the plates. Compute the value of K for mica from these data.

8. A 5.00-μf capacitor is charged to a PD of 15.0 volts. The capacitor is then connected to a ballistic galvanometer and a deflection of 22.4 cm is observed. Calculate the charge sensitivity of the galvanometer. A second capacitor is charged by the same source, then connected to the galvanometer, and a deflection of 12.5 cm is observed. What is the capacitance of this capacitor?

9. Describe the differences between the use of a galvanometer as a current-measuring instrument and as a ballistic galvanometer for the measurement of charge.

10. A capacitor has 33 plates alternately connected in parallel. The plates are separated by mica ($K = 5.82$) sheets 0.250 mm thick. The area of each plate is 225 cm². What is the capacitance of this capacitor?

11. Three capacitors of 0.125, 0.250, and 0.375 μf are connected in parallel. This group is connected in parallel with another group of three capacitors in series, with capacitances of 5.00, 4.00, and 3.00 μf. The combination is connected to a 220-volt source. Calculate the charge in the system and in each capacitor.

CHAPTER 38

ALTERNATING CURRENTS

In an a-c circuit containing only pure resistance (that is, no inductance or capacitance) the current and voltage are always *in phase*. This means that both are zero at the same instant, both pass through their maximum values at the same instant, and always have similar time relationships. The *phase diagram* for such a circuit, Fig. 38.1, has the IR drop laid off along the same direction (phase) as the current. In a pure resistor the effective value of the current I is simply the ratio of the effective value of the voltage V to

FIG. 38.1   Voltage across pure resistor in phase with current

the resistance R or $I = V/R$. Alternating-current ammeters and voltmeters are designed to read effective values.

In an inductive circuit the current lags behind the voltage, that is, the current does not reach its maximum value until some time after the voltage is a maximum. In a circuit containing only inductance (that is, no resistance or capacitance) the angle of lag of current behind voltage is 90°. In general, however, circuits contain both inductance and resistance. In such cases the angle of lag is less than 90°. In other words the phase angle for circuits containing resistance and inductance may vary from 0° for a circuit with resistance only to 90° for a circuit with inductance only.

In circuits in which capacitance predominates the current *leads* the voltage, that is, the current is always ahead of the voltage in phase. This phase angle may vary from 90° for a circuit containing only capacitance to 0° for a circuit containing resistance only. The most general case is for circuits containing resistance, inductance, and capacitance. These may be present in all proportions so that the current may either lag or lead the voltage by 90° or less.

Reactance and Impedance.   Whenever there is inductance in a circuit, an induced emf is set up as the current changes. The ratio of the voltage V_L required to maintain a current in an inductor to the current is called the *inductive reactance X_L*

$$X_L = V_L/I \tag{1}$$

Inductive reactance may be expressed in terms of the self-inductance L and the frequency f by the relation

$$X_L = 2\pi fL \tag{2}$$

Reactance is in ohms when f is in cycles per second and L is in henrys.

The portion of the applied voltage which must be used because of this induced emf is called the *reactance drop in potential*. It is equal to the product of the current and the inductive reactance and may be written as

$$V_L = IX_L \tag{3}$$

259

In this discussion of a-c circuits we shall discuss series circuits only. In such circuits the current is the same in all parts. Therefore we refer phases to the common current phase. Thus the current is ahead of or behind the various voltages in phase.

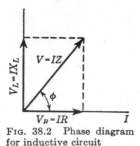

FIG. 38.2 Phase diagram for inductive circuit

The voltage and phase relations in a circuit containing resistance and inductance in series are shown in the phase diagram of Fig. 38.2. The resistance drop (IR) is in phase with the current and is laid off along the positive direction of the X axis. The reactance drop (IX_L) leads the current by 90° and is laid off along the Y axis. The impressed voltage V is the vector sum of IR and IX_L. The angle ϕ is the phase angle.

The joint effect of resistance and reactance in an a-c circuit is known as impedance; it is designated by the symbol Z. *Impedance* is defined as the ratio of the effective voltage to the effective current. The defining equation is

$$Z = V/I \tag{4}$$

Impedance is measured in ohms, when V is in volts and I in amperes. From Fig. 38.2 it is clear that, for a circuit containing resistance and inductance

$$Z^2 = R^2 + X_L^2 \tag{5}$$

The current in such a circuit is given by

$$I = \frac{V}{Z} = \frac{V}{\sqrt{R^2 + X_L^2}} \tag{6}$$

An application of this equation offers a very direct method for measuring X_L. When X_L is known, L can be computed by Eq. (2). This is done as follows: The current in an inductive circuit is measured when a known voltage is impressed. The resistance R is easily measured and, if the frequency is known, every factor in Eqs. (2) and (6) is known except L.

The conventional wiring diagram for a circuit containing an inductor is shown in Fig. 38.3. In this diagram the resistance portion of the inductive circuit is conventionally represented as separate from the inductive portion; in practice this can only be approximately true. In the phase diagram of Fig. 38.2 the voltage V_L represents the portion of the total voltage used to maintain the current in the strictly inductive portion

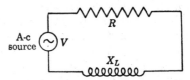

FIG. 38.3 Conventional wiring diagram of inductive circuit

of the circuit, and V_R represents the voltage across the resistance of the circuit.

When an a-c source is connected to a capacitor, charges continually flow into and then out of the plates as the voltage alternately rises and falls. Hence an a-c ammeter in the circuit shows a continuous reading. The charge and discharge are associated with a *capacitive drop* in potential that is 90° behind the current in phase. The *capacitive reactance* X_C is the ratio of the effective value of the capacitive drop in potential V_C to the effective value of the current

$$X_C = V_C/I \qquad (7)$$

The value of X_C may be obtained from

$$X_C = 1/2\pi fC \qquad (8)$$

The reactance X_C is in ohms when C is in farads.

The impedance of a circuit containing resistance and capacitive reactance is given by the equation

$$Z^2 = R^2 + X_C{}^2 \qquad (9)$$

The current in a circuit containing resistance and capacitive reactance is expressed by the equation

$$I = \frac{V}{Z} = \frac{V}{\sqrt{R^2 + X_C{}^2}} \qquad (10)$$

The conventional wiring diagram for a circuit containing a resistor and a capacitor in series is illustrated in Fig. 38.4. The corresponding phase diagram is

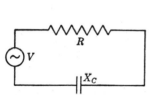

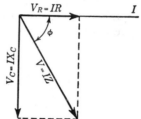

FIG. 38.4 Conventional wiring diagram of capacitive circuit

FIG. 38.5 Phase diagram for capacitive circuit

shown in Fig. 38.5. In actual practice a good-quality capacitor connected directly to an a-c power source gives a phase angle very close to 90°.

Series Circuits Containing Resistance, Inductance, and Capacitance. We will first consider an ideal case of a series circuit (Fig. 38.6) containing resistance (without inductive or capacitive effects), inductance (without resistance or capacitive effects), and capacitance (without resistance or inductive effects). The voltages in this circuit are shown in the "clock" diagram of Fig. 38.7. In Fig. 38.7a the voltage V_R across the resistor is laid off in

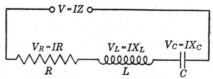

FIG. 38.6 Series circuit containing resistance, inductance, and capacitance

phase with the current. The voltage V_L across the inductive coil is 90° ahead of the current. The voltage V_C across the capacitor lags the current by 90°. Hence the net reactive voltage is $V_L - V_C$ and is designated V_X in Fig. 38.7b. From

$$V_X = IX = IX_L - IX_C = I(X_L - X_C)$$

it follows that the net reactance X of the series circuit is $X = X_L - X_C$. The impedance Z is given by

$$Z = \sqrt{R^2 + X^2} = \sqrt{R^2 + (X_L - X_C)^2}$$

$$Z = \sqrt{R^2 + \left(2\pi fL - \frac{1}{2\pi fC}\right)^2} \tag{11}$$

We may now write what is one of the most useful equations in a-c circuits in the following expression for the current in a series circuit containing resistance, inductance, and capacitance, namely,

$$I = \frac{V}{\sqrt{R^2 + \left(2\pi fL - \frac{1}{2\pi fC}\right)^2}} \tag{12}$$

The ideal case represented by Fig. 38.6 never exists in practice. The resistor usually contains more or less inductance; the reactor necessarily includes some

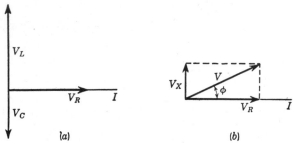

(a)   (b)

FIG. 38.7 Phase diagram of series circuit containing resistance, inductance, and capacitance; inductance reactance is predominant

resistance; the capacitor may have sufficient losses to offer appreciable equivalent resistance. However, the phase diagram of the voltages in such a circuit may be reduced to one like that in Fig. 38.7, where in this case V_R represents the voltage across all the resistance in the circuit, V_L the total inductive voltage of the circuit, and V_C the total capacitive voltage. As before, V_X represents the net reactive voltage and V the voltage of the entire circuit.

In practice it is common to have a-c series circuits containing a nonreactive resistor, a pure capacitor, and an inductor that has both resistance and inductive reactance. The voltages across the parts of such a circuit are easily measured. Let V_1 represent the voltage across the resistor, V_2 the voltage across the inductor, V_3 the voltage measured across both the resistor and inductor, V_4 the voltage across the capacitor, and V the source voltage. The phase diagrams for such circuits, of which those in Fig. 38.8 are typical, may be constructed as follows: Use the positive direction of the X axis as the phase of the current I. Choose a convenient scale for the voltage. Lay off in the direction of I the line OA to represent the voltage V_1 across the resistor. From the end A of this line describe in the first quadrant an arc of radius equal to the voltage V_2 across the inductor. Swing a second arc from O of radius equal to the voltage V_3 across both the resistor and the inductor. Draw lines OB and AB to the intersection B of the two arcs.

These lines represent in magnitude and phase V_3 and V_2, respectively. The phase angle α is the angle by which the current lags the voltage in the inductor. Draw a line BC from B vertically downward to represent voltage V_4 across the capacitor. It is assumed that this voltage is 90° out of phase with the current. Finally, draw the line OC which closes the polygon of voltages. This line represents the voltage V of the source. The circuit phase angle ϕ is the angle by which the current lags the source voltage in Fig. 38.8a and leads the source voltage in Fig. 38.8b. The voltage represented by AD is the drop of potential caused by the equivalent resistance of the inductor. The voltage represented by DB is the component of the inductor voltage due to inductive reactance.

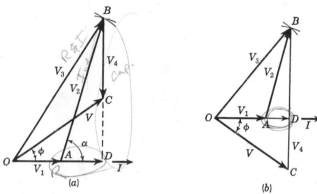

Fig. 38.8   Voltage relationships in series circuit containing a resistor, an inductor, and a capacitor. (a) inductive reactance predominates; (b) capacitive reactance predominates

Electric Resonance. It is apparent from inspection of the reactance term $(2\pi fL - 1/2\pi fC)$ in Eq. (12) that it is possible for the inductive reactance to be just equal to the capacitive reactance. In this special case, when $X_L = X_C$, the net reactance is zero and the current is given by $I = V/R$. The current in such a circuit is limited only by the resistance of the circuit and the circuit phase angle is zero, that is, the current and source voltage are in phase. In the phase diagram of such a circuit the point C falls exactly on the current axis at D. This condition is called *electric resonance*.

Power in A-C Circuits. In d-c circuits the power is given by $P = VI$. In such circuits both the current and voltage are assumed to be steady and in phase. In a-c circuits this is not the case. During half of the cycle, energy is supplied to the reactive component of the circuit, but this energy is returned to the source during the other half of the cycle. Hence no power is required to maintain the current in the part of the circuit which is purely reactive. All the power is used in the resistance portion of the circuit. From Fig. 38.7(b) $P = IV_R$. But $V_R = V \cos \phi$ and therefore

$$P = VI \cos \phi \tag{13}$$

where P is the average power in watts when V is the effective value of the voltage and I is the effective value of the current. The angle ϕ is the angle of lag of the current behind the voltage. The quantity $\cos \phi$ is called the *power factor* of the circuit. Note that the power factor can vary anywhere from zero for a purely reactive circuit to unity for a pure resistance. From the derivation of Eq. (11) and Fig. 38.7b, it may be seen that the phase angle ϕ is defined by the equation

$$\cos \phi = R/Z \tag{14}$$

There are power losses in inductors other than the usual I^2R copper losses. Such losses are due to eddy currents and hysteresis in the core of the inductor, especially if the coil is wound on an iron core. Hence the equivalent resistance of such a coil is considerably larger than the ordinary ("ohmic") resistance as measured by the use of direct current. The *equivalent resistance* of an inductor may be obtained from the relation

$$R_{\text{equiv.}} = P/I^2 \tag{15}$$

where P is the power loss and I is the current in the coil. A capacitor may have a power loss, caused by leakage resistance or dielectric hysteresis. The equivalent resistance of a capacitor can be calculated from an equation similar to Eq. (15).

Experiment 38.1

A-C SERIES CIRCUITS

Object: To study the electrical characteristics of an a-c series circuit containing a resistor, an inductor, and a capacitor; and to observe the phenomenon of electric resonance.

Method: A rheostat, an iron-cored coil, and a mica capacitor are connected in series and joined to a source of 60-cycle power. The current in the circuit and the voltages across the resistor, coil, resistor and coil, and the capacitor are measured. Phase diagrams are drawn and analyzed for each case. The reluctance of the magnetic circuit of the coil is varied to change the inductive reactance, while the resistance and capacitance are kept constant, in order to demonstrate the phenomena of electric resonance.

Apparatus: Electric-resonance apparatus; a-c ammeter, ranges 0.5 and 1 amp; high-resistance a-c voltmeter, ranges 30, 120, and 300 volts; wattmeter, ranges 75 and 150 watts; DPDT switch connected to the 120-volt a-c and d-c power sources; d-c voltmeter, ranges 15 volts and 150 volts; 150-ohm rheostat; two pairs of voltmeter leads, with clip connectors; protractor; ruler; compass.

The electric-resonance apparatus, Fig. 38.9, consists of a rheostat (replacing the lamp shown in the illustration), a variable inductive coil, and a mica capacitor, connected in series. A switch is provided to short-circuit the capacitor when it is desired to have only the resistor and inductor in the circuit. The inductor consists of a fixed coil, wound on a closed iron core that includes an adjustable portion to vary the reluctance of the magnetic circuit. This adjustment is made by a screw-drive mechanism that raises or lowers an iron yoke which opens or closes the magnetic circuit. A millimeter scale and indicator attached to the movable yoke is used as an arbitrary measure of the magnetic reluctance which controls the inductance of the coil. The iron core, although laminated to minimize eddy currents, has material and variable iron losses which result in appreciable equivalent resistances dependent upon the currents. Binding posts, provided for each part of the assembly, facilitate separate voltage measurements.

Procedure: I. *Observations with D-C Source.* 1. Short-circuit the capacitor by closing the switch at its terminals, thus leaving only the resistor and inductor in the circuit. Use the following procedure to compare: (*a*) the sum of the voltages across the rheostat and coil with the source voltage and (*b*) the sum of the power in the rheostat and the power in the coil with the power taken from the source.

2. Connect in series the coil, rheostat set for maximum resistance, high range of ammeter, and the current coil of the wattmeter. Join the terminals of this circuit to the center posts of the DPDT switch, after being certain that the switch is open and that the wires from the other posts on the switch are not plugged into the power source. Have the instructor check the wiring and supervise the preliminary observations.

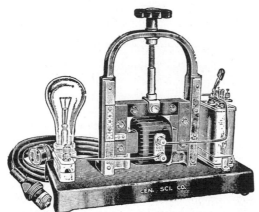

FIG. 38.9 Electric-resonance apparatus

3. Close the circuits to the 115-volt d-c power line. Adjust the rheostat to give a full-scale (1.00 amp) reading of the ammeter. Use the d-c voltmeter for the voltage observations and the wattmeter for the power measurements. Measure and record the following: current, source voltage, rheostat voltage, coil voltage, power in coil, power in rheostat, and power in entire circuit. Complete Part 1 of the Computations and Analysis before proceeding with the experiment.

II. *A-C Circuit Containing Resistor and Inductor.* 4. Repeat the observations of Steps 1 to 3, except that the a-c source and the a-c voltmeter should be used. Throw the DPDT switch so that the circuit used in Step 1 is connected to the 115-volt a-c power line. Adjust the iron yoke so that there is a large air gap in the magnetic circuit of the iron core (scale setting 40). Adjust the rheostat until a current, of about 1 amp is obtained. Measure and record the following: current, source voltage, rheostat voltage, coil voltage, power in coil, power in rheostat, and power in entire circuit.

5. Repeat the observations of Step 4, except that a small air gap (scale setting 20) is used and the rheostat is adjusted to produce a current of about 0.2 amp. Complete Part 2 of the Computations and Analysis before proceeding with the experiment.

III. *Electric Resonance.* 6. Open the capacitor switch and thus add the capacitor to the series circuit. Study the variation of the current with coil inductance when all other factors are kept constant.   Proceed as follows: Adjust the rheostat until the ammeter reads full scale (1.00 amp) for resonant conditions.   Resonance is obtained by varying the reluctance of the magnetic circuit until the current is a maximum.   After the circuit is properly adjusted for resonance, do not change the rheostat or capacitor.   Turn the screw until the scale reads zero and record the current and the scale setting.

Change the setting by 5-mm steps until resonance is approached.   Near the resonant conditions change the setting by 2.5-mm intervals.   Read and record the current and scale setting until the air gap is a maximum.   Interpret these data in Part 3 of the Analysis.

IV. *Phase Diagrams for Circuits Containing R, L, and C.*   7. Examine as follows the relationships between the various voltages in the circuit containing *R*, *L*, and *C*: Set the air gap for large reluctance, with the scale near 35 or 40. Adjust the rheostat to give a current of about 0.7 amp.   Observe and record the following: scale setting, current, source voltage, rheostat voltage, coil voltage, voltage across both coil and rheostat, and capacitor voltage.   Repeat these observations for a low reluctance in the magnetic circuit, with the scale set at 10 and the rheostat adjusted to give a current of about 0.22 amp.   Change the voltmeter terminals as necessary in order to select the scale which can be read most accurately.

Computations and Analysis: 1. Using the data of Steps 2 and 3 compare the sum of the voltages across the rheostat and coil with the source voltage.   Comment on this comparison.   Calculate the resistance R_{dc} of the coil and the power loss $I^2 R_{dc}$ in the coil.   Compare this power with that observed from the watt-meter.   Compare the sum of the power in the coil and in the rheostat with that for the entire circuit.   Comment on these power observations.

2. Considering the data of Steps 4 and 5, answer the following questions and perform the analyses indicated:

a. Does the arithmetic sum of the voltages across the rheostat and the coil equal the source voltage, as in Part I?   Explain.   Does the sum of the power in the rheostat and the coil equal the power in the circuit?   Try to account for any difference.

b. Note the fact that the power in the coil in the a-c circuit is not equal to $I^2 R_{dc}$ where R_{dc} is the resistance as measured from the d-c observations.   Why is this true?   Calculate the equivalent resistance of the coil from

$$R_{equiv.} = P/I^2$$

and compare this equivalent resistance with that observed in Part I.   Explain reasons for any difference found in these resistances.

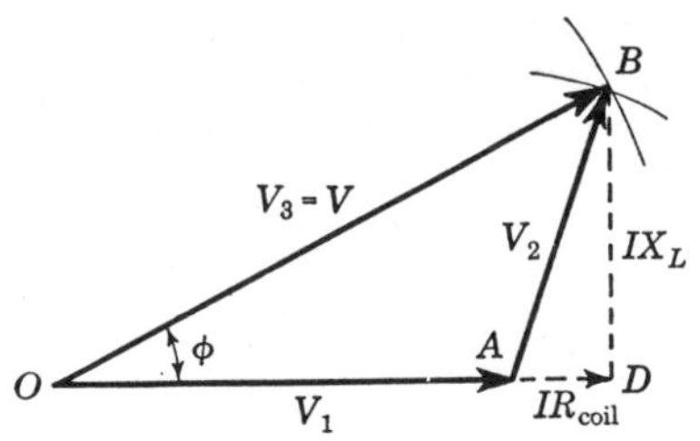

c. Construct a phase diagram of the various voltages for each of these two a-c cases. Choose a convenient scale and lay off the rheostat voltage V_1 in phase with the cur-rent, along the X axis (line OA in Fig. 38.10).   Swing an arc of length AB, scaled to represent the voltage V_2 across the coil, using A as the center of the arc.   Then swing another arc of length OB, with O as the center, scaled to represent the

FIG. 38.10   Phase diagram for circuit containing rheostat and inductor

source voltage V. The intersection of these two arcs determines the location of B and hence the phase angle ϕ. Comment on the differences in the two phase diagrams.

d. From each of these phase diagrams calculate the drop of potential due to the equivalent resistance of the coil. This is the X component AD of the coil voltage. Calculate the reactive component IX_L of the coil voltage. This is the Y component BD of V_2. Compute the impedance of the coil from $Z = V_2/I$. Compare this impedance with the d-c resistance of the coil. How does the impedance of the coil in the case of maximum reluctance of the magnetic circuit compare with the impedance when the coil has a minimum reluctance in its magnetic circuit?

e. From the two values of IX_L obtained from the phase diagrams calculate X_L and L. Explain why the two values of L are so different.

f. Compute the power factor in each of these two cases. Why are the values so different?

3. Use the data of Step 6 to plot a curve to show the variation of current with the setting of the air gap. Explain reasons for the shape of the curve.

4. From the data of Step 7 draw a phase diagram for each of the series of observations. Proceed as indicated in the theory given above. These diagrams will be similar to those in Fig. 38.8. Compare the values of V as observed from the phase diagrams with those measured in the experiments. Calculate X_C and C from the observed currents and capacitor voltages. How do the two values of C compare?

Review Questions: 1. Explain what is meant by the phase relation between current and voltage in an a-c circuit. What phase relation exists in a circuit containing resistance only? inductance only? capacitance only? resistance and inductance? resistance and capacitance? 2. Define inductive reactance, give the defining equation, and state the unit of each factor. 3. Draw a phase diagram for a circuit containing resistance and inductance. 4. Define impedance, give the defining equation, and state the unit of each factor. 5. Write the equation for the current in a circuit containing resistance and inductance. Show how X_L and L can be determined by the use of this equation. 6. Define capacitive reactance, give the defining equation, and state the unit of each factor. 7. Write the equation for the current in a circuit containing resistance and capacitance. Show how X_C and C can be determined by the use of this equation. 8. Write the equation for the current in a circuit containing R, L, and C. Show how the phase diagram for such a circuit is constructed. 9. What is meant by electric resonance? Under what conditions is resonance obtained? 10. State the equation for power in an a-c circuit. What is power factor? What values may it have? 11. Explain what is meant by the equivalent resistance of an inductor; of a capacitor.

Questions and Problems: **1.** A d-c source of 115 volts is connected to an iron-cored solenoid and the current is measured. The solenoid is then connected to a 60-cycle a-c source of 115 volts and the current measured. How would the two currents be expected to compare? State reasons.

2. Answer question 1 for the case of a capacitor instead of a solenoid.

3. The a-c voltage across an air-cored solenoid is varied and the currents noted. Plot a curve of current against voltage. An iron core is inserted into the solenoid and the observations are repeated. Sketch the sort of curve of current against voltage which would be expected.

4. A noninductive resistor and an inductor are connected in series. When a d-c voltage of 120 volts is applied to the series arrangement, the voltage is measured across the resistor and also across the inductor.   How would one expect the sum of these individual voltages to compare with that of the 120-volt source?   Answer this question also for the case where the source is an a-c one.

5. A 60-cycle power line is connected first to an inductor and then to a capacitor, each of negligible equivalent resistance.   The current in each case is observed to be 10 amp.   What would the currents be in each case if the same voltage at 400 cycles/sec were applied?

6. A coil having a resistance of 22.4 ohms and an inductive reactance of 20.0 ohms is connected to a 120-volt a-c source.   What is the current and the power in the coil? Sketch a phase diagram for this case.

7. An inductor connected to a 120-volt d-c power source takes a power of 500 watts. When this coil is connected to a 120-volt a-c source a power of 275 watts and a current of 2.50 amp are observed.   Calculate the actual and the equivalent resistances of the inductor.   What is the inductance of the circuit?

8. A 100-μf capacitor is connected to a 60-cycle 120-volt source.   What current will result?   A 25.0-ohm resistor is then inserted in series with the capacitor.   Calculate the current for this case.   What is the power taken from the source?   Draw a phase diagram of the voltages.

9. A series circuit consists of a 115-volt 60-cycle source, a 6.00-ohm resistor, a capacitor of reactance 5.00 ohms, and an inductor having negligible resistance and reactance 12.5 ohms.   Calculate the power.   Construct a phase diagram of the voltages.

10. A 125-volt 60-cycle a-c source is connected to a series circuit consisting of a 75.0-ohm resistor, a capacitor of negligible resistance and reactance of 100.0 ohms, and an inductor having a reactance of 50.0 ohms and a resistance of 30.0 ohms.   What is the current and the power in the circuit?   Construct a phase diagram of the various voltages.

11. A coil of inductance 80 mh (resistance zero), a 7.0-ohm resistor, and a 25-μf capacitor are connected in series across 110-volt a-c supply mains of variable frequency.   For what frequency is the current a maximum?   What is that maximum value?   When the current has this value what is the voltage across the inductor? across the resistor? across the capacitor?

12. An alternating voltage of 115 volts and frequency 60 cycles/sec is impressed upon a circuit consisting of an electric lamp of 12 ohms resistance, an inductor having negligible resistance and an inductance of 2.50 henrys, and a capacitor of negligible resistance, all connected in series.   What must be the value of the capacitance for the circuit to be in resonance?   What is the value of the current under these circumstances?

PART V—LIGHT

CHAPTER 39

PHOTOMETRY

The art of comparing the luminous intensity of light sources is called *photometry*. An instrument used for this purpose is known as a *photometer*.

The luminous intensity I of a source of light is usually referred to the *standard candle* as an arbitrarily selected unit. This rather unsatisfactory source has been replaced for practical purposes by standardized electric lamps. These lamps are calibrated with reference to the standard international candle. A 100-watt tungsten electric lamp, for example, produces a luminous intensity of more than 90 candles.

The portion of the radiant energy per unit time from an incandescent source that is effective in producing the sensation of sight is known as *luminous flux F*. The *luminous intensity I* of a point source is defined as the flux emitted per unit solid angle ω, $I = F/\omega$. The unit of luminous flux is defined from the standard candle as a source. The *lumen* is the luminous flux that streams into a unit solid angle (1 steradian) from a point source of one candle at the apex of the angle. Since there are 4π steradians in all space about a point, it follows that the total flux emitted by a point source is given by

$$F = 4\pi I \tag{1}$$

The flux is given in lumens when I is in candles.

When visible radiation falls upon a surface, we say that the surface is illuminated. The amount of the illumination is called illuminance. The *illuminance E* of a surface is defined as the luminous flux per unit area. In symbols

$$E = F/A \tag{2}$$

When the flux is 1 lumen and the area is 1 ft², the illuminance is 1 lumen/ft². This illuminance is commonly called a *foot-candle*.

For light from a point source the geometry of the case is such that the illuminance of a surface normal to the flux varies inversely with the square of the distance from the source. For the special case of a point source, when the surface is perpendicular to the light beam,

$$E = I/s^2 \tag{3}$$

An electric lamp bulb is not strictly a point source. However, the equations for point sources apply fairly well if the distance from lamp to screen is large in comparison with the dimensions of the filament.

The comparison of the luminous intensities of two sources by a photometer involves the use of a standard source and an unknown source. These sources are arranged to illuminate two adjacent surfaces normal to the flux, which are visually compared. The relative distances of the sources from the illuminated surfaces is adjusted until a match of illuminance is obtained. When the two surfaces have equal illuminance, $E_1 = E_2$, and from Eq. (3) it follows that

$$\frac{I_1}{s_1{}^2} = \frac{I_2}{s_2{}^2} \quad \text{or} \quad \frac{I_1}{I_2} = \frac{s_1{}^2}{s_2{}^2} \tag{4}$$

where I_1 and I_2 are the luminous intensities of the two sources and s_1 and s_2 are their respective distances from the screen. When one source is a standard lamp of known intensity, the luminous intensity of an unknown lamp may be found by comparison, using Eq. (4), the working equation of the photometer. One form of photometer is shown in Fig. 39.1.

FIG. 39.1 Bench-type photometer

For the present purpose the *luminous efficiency* of a source will be defined as the number of lumens of light flux produced per watt of power supplied to the source. The luminous efficiencies of some typical light sources are given in the accompanying table.

LUMINOUS EFFICIENCIES OF ELECTRIC LAMPS

Lamp source		Luminous intensity	Luminous efficiency
Rated power	Type		
watts		candles	lumens/watt
100	Carbon	24	3
40	Tungsten	34	11
100	Tungsten	120	15
500	Tungsten	800	20
250	Photoflood	710	36
15	Daylight fluorescent	40	33
15	Green fluorescent	70	60
250	Mercury vapor	800	40

The luminous efficiency of an electric lamp may be measured by using a photometer to determine the luminous intensity and a wattmeter, or a voltmeter and an ammeter, to measure the power input P. From the definition of luminous efficiency and Eq. (1)

$$\text{Luminous efficiency} = 4\pi I/P \tag{5}$$

It should be noted that this measure of efficiency is not the usual percentage value, since the numerator and denominator of Eq. (5) are expressed in quite different units.

Weston Foot-candle Meter. This device, Fig. 39.2, makes use of the photo-electric effect whereby light falling on a sensitized surface produces an emf that may be used to operate a galvanometer. The scale of the meter can be calibrated to read the illumi-nance directly in foot-candles.

Experiment 39.1

THE PHOTOMETER

Object: (1) To measure by means of a photometer the luminous intensities of several electric lamps. (2) To study the variation in the luminous efficiency of an electric lamp as a function of the voltage applied to it.

Fig. 39.2 Foot-candle meter

Method: A standard lamp and a test lamp are placed at opposite ends of a photometer bench. The position of the photometer box is adjusted until the screen is equally illuminated from the two sources. The luminous intensity of the test lamp is determined from the photometer equation. This procedure is repeated for various voltages on the test lamp in order to determine the relationship between the luminous efficiency and the voltage.

Apparatus: A two- or three-meter photometer bench, with a Lummer-Brodhun prism or a Bunsen photometer box; standard lamp; assortment of test lamps, including a carbon-filament lamp and tungsten lamps of 25-, 40-, 60-, 75-, 100-, and 150-watt sizes; 125-volt voltmeter; voltmeter with 60- and 120-volt ranges; 20-ohm rheo-stat; 180-ohm rheostat; ammeter with ranges of 0.5, 1, and 2 amp; foot-candle meter.

Lummer-Brodhun Photometer. In order to match the illuminances on the two sides of a photometer screen accurately, both sides must be made visible to the observer at the same time. One method of accomplishing this purpose is to use the *Lummer-Brodhun photometer head*, or sight box (illustrated in Fig. 39.3). This unit has an optical system designed to present the eye with a pattern

Fig. 39.3 Lummer-Brodhun pho-tometer head

in which the illuminances from the two sources are sharply contrasted, except when the screen is in the balanced position. A schematic diagram of the Lummer-Brodhun head is shown in Fig. 39.4. Light from the two sources of intensities I_1 and I_2 falls on a chalky white diffusing screen P at the back of the box. Some of the light which is scattered by the screen is reflected by the prisms T_1 and T_2 into the comparison glass "cube" C. The totally reflecting prisms and the cube enable the observer to view simultaneously the two sides of the screen P. The cube C consists of two 90° prisms with a figure etched on the hypotenuse face of one of the prisms. If the two sides of P are unequally illuminated, the etched figure will be visible, but when equally illuminated the figure disappears. Since, however, it is easier to judge equality of contrast than equality of brightness, the glass plates G_1 and G_2 are interposed. With these plates in place about 8 per cent of the light in certain regions of the cube is absorbed and the figure is always visible. Figure 39.5 shows the appearance of the field as seen by an observer at

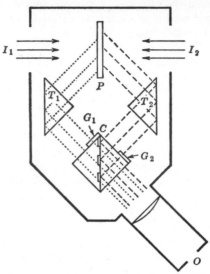

Fig. 39.4   Schematic diagram of rays in Lummer-Brodhun box

O (with the glass plates in position) for the three different conditions of illumination. Actually with this apparatus it is the brightness of the two sides of screen P that are compared. If the brightness of the two surfaces is the same, the illuminance E_1 is equal to the illuminance E_2, provided the reflectivities are equal. To correct for small differences in reflectivity, the photometer is so mounted

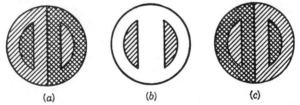

Fig. 39.5   Field in a Lummer-Brodhun photometer when (a) the right side of the screen is brighter, (b) the two sides are equally bright, and (c) when the left side is brighter

that it may be rotated about a horizontal axis. Half of the readings are taken in each position.

The Bunsen Photometer Box. This screen, Figs. 39.6 and 39.7, is simpler than the Lummer-Brodhun but offers less accuracy. The screen is a white paper P that has a grease spot or thin gauze in one area. This area will appear darker than the surrounding paper if viewed from the side of greater illuminance but brighter than the surroundings from the side of lesser illuminance. When the two sides are equally illuminated, the spot will be slightly darker than its surroundings but the contrast will be the same on the two sides. The mirrors M_1 and M_2 are placed so that an observer at O may see both sides of the screen simultaneously.

Photometer experiments should be performed in a "darkroom" with nonreflecting walls. If such a room is not available, a piece of black cloth should be placed behind and at the sides of each lamp. The photometer box should also be screened from extraneous light. The distances from the lamps to the screen should be large in comparison with the dimensions of the lamps.

Since this experiment utilizes the commercial 115-volt power line, it is essential that the student be exceedingly careful to make the proper connections. The circuits should be checked by the instructor before the plugs are inserted into the power outlets.

Fig. 39.6   Bunsen photometer box

Procedure: 1. Arrange the apparatus as shown in Fig. 39.7. Insert the 20-ohm rheostat in series with the standard lamp L_1. This rheostat provides for the adjustment of the standard lamp to the voltage for which its source intensity is known. Place this lamp at the zero position on the bench. Record the source intensity of the standard lamp.

2. For the test lamp L_2 select a modern-type lamp of about the same source intensity as the standard lamp. Connect the 180-ohm rheostat and the ammeter in series with this lamp. Place the test lamp at the end of the bench opposite the standard lamp. The center of each lamp and the photometer screen should be placed in the same horizontal plane. Have the wiring and arrangements checked by the instructor.

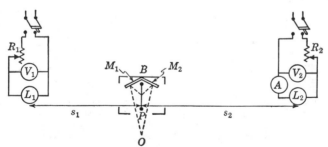

Fig. 39.7   Arrangement of apparatus for Bunsen photometer

3. Adjust the standard lamp to its rated voltage and keep it at this value throughout the experiment. Adjust the test lamp to its rated voltage and record the voltage and current. Move the viewing screen horizontally until both sides of the screen seem to be equally illuminated, as in Fig. 39.5. The observer should adjust the screen to approach the final position first from the right and then from the left. Record the distances of the test lamp and the standard lamp from the screen, the test lamp voltage, and the current.

4. For each of the lamps available, including a carbon-filament type, take one set of observations as in Step 3, with the lamps adjusted to their rated voltages.

5. Use the 150-watt test lamp to take a series of about six sets of observations with the lamp voltage adjusted to values ranging from slightly above the rated voltage down to a value such that the readings become uncertain because of the difference in color of the two sources. Note the value of the voltage for which the filament is barely visible.

6. Use the Weston foot-candle meter to perform exercises from Exp. 39.2 as selected by the instructor.

Computations and Analysis: 1. From the data of Step 3 calculate the luminous intensity, in candles, of the test lamp at its rated voltage. Compute the luminous efficiency of this lamp, by the use of Eq. (5).

2. Calculate the luminous efficiency of each of the additional lamps used in Step 4.

3. Use the data of Step 5 to determine the luminous intensity and the luminous efficiency of the 150-watt lamp at various voltages. Plot a curve to show the variation of the luminous efficiency with the lamp voltage. (Begin both axes at zero.) Give reasons for the shape of this curve.

4. Comment on the relative values of the efficiencies of the lamps.

Experiment 39.2

ILLUMINATION STUDIES WITH A FOOT-CANDLE METER

Object: To study the illumination produced by electric lamps at various distances from the lamps and at different orientations about the lamps.

Method: A foot-candle meter is used to measure the illuminance produced at a fixed position by several types of electric lamps, including a standard lamp of known luminous intensity. The intensities of the other lamps are determined from the readings of the meter and the intensity of the standard source. The illuminance at different distances from a lamp is measured and compared with the values to be expected from the inverse-square law. The effect of bare lamps is compared with that of shaded lamps. Other tests are made of the way in which the illuminance changes with angle around the lamp.

Apparatus: Foot-candle meter (Fig. 39.2); optical bench (or ruled bar); standard lamp; various test lamps, including a carbon-filament lamp and tungsten-filament lamps of 25-, 40-, 60-, 75-, 100-, and 150-watt sizes; rods

FIG. 39.8 Lamp socket, with adjustable circular scales

and clamps, or other arrangement for supporting the lamp bulb above the table; reflector shade for bulb; diffuser shade for bulb; meterstick; 115-volt power source, with switch; special lamp socket, with circular scales (Fig. 39.8).

Procedure: 1. *Luminous Intensities of Various Sources.* Arrange the apparatus in a darkroom where other light sources will not interfere with the experiment. Place the lamp socket at the zero end of the bench. Mount the foot-candle meter in a holder so that it can be moved along the bench. With the largest bulb available in the socket move the meter until the reading is full scale. Keep the meter in this position and insert successively in the socket the various bulbs available, including the standard lamp. (If necessary the voltage across the standard lamp can be adjusted to the rated value by the use of a rheostat in series with the lamp.) Record the readings of the meter for the various bulbs.

2. *Illuminance as a Function of Distance from Source.* Insert the 60-watt bulb in the socket. Place the meter at a position which gives a full-scale reading. Record the positions and readings as the meter is moved at 10-cm intervals until the reading becomes too small to be read accurately.

3. *Luminous Intensity in Different Directions.* Arrange the lamp and meter to give a full-scale deflection on the meter when the lamp is rotated around a vertical axis to the position which produces a maximum intensity. Adjust the circular scale to 0° for this condition. Rotate the bulb around a vertical axis in 15° increments from 0 to 360° and record the positions and meter readings. Repeat these observations for a bulb of different design. Repeat a similar series of observations for a rotation of the bulb around a horizontal axis.

4. *Illuminance at Different Places Near a Lamp.* Arrange a large rectangular table with its center directly below a bare lamp bulb suspended above the table. Draw two lines from the opposite corners of the table, intersecting at the middle. Show the dimensions of this figure on the record sheet. Select a lamp of such size that when the lamp is at a convenient height above the table a full-scale deflection of the meter is obtained at the center. Record the rating of the lamp and its distance above the center of the table. Record the illuminance at the center of the table, the corners, and two stations along the diagonals. Repeat these observations with a reflector shade on the bulb. Repeat also with a diffusing shade on the bulb. Measure the illuminance at a number of places around the laboratory, particularly near windows, the center of the room, and in the darker corners.

Computations and Analysis: 1. Construct the straight-line curve of luminous intensity against illuminance, for the arrangement used in Step 1. Use the data for the standard lamp for one of the points and the origin for the other point. Since this curve is a straight line, the source intensity of the various lamps can be read directly from the curve at the points of measured illuminance.

2. From the data of Step 2 plot a curve to show the illuminance as ordinates against the reciprocal of distance squared as abscissas. Explain the significance of the shape of this curve.

3. Plot on polar coordinate paper the illuminance observed in Step 3 as a function of angle, when the bulb was rotated around a vertical axis. Show on the same paper the curve for the other lamp of different design. Plot on another sheet the similar curves for the cases in which the lamp was rotated around a horizontal axis. Discuss the significance of these curves.

4. Discuss the significance of the data observed in Step 4. Compare the illuminances observed with those ordinarily desired for close work (see table in a textbook or a handbook).

Review Questions: 1. Explain what is meant by the term photometry; by photometer. 2. How is the luminous intensity of a lamp usually specified? 3. Explain what is meant by the term luminous flux. Define the lumen. 4. Define illuminance. What is the usual unit of illuminance? 5. Describe the photometer principle for the comparison of luminous intensities. Derive the working equation for the photometer. 6. Define luminous efficiency. What is the unit? Show how the luminous efficiency of a lamp may be measured. 7. By the aid of a sketch describe the principles and use of the photometer that you used. 8. Describe the Weston foot-candle

meter and its use.   9. How does illuminance vary with distance from a point source?
10. Show why the luminous flux is different in various directions around an electric
lamp bulb.

Questions and Problems: 1. Justify the fact that the total luminous flux emitted
by a source is given by $F = 4\pi I$.

2. Show why the illuminance on a surface normal to the flux from a point source
is given by $E = I/s^2$.

3. The working equation for the photometer, Eq. (4), is a *direct*-square relation.
Show why this equation is not inconsistent with, but is a result of, the *inverse*-square
law.

4. A skinflint landlord replaced the lamps on his 115-volt circuit with lamps
designed for 120-volt use.   What effect was obtained?   Would the life of the 120-volt
lamps be longer?   Why?   If sufficient light was provided for equal illumination with
the second set of lamps, was the move an economical one?   Explain.

5. Sketch and describe some type of photometer other than the instrument used
in your experiment.

6. Why is a Photoflood bulb superior to an ordinary lamp for photography?
Explain why this lamp has such a high luminous efficiency.

7. Describe the operation of a flash bulb outfit in photography.   Why is this
widely used?

8. Would you expect more, less, or an equal amount of light from one 100-watt
bulb as compared with two 50-watt bulbs?

9. In a typical photometer experiment it was observed that two lamps produced
equal illuminances when the screen was 40 cm from one lamp and 160 cm from the
other.   The first lamp was rated at 8.0 candles.   What was the luminous intensity
of the second lamp?   If the second lamp was rated at 150 watts, what was its luminous
efficiency?   What was the illuminance at the screen?

10. A 40-candle lamp is placed 100 cm from a photometer screen.   An unknown
lamp is 200 cm from this screen when the two lamps produce equal illuminances at
the screen.   The unknown lamp takes a current of 1.20 amp at 110 volts.   What is
the luminous source intensity of the unknown lamp?   What is its luminous efficiency?
What is the illuminance at the screen?

11. A standard lamp is rated at 17.5 candles at 115 volts.   The standard lamp is
placed 120 cm from a photometer screen.   An unknown lamp, operating at 115
volts and 0.90 amp, must be placed 180 cm from the screen to produce equal illuminance.   What is the luminous efficiency of the unknown lamp?

CHAPTER 40

REFLECTION AND REFRACTION OF LIGHT

In geometrical optics the propagation of light is studied in terms of *rays*, which are straight lines drawn in the direction the waves are moving. When light strikes a plane surface of a different medium, part of it may be absorbed, part reflected, and part refracted.

Reflection of Light. The angle between an incident ray and the normal to the surface is called the *angle of incidence i*. The angle between the reflected ray and the normal is called the *angle of reflection r*. By the *law of reflection* the angle of incidence is equal to the angle of reflection.

The position of an image formed by a plane mirror may be readily located by a ray diagram, following the law of reflection, Fig. 40.1. The image is virtual, as far behind the mirror as the object is in front of the mirror, and the same size as the object. A plane mirror does not change the curvature of the light incident upon it.

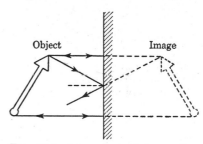

FIG. 40.1 Image formed by a plane mirror

If the reflecting surface is curved, the same law of reflection holds as for a plane mirror, but the size and position of the image are quite different. Spherical mirrors are classified as *concave* or *convex* when the reflecting surface is on the inside or outside, respectively, of the spherical shell, Fig. 40.2. The center of

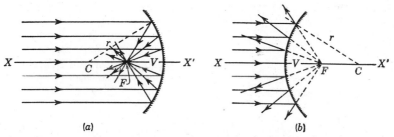

FIG. 40.2 Reflection of parallel rays: (a) from concave mirror and (b) from convex mirror

curvature *C* is the center of the sphere. The *radius of curvature r* is the radius of the sphere. A line connecting the vertex *V* and the center of curvature *C* is called the *principal axis*. In Fig. 40.2 there are also shown incident light rays parallel to the principal axis of the mirrors. From the law of reflection it follows that these rays will converge through a common point *F* after reflection from a concave mirror, or will diverge after reflection from a convex mirror, as though

they originated from a common point F behind the mirror.   Such a point is called
the *principal focus* of the mirror.   The distance f of the principal focus from the
mirror is called the *focal length*.   It may be shown that $r = 2f$.

Simple geometrical constructions are available for locating images formed by
spherical mirrors, as shown in Fig. 40.3.   Two rays from any point O, whose

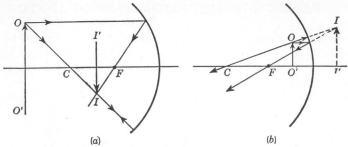

(a)   (b)

FIG. 40.3   Location of images formed by concave mirror

directions after reflection can readily be predicted, are drawn: first, the ray parallel to the principal axis, which after reflection passes through F; and second, a
ray from O in the direction OC through the center of curvature, which strikes the
mirror normally and is reflected back upon itself.   The intersection of these two

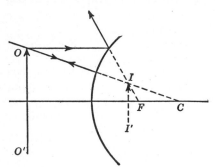

FIG. 40.4 Image formed by convex
mirror

rays at I locates the image of O.   Another similar pair of rays could be drawn
from O' to locate I'.   The image in Fig.
40.3a is *real*, since it is formed by converging rays that could form an image on a
screen at II'.   In Fig. 40.2b the reflected
rays are diverging as if they came from
II'; such an image is said to be *virtual*.
The image of any real object formed
by a *convex* mirror (Fig. 40.4) is always
virtual, erect, and diminished.

There is a simple relation between the
distance p of an object from the mirror,
the distance q of the image from the

mirror, and the focal length f; this relationship makes it easy to locate images
in spherical mirrors.   This *mirror equation* is

$$\frac{1}{p} + \frac{1}{q} = \frac{1}{f} \tag{1}$$

The convention with regard to the algebraic signs of the quantities in Eq. (1) is as
follows: f is positive for concave mirrors, negative for convex mirrors; p is positive
for real objects and negative for virtual objects; q is positive for real images and
negative for virtual images.   Only the rays which are parallel to the principal
axis and which are close to the axis are focused at the principal focus.   The
extreme rays farther from the axis in a mirror of large aperture (Fig. 40.5) cross
the axis closer to the mirror than do the rays reflected nearer the center.   This

imperfection of spherical mirrors is called *spherical aberration*. It results in a blurring of the image. The trace of the surface formed by the intersecting rays is called the *caustic curve* of the mirror. Spherical aberration can be minimized by using a diaphragm in front of the mirror to block off the rays far from the axis.

Refraction of Light. When light passes from one medium to another, it undergoes an abrupt change in direction if the speed of light in the second medium differs from that of the first. This bending of the light path is called *refraction*.

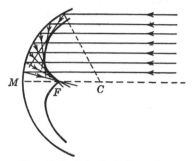

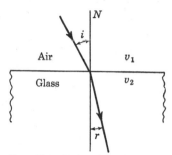

Fig. 40.5   Spherical aberration

Fig. 40.6   Refraction at a plane surface

If the speed of light in the second medium is less than that in the first medium (that is, the second medium has a greater *optical density*), the ray is bent *toward* the normal. When light travels from an optically more dense medium into one of lesser optical density, the ray is bent *away from* the normal.

In Fig. 40.6 a ray of light is shown passing from air to glass. The angle of refraction r is less than the angle of incidence i. There is an experimental relationship between these two angles known as Snell's law, namely: The ratio of sin i to sin r is a constant (for a given kind of light and pair of substances); this constant is known as the *index of refraction n*. Expressed in equation form

$$\frac{\sin i}{\sin r} = n \tag{2}$$

It may also be shown that

$$n = v_1/v_2 \tag{3}$$

where v_1 and v_2 are the speeds of the light in the first and second mediums, respectively.

When light passes from an optically denser medium into a medium of lesser optical density such as air, as in Fig. 40.7, the angle of refraction r is greater than the angle of incidence i, and r increases more rapidly than does i. The value of r approaches the limiting value of 90° beyond which no light is refracted into the air. The angle of incidence i_c in the denser medium which makes the angle of refraction 90° is called the *critical angle* of incidence. From the law of refraction

$$\sin i_c = 1/n \tag{4}$$

When the angle of incidence is increased beyond its critical value, the light is *totally reflected*, making the angle of reflection r' equal to the angle of incidence i'. Total reflection can take place only when the light in the denser medium is incident on the surface separating it from the less dense medium and the angle of incidence is greater than the critical angle.

When light passes through a glass prism in air, it is bent toward the thicker part of the prism. In Fig. 40.8 the prism angle is A and the angle of deviation is D.

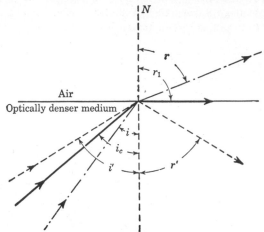

FIG. 40.7   Total internal reflection

Measurements of refraction of light show that the amount of refraction depends upon the wavelength of the light and the nature of the refracting substance. Blue-producing light is refracted more than the red-producing light.

Lenses. A transparent body with regularly curved surfaces ordinarily produces changes in the path of light; such a body is called a *lens*. The most common forms of lenses are those in which at least one surface is a part of a sphere. As shown in Fig. 40.9a rays parallel to the principal axis, after passing through a convex lens, will converge to the principal focus F; rays parallel to the principal axis, after passing through a concave lens, diverge from the

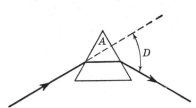

FIG. 40.8   Refraction by a prism

principal focus, at which a *virtual image* is formed, Fig. 40.9b. For both mirrors

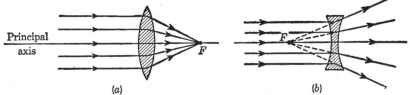

FIG. 40.9   Focusing of light: (a) by converging lens and (b) virtual focus of diverging lens

and lenses it may be shown from the graphical constructions that

$$\frac{\text{Size of image}}{\text{Size of object}} = \frac{\text{distance of image}}{\text{distance of object}}$$

The first ratio is called the *linear magnification M*. Thus

$$M = q/p \qquad (5)$$

As in the case of mirrors, rays not near the principal axis are focused closer to the lens than are the central rays. This *spherical aberration* is minimized by using a diaphragm to decrease the aperture of the lens. This small aperture produces a sharper image but reduces its brightness.

Experiment 40.1

REFLECTION AND REFRACTION BY USE OF THE OPTICAL DISK

Object: To study by means of the optical disk the fundamental principles of the reflection and refraction of light.

Method: An illuminator sends parallel rays of light against the surface of plane and curved objects on an optical disk. The basic phenomena of reflection

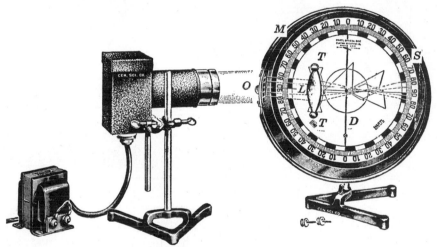

Fig. 40.10   Optical disk and light source

and refraction are studied by observing the paths of the rays when they strike the surfaces of various kinds of mirrors and lenses.

Apparatus: Hartl optical disk with accessories; illuminator, arc light or electric bulb form; ruler; protractor; compass; cross-ruled paper.

The optical disk, Fig. 40.10, consists of a metal disk D upon which is etched a circular graduated scale S. The disk is partly shielded by a semicylindrical sheet-metal screen M. The disk and screen turn independently upon the same horizontal axis; the friction is sufficient to hold them in any desired position. A square opening O in the screen is provided with guides to hold slotted plates. Thumbscrews T are used to hold different objects, such as a convex lens L, at positions outlined on the disk. The accessories used with the optical disk are shown in Fig. 40.11. A chart showing sketches of experiments that may be performed with the optical disk is provided with the apparatus.

Procedure: I. *Preliminary Adjustments.* 1. Place the disk D so that a beam of parallel rays from the illuminator strikes the edge of the disk. Turn the screen M so that it is between the illuminator and the disk. Adjust the illuminator until the beam covers most of the opening O in the screen and traces its path across the face of the disk. Put the three-slot plate in place and cover two of the slots. Adjust the disk and screen until the beam of light crosses along the zero axis of the disk.

II. *Plane Mirrors.* 2. Fasten the plane mirror to the disk so that the face of the mirror coincides with the diameter 90-90. Pass the ray along axis 0-0 and

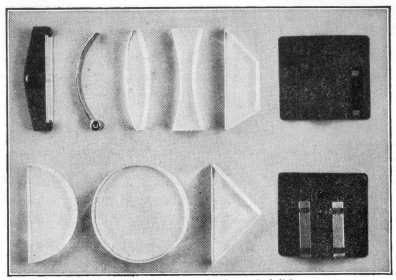

Fig. 40.11 Accessories for optical disk

turn the screen on its horizontal axis until the ray strikes the mirror exactly at the center of the disk. (See item 1 on the chart provided with the apparatus.) Leave the screen stationary and turn the disk to several different positions, increasing the angle of incidence. Record the angles of incidence and the corresponding angles of reflection. Note whether the law of reflection applies for the various angles of incidence.

3. Substitute the seven-slot plate and check to see if parallel rays are still parallel after reflection from a plane mirror (item 2 on chart).

III. *Spherical Mirrors.* 4. Replace the plane mirror by the concave mirror. (It is sometimes desirable to put a piece of paper under the back edge of the mirror to tip it slightly forward.) Rotate the disk until the single light ray crosses the center of the disk in the axis of the mirror. Leave the disk stationary and rotate the screen slightly backward and forward to move the ray parallel to itself across the face of the mirror. The reflected ray will be seen always to pass through one point on the axis of the disk; this is the principal focus F of the mirror (item 3).

5. Attach a piece of paper to the disk with masking tape and sketch the mirror, rays, principal axis, and principal focus. Measure f and compare it with the radius of curvature of the mirror as determined by a compass and ruler.

6. Adjust the screen and disk so that the ray passes through the center of curvature of the mirror. Record the way in which this ray is reflected.

7. Use the seven-slot plate to check the location of the principal focus of the mirror (item 5).

8. Remove the slotted plate and use the full opening. Note the spherical aberration. Trace the caustic curve on a sheet of paper.

9. Turn the disk so that the convex side of the mirror faces the opening O and repeat the appropriate observations (item 4). Use the seven-slot plate to trace with dotted lines the reflected rays back to the virtual principal focus (item 6).

10. If a diverging-ray attachment is available, perform the experiments suggested by items 21 and 22 on the chart. What conclusions may be drawn from these experiments?

IV. *Refraction of Light.* 11. Attach the semicylindrical glass plate to the disk so that the straight edge coincides with the 90-90 diameter (item 7). Adjust the screen so that the single ray touches the flat edge at the 0-0 axis. Sketch the reflected and refracted rays. Compare the angle of incidence with the angle of reflection and the angle of refraction. Read the angles of refraction for several angles of incidence, such as 30°, 40°, and 55°. Compute the corresponding values of the index of refraction and comment on the results.

12. Rotate the disk through 180° and send the light through the glass in a direction opposite to that of Step 11 (see item 8). Note the reflected and refracted rays. Compare the angle of incidence with the angle of reflection; with the angle of refraction. What difference is seen as compared with Step 11? Note the new angles of refraction when the same angles of incidence are used as were observed in Step 11 for the angles of *refraction*. Do the observations indicate that the light paths are reversible?

V. *Total Internal Reflection.* 13. Vary angle i and note that as i is increased the intensity of the refracted ray is decreased. Turn the disk carefully near an angle of incidence of about 41.5° and note the value of the critical angle for which the refracted ray is parallel to the surface (item 9). Measure the critical angle for white, red, and blue light, obtained by covering the slit with colored glass. Comment on the values of the index of refraction.

14. Insert the 90° prism so that the long face is on the vertical 90-90 axis, with the apex of the prism toward the right. Use two differently colored rays to show the paths of the rays that are totally internally reflected. Turn the disk through 45° so that the rays fall perpendicularly on one leg face and trace the totally internal-reflected rays. This is an example of the use of such a prism in modern field glasses.

VI. *Refraction through a Parallel Plate.* 15. Arrange the trapezoidal glass plate on the disk in such a position that a ray enters one of the parallel surfaces and emerges at the other (item 10). Note the relative directions of the incident and emergent beams. Repeat for several angles of incidence and comment on the results.

VII. *Refraction by a Prism.* 16. Rearrange the trapezoidal prism so that either the 45° or the 60° angle is bisected by the 90-90 diameter (items 11 and 12). Use a colored beam to avoid dispersion. Let the single beam fall on the edge of the glass so that half the beam passes through the glass and the other half passes by the edge undeviated. Read the deflection caused by the glass. Record the deviation both for the 45° angle and the 60° angle. Comment on the variation of the deviation with the refracting angle of the prism.

17. With the single beam of white light hold a prism against the disk behind the square opening of the screen. Note the dispersion of the white light into a prismatic spectrum. Record the colors in their order of deviation.

VIII. *Refraction by Lenses.* 18. Fasten the convex lens to the disk parallel to the 90-90 diameter, as in item 15. Send a single beam along the 0-0 axis passing through the center of the lens. Rotate the disk through a small angle on either side of the initial position and observe the incident and emergent rays. When such a ray passes through the optical center of a lens, how does it emerge?

19. Turn the disk so that the 0-0 diameter (axis of lens) is parallel to the incident ray. Rotate the screen about its horizontal axis and note that the refracted ray turns about one point on the disk. Record this location of the principal focus of the lens (item 14). Show the location of F by using the plate with seven slits (item 16). Remove the slotted plate and shine the light through the square opening. Is the light sharply focused? Sketch the caustic curve (item 17).

20. Attach the concave lens in place of the convex lens. Use the seven parallel rays to trace the paths of the diverging rays and to locate the virtual principal focus of the lens (item 18).

21. *Optional Experiments.* Show the passage of light through a combination of lenses, as in item 19. If a diverging-ray attachment is available, the experiments shown in items 23 to 25 may be performed.

Experiment 40.2

REFLECTION BY SPHERICAL MIRRORS

Object: To study the phenomena of the reflection of light by spherical mirrors; in particular, to measure the focal length and radius of curvature of a concave and a convex mirror and to demonstrate the mirror equation.

Fig. 40.12 Optical bench and accessories

Method: Images are formed by concave and convex spherical mirrors and from the measured object and image distances the optical constants of the mirrors are determined and the mirror equation is demonstrated.

Apparatus: Optical bench and accessories; illuminated object-image screen; concave mirror; convex mirror; convex lens, about 6 in. focal length; hooded screen.

The optical bench, Fig. 40.12, consists essentially of a horizontal lathe-bed type of mounting for sliding carriages equipped with upright clamps in which illuminators, mirrors, lenses, and screens may be inserted. The bench is provided with a metric scale on which the necessary distances are measured. The illuminated object, Fig. 40.13a is an arrow cut in a black plate. The same plate has a white screen arranged just above the arrow so that object and image may be superimposed when desired.

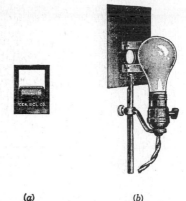

Procedure: I. *Focal Length of Concave Mirror.* 1. *Coincidence method.* Place the concave mirror N in a holder on the bench, with the concave surface of the mirror toward the illuminated object O, Fig. 40.14. Slide the mirror away from the object until a sharp image of the arrow is formed on the screen. Adjust the mirror back and forth around the proper position so that an average distance

(a)   (b)

Fig. 40.13   Object-image screen *a* for use in illuminated holder *b*

is obtained. Record the distance from A to B. Show from the mirror equation why this distance is equal to the radius of curvature r of the mirror.

2. *Mirror-equation method.* Place a small white screen I (a white card may be used) in a holder at C between the object O and the concave mirror N, Fig. 40.15.

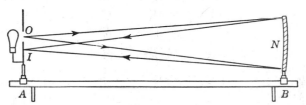

Fig. 40.14   Image I of object O placed at the center of a concave mirror

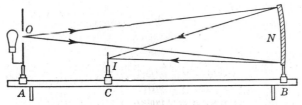

Fig. 40.15   Object and image formed by concave mirror

The top of the white screen should be just below the bottom of the object. Locate the position of the white screen for a sharp image when the mirror is 50 cm from the object. Record the object and image distances. From the mirror equation compute r and f for the mirror. Repeat this series of measurements with the mirror at 100 cm from the object and again calculate r and f.

II. *Focal Length of Convex Mirror.* 3. *Coincidence method.* Mount a convex lens L at D near the combined object-image screen at A, Fig. 40.16. Leave the holder at C temporarily empty. Place the hooded screen S about 80 cm from O and adjust its position until a sharp image of the object is obtained. Now place the convex mirror N' at C between the lens and the hooded screen. Adjust its position until a sharp image I is formed just beside the object O. If the radius of curvature of the mirror is greater than the distance from the hooded screen to the lens, no sharp image is obtainable. In this case the distance from the object to

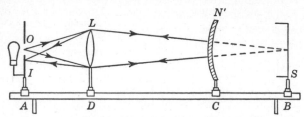

FIG. 40.16 Coincidence method for finding radius of curvature of convex mirror

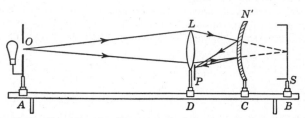

FIG. 40.17 Focal length of convex mirror by use of mirror equation

the hooded screen should be increased and the system refocused. Measure the distance from the mirror to the hooded screen. Show why this distance is the radius of curvature r of the mirror. Place the screen at a different position and obtain a second set of readings. Again measure r. Calculate f from the average of the observed values of r.

4. *Mirror-equation method.* Place the convex lens L about 45 cm from the object O, Fig. 40.17. Locate the image by adjusting the position of the hooded screen S. Note in this case that the image rays are highly convergent. Insert the convex mirror N' about 10 cm in front of the hooded screen. Adjust its position until the image I is formed on a piece of white paper P held on and partly covering the nearer face of the convex lens. Note and record the lens-to-mirror distance q, the mirror-to-hooded-screen distance p, and from these compute f and r.

Review Questions: 1. What may happen to a beam of light striking the surface of a different substance? 2. Explain what is meant by angles of incidence and reflection. State the law of reflection. 3. Describe the image formed by a plane mirror. 4. Explain the meanings of the terms: concave mirror; convex mirror; center of curvature; principal axis; principal focus; focal length. Distinguish between real and virtual images. 5. What relation exists between the focal length and radius of curvature of a spherical mirror? 6. Describe and illustrate a ray-diagram method for locating an image formed by a spherical mirror. 7. State the mirror equation and the convention for the algebraic signs of the quantities in this equation.   8. Explain by the aid of a diagram the meaning of spherical aberration.   What is a caustic curve?   9. Describe the phenomenon of refraction of light waves.   State Snell's law.   Define index of refraction.   10. Under what circumstances does total internal reflection occur?   What is the critical angle?   State the relation between index of refraction and critical angle.   Sketch a ray through a totally reflecting prism. 11. What is a lens?   Distinguish between concave and convex lenses.   Define principal axis and principal focus of a lens.   12. Define magnification and state its relation to the distances frequently used in mirrors and lenses.

Questions and Problems: **1.** Show geometrically that the image of an object formed by a plane mirror is as far behind the mirror as the object is in front of the mirror.   Use the spherical-mirror equation to justify this same fact.

2. If light waves are to converge to a point after reflection from a plane mirror, what must be their form before reflection?   Explain by a sketch.

3. Justify geometrically the fact that when a plane mirror is rotated the beam of light reflected from it rotates through twice the angle turned by the mirror.

4. Draw a curve to portray the variation of image distance with object distance for a concave mirror as the object distance is varied from zero to infinity.   Describe the variation in size of these images.

5. A person 5 ft 8 in. tall stands 3 ft in front of a plane mirror.   What size mirror is required to enable the person to obtain a full-length image?   How must the mirror be arranged?   Explain reasoning by a sketch.   Answer the same question for a distance of 10 ft from the mirror.

6. An object is 150 mm in front of a concave mirror and the image is located 600 mm behind the mirror.   Describe the characteristics of the image.   What is the magnification?   Calculate r and f for the mirror.

7. A man wishes to use a shaving mirror to give an image with a twofold magnification.   If the distance from the eye to the image is 25 cm, the distance of comfortable vision, what must be the radius of curvature of the mirror?   Draw a ray diagram for this case.

8. An object 10.0 cm high is placed 15.0 cm from a convex mirror of radius of curvature 6.00 cm.   Where is the image and what are its size and characteristics? Illustrate by a ray diagram.

9. A light ray in air strikes water at an angle of incidence of 40°.   The index of refraction for water is 1.33.   What is the angle of refraction?   What is the critical angle for the water?

10. A ray of light is incident normally upon a prism of refracting angle 60°.   The index of refraction is 1.65.   By what angle is the emergent ray deviated from the incident ray?

11. What type of lens is a glass full of water?   Does this lens have more or less spherical aberration than the common magnifying glass?   Why?

CHAPTER 41

THIN LENSES; OPTICAL INSTRUMENTS

A lens produces a change in the curvature of a wave front passing through it. For a converging (positive) lens, rays parallel to the principal axis are brought to a focus called the *principal focus F*, Fig. 41.1. The distance from the center of the lens to F is called the *focal length f* of the lens. For a diverging (negative) lens,

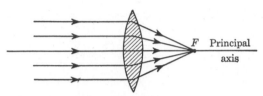

FIG. 41.1   Principal focus of a converging lens

rays parallel to the principal axis diverge as they leave the lens from the *virtual* principal focus, Fig. 41.2.

Only from a very distant source are the rays reaching a lens parallel. For nearby sources the rays leaving an object diverge. It is possible to determine the

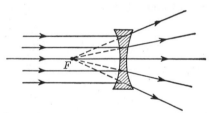

FIG. 41.2   Principal focus (virtual) of a diverging lens

position and size of an image formed by a thin lens by a simple geometrical construction. This is done by drawing two rays whose complete paths we know, starting from an object point and focussing at the corresponding image point. In Fig. 41.3a one ray is shown leaving the tip of the arrow object OO' and directed parallel to the principal axis. After refraction by the lens this ray passes through F. Another ray also from the tip of the arrow is drawn through the optical center of the lens. For a thin lens this ray is undeviated. The intersection of the two rays at I locates the image point which corresponds to the object point O. The other image points corresponding to additional object points may be located by similar constructions, thus giving the complete image II'.

The location of the image for an object placed closer to a converging lens than the principal focus is shown in Fig. 41.3b. It is seen that the rays diverge after passing through the lens. If the refracted rays are traced backward, they intersect at a *virtual* focus. The entire virtual image is represented conventionally by the dotted arrow. Such a virtual image cannot be formed on a screen but it may be viewed by looking into the lens, from the right in the figure.

The image of an object formed by a diverging (negative) lens is found by a similar construction, as in Fig. 41.3c. Here the ray that is parallel to the prin-

288

cipal axis diverges from the virtual focus F after passing through the lens. The image is seen to be virtual in this case.

Magnification. It will be seen from all of the ray diagrams of Fig. 41.3 that the angle subtended at the lens by the image is always equal to the angle subtended

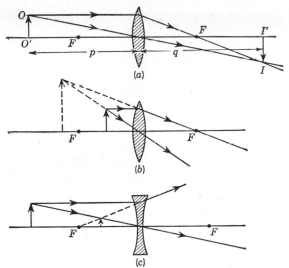

FIG. 41.3 Location of images by ray diagrams

by the object. Hence we see from the graphical construction that

$$\frac{\text{Size of image}}{\text{Size of object}} = \frac{\text{distance of image from lens}}{\text{distance of object from lens}}$$

The first ratio is called the *linear magnification*, or simply the magnification. Hence, in symbols

$$M = q/p \tag{1}$$

where p is the distance of the object from the lens and q is the distance of the image.

The Thin-lens Equation. It is possible to find the location and size of an image by algebraic means as well as by the graphical method already outlined. Analysis shows that the focal length f of a thin lens, the distance p of the object from the lens, and the distance q of the image are related by

$$\frac{1}{p} + \frac{1}{q} = \frac{1}{f} \tag{2}$$

This relation holds for any case of image formation by either a converging or diverging lens, provided that the following conventions are observed:

1. Consider f positive for a converging lens and negative for a diverging lens.

2. Object and image distances are taken as positive for real objects and images, negative for virtual objects and images. The normal arrangement is taken to be object, lens, and image, going from left to right in the diagram. If q is negative,

this means that the image lies *to the left* of the lens, rather than to the right, and is therefore virtual.

Experiment 41.1

THIN LENSES

Object: To measure the focal lengths of some positive and negative thin lenses and the equivalent focal lengths of some lens combinations by the use of the thin-lens equation.

Method: A thin convex lens mounted on an optical bench is used to form an image of a distant object at the principal focus of the lens. The lens is also used to form an image on a screen of a nearby object. The focal length is calculated from the observed object and image distances, by the use of the thin-lens equation. These values of f are checked by the use of the conjugate-position method. The magnification produced by the lens is measured from the ratio of the image to object size and compared with the value calculated from the ratio of the image distance to the object distance. The equivalent focal length of two thin lenses in contact is measured and compared with the calculated value. The focal length of a diverging lens is measured by using it in conjunction with a converging lens.

Theory: *Measurement of the Focal Length of a Converging Lens.* Three methods for measuring the focal length of a positive lens are available:

1. *Parallel-ray method.* Light from a distant landscape, or an illuminator giving parallel rays, is focused on a screen by a convex lens, as in Fig. 41.1. Since for this case p may be considered infinite, from Eq. (2) it follows that $q = f$. Hence the distance from the lens to the screen gives directly an approximate value of f.

2. *Thin-lens-equation method.* A measured image distance q corresponding to a known object distance p, Fig. 41.3a, is used to substitute into the thin-lens equation, Eq. (2), thus enabling the value of f to be determined.

3. *Conjugate-position method.* For any thin lens the values of p and q are interchangeable. Such a pair of values are called *conjugate* distances, because, if the object distance is changed from value p to value q, the image distance changes from value q to value p. It is more convenient experimentally to change the position of the lens to achieve this change of p and q than it is to change the position of the object and image. First the lens is placed in position L_1 (Fig. 41.4) a

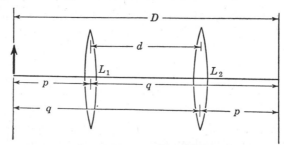

Fig. 41.4   Double-position method for measuring focal length

distance p from the object. The image is formed at distance q. Next, without moving the object or the screen, the lens is placed at position L_2, a distance q from the object. The second image distance is found to be equal to the first object distance p. Let D denote the distance from screen to object and d the distance between the two symmetrical positions of the lens with respect to object and screen. Then

$$D = p + q \qquad\qquad d = q - p$$
$$p = D - q \qquad\qquad q = d + p$$
$$ = D - (d + p) \qquad = d + (D - q)$$
$$ = \frac{D - d}{2} \qquad\qquad = \frac{d + D}{2}$$

Substituting these values of p and q into the thin-lens formula, Eq. (2), gives

$$f = \frac{pq}{p + q} = \frac{\dfrac{D - d}{2} \times \dfrac{D + d}{2}}{\dfrac{D - d}{2} + \dfrac{d + D}{2}} = \frac{D^2 - d^2}{4D} \qquad (3)$$

Since D and d may be accurately and easily measured, this procedure offers a convenient method for the determination of focal lengths.

Equivalent Focal Length of Two Thin Lenses in Contact. This equivalent focal length may be found experimentally by considering the lens combination as a single lens and proceeding as in any of the methods previously described. If the individual focal lengths are known, the equivalent focal length f may be calculated from the equation

$$\frac{1}{f} = \frac{1}{f_1} + \frac{1}{f_2} \qquad (4)$$

where f_1 and f_2 are the individual focal lengths. Proper regard must be observed for the algebraic signs of these focal lengths.

Focal Length of a Concave Lens. Since a concave lens makes a wave front more divergent, it cannot form a real image of a real object, and therefore the focal length of a concave lens cannot be determined by the methods just described for convex lenses. The focal length of a negative lens may, however, be obtained by placing it in contact with a positive lens of shorter and known focal length, measuring the focal length of the combination, and using Eq. (4) for the equivalent focal length of thin lenses in contact.

Another method of determining the focal length of a negative lens is to use it in conjunction, but not in contact, with an auxiliary positive lens. Let O, Fig. 41.5,

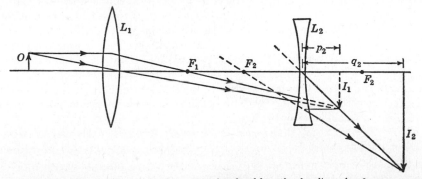

Fig. 41.5 Arrangement for measuring focal length of a diverging lens

be the object and I_1 the image which would be formed by the converging lens L_1 (if L_2 were not present). If now a diverging lens L_2 be placed between L_1 and I_1, the rays entering L_2 are converging toward I_1. Hence I_1 is a *virtual* object for L_2. If I_1 is between L_2 and its principal focus F_2, the rays leaving L_2 will be less convergent than those entering the lens but will converge to form a real image at I_2. By measuring p_2 (considered negative) and q_2 and substituting the values in Eq. (2), the focal length of the diverging lens can be determined. This value of f_2 can then be checked by placing L_2 in contact with a converging lens of known focal length and finding the equivalent focal length of the combination as described above.

Apparatus: Optical bench and accessories, Fig. 41.6, including: object box with illuminated screen;* four lens holders; screen (white board or frosted glass);

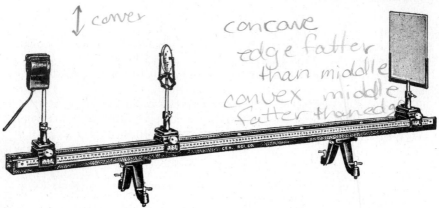

FIG. 41.6   Optical bench and accessories

convex lenses, 5-, 10-, and 15-cm focal lengths; concave lens, 15-cm focal length; vernier caliper; illuminator, arranged to give parallel rays.

One form of optical bench is the lathe-bed type, consisting essentially of a horizontal bed with two rails on which ride carriages for lens holders and other devices. The bed is provided with a meter scale for the measurement of distances.

In addition to the apparatus provided the student should bring a ruler and a compass.

This experiment should be performed in a room with subdued light.

Procedure: I, *Focal Length of Converging Lenses.* 1. *Use of distant landscape as a source.* Select the thinnest convex lens available (focal length about 15 cm). Mount it in the lens holder on the optical bench and place it near the screen. Point the optical bench through an open window toward a distant building. Place the screen at the zero of the scale on the bench and adjust the position of the lens until the central part of the image of the building is sharply outlined on the screen. Record the distance from the lens to the screen. Show why this distance is the focal length of the lens.

2. *Use of parallel-ray illuminator as a source.* Arrange an illuminator at one end of the laboratory to give approximately parallel rays directed toward the

* Instead of the illuminated object shown in Fig. 41.6 a good object which can be used is the filament of an unfrosted lamp bulb, with the bulb mounted horizontally.

screen of the optical bench. Place the lens on the bench and adjust it to focus the beam of light sharply. Record the distance from the lens to the screen and show why this distance is the focal length of the lens.

3. *Use of thin-lens equation.* Place the illuminated object at the zero position on the optical bench and arrange the screen at a distance of about five times the focal length from the object, as in Fig. 41.4. Mount the lens near the object at position L_1 and find the position of the screen for which the image is sharply defined. Choose a position of the lens which will give a sharply outlined image of moderate size (covering about half the screen). In determining the position of the screen move it from left to right and then from right to left and average the two readings for the final value. Measure p_1 and q_1 and calculate f from Eq. (2). Draw a ray diagram, to scale, for this case. Make the diagram to as large a scale as possible. *rough sketch*

4. *Conjugate-position method.* With the same arrangement as used in Step 3 and keeping the position of the screen fixed, move the lens toward the screen to position L_2 until an image is again sharply defined, Fig. 41.4. Observe and record the values of D and d. Calculate f from Eq. (3). Note the percentage difference between the indivudual values of f and the average of the four values found in Steps 1 to 4.

5. Measure the focal length of the other convex lenses by the double-position method as in Step 4.

II. *Magnification.* 6. Using one of the arrangements of Step 5 where the image is reasonably large, measure with a vernier caliper the sizes of the image and object. Note the percentage difference between this magnification II'/OO' and the value obtained from q/p.

III. *Focal Length of Thin Lenses in Contact.* 7. Insert the two thinner convex lenses in contact in a single holder and measure their equivalent focal length f by the double-position method, as in Step 4. Note the percentage difference between this observed value of f and that obtained from the measured values of the focal lengths f_1 and f_2, by the use of Eq. (4).

IV. *Focal Length of a Concave Lens.* 8. Place the screen near the 30-cm position. Mount the 5-cm (thickest) convex lens L_1 near the object. Adjust L_1 until the image I_1 is sharp. Then place the concave lens L_2 (focal length about 15 cm) near the 17-cm mark on the bench, as in Fig. 41.5. Note that the rays now focus farther away at I_2. Locate and record the position of I_2. The distance from the concave lens to I_1 is taken as the object distance (with a negative sign) for the concave lens. The image distance (positive) for this lens is the distance from the concave lens to I_2. Record these values of p_2 and q_2 and solve for the focal length f_2 of the concave lens, using Eq. (2).

9. Obtain an independent check on the focal length of the concave lens by mounting it in contact with the 5-cm convex lens and proceeding as in Step 7.

Experiment 41.2

OPTICAL PRINCIPLES OF THE EYE

Object: To study the optical principles of the eye by means of a model.

Method: A model, built to conform roughly with the shape of the eye, consists of a tank of water containing a movable white screen which serves as the retina.   A window in the tank is covered with a meniscus lens which performs the function of the cornea.   One of a series of interchangeable lenses takes the place of the crystalline lens of the eye.   The image of an illuminated object is projected on the retina and its appearance observed.   The accommodation of the eye to different object distances is demonstrated by interchanging lenses.   A movable retina is used to show farsightedness and nearsightedness.   Astigmatism is produced by the use of a cylindrical lens.   The correction of optical defects is accomplished by the use of auxiliary lenses.

Theory: The human eye is a most ingenious optical mechanism.   A schematic sketch of the eye is shown in Fig. 41.7.   The eye consists of a lightproof chamber

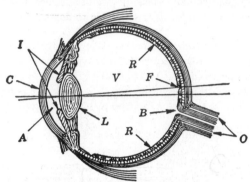

Fig. 41.7   Diagram of the human eye. R, retina; I, iris; C, cornea; V, vitreous humor; F, fovea centralis; B, blind spot; A, aqueous humor; L, crystalline lens; O, optic nerve

into which light from an object enters through the *cornea C* and the *crystalline lens L*.   This lens forms an inverted image of the object upon the *retina R* at the back part of the black-coated chamber.   The *optic nerve O* receives an impression of this image and transmits it to the brain, where the sensation of sight is obtained.

The amount of light admitted to the eye is controlled by an opening known as the *iris I*.   Its aperture is adjusted to varying light intensities by muscular action. The major portion of the eye is filled with transparent fluids, that portion behind the lens being called the *vitreous humor V* and the part in front of the lens being designated the *aqueous humor A*.

At the place where the optic nerve leaves the eye there is an insensitive region known as the *blind spot B*.   The region of the retina of greatest sensitivity is the *fovea centralis F*.   As the eye scans an object, it rotates in its socket to bring successive parts of the image onto the fovea.

In order that images of all objects at varying distances may focus at the same place, the crystalline lens changes its curvature by unconscious muscular control. This action is known as *accommodation*.

An optical defect known as *nearsightedness* or *myopia* is caused by the lens not being sufficiently flattened, so that images are formed in front of the retina.   Such a person unconsciously brings an object near his eye, so that the image recedes from the lens as near the retina as possible.   This defect is remedied by spectacles having diverging lenses.

A common optical defect, especially among older people, is the inability of the crystalline lens to contract sufficiently properly to focus light from nearby objects. This is known as *farsightedness* or *hyperopia* and is corrected by wearing spectacles with positive (converging) lenses.

Another very common optical defect is *astigmatism*. This is the phenomenon whereby the rays of light from different planes in an object do not all focus at the same point on the retina. Because of this effect, lines in some directions are seen clearly while those in other directions are blurred. Astigmatism is caused by differences in the curvature of the cornea or crystalline lens at different diameters. It is corrected by the use of *cylindrical* spectacle lenses, with the axis of the cylinder properly chosen so as just to neutralize the cylindrical curvature of the cornea or crystalline lens. The corrective cylindrical lenses are also combined in the spectacles with the proper spherical lenses to correct for nearsightedness or farsightedness.

The *power* of a lens is a term used by opticians to designate the ability of the lens to change the curvature of a wave incident upon it. Since a lens changes the curvature of a wave by an amount $1/f$, this is the measure of its power. The common unit of lens power is the *diopter*, which is defined as the power of a lens having a focal length of 1 m. Hence a $+3$-diopter lens would be a converging lens of

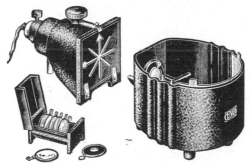

FIG. 41.8   Eye model and accessories

$\frac{1}{3}$-m focal length; a diverging lens of focal length 50 cm would have a power of $-1/0.5 = -2$ diopters. These two lenses in contact would have a combined power of $3 - 2$ or 1 diopter; their equivalent focal length would be 1 m.

Apparatus: Eye model and accessories, Fig. 41.8, including a set of lenses and an illuminated object box.

The model eye consists of a metal tank, Fig. 41.9, shaped roughly like a horizontal section of the eyeball. A window in one side of the tank is covered with a meniscus lens C which serves as the cornea. The tank is filled with water which takes the place of the aqueous and vitreous humors. The interchangeable crystalline lens L is mounted in a septum which marks the boundary between the humors. Two supports G_1 and G_2 are provided for the insertion of additional lenses and a diaphragm. In front of the cornea are two additional supports S_1 and S_2 for spectacle lenses. The retina is represented by a circular white area on a removable curved screen R which may be located also at R_m for a myopic eye and at R_h to simulate a hyperopic eye. The blind spot is represented by a black spot painted on the retina.

A diaphragm and the following set of six lenses are mounted individually in a metal holder.

1. Spherical convergent (+7.00 d)
2. Spherical convergent (+20.00 d)
3. Spherical convergent (+2.00 d)
4. Spherical divergent (−1.75 d)
5. Cylindrical divergent (−5.50 d)
6. Cylindrical convergent (+1.75 d)

In the case of the cylindrical lenses the axis of the cylinder is indicated on the mount.   Numbers 1 and 2 of these lenses serve in turn as crystalline lenses.   The

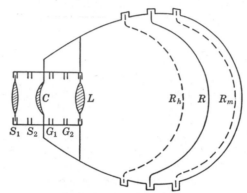

FIG. 41.9   Plan of eye model

others are used to simulate defects of vision or as correcting eyeglass lenses in front of the cornea.

The study of the optical principles of the eye by the aid of the eye model depends upon the examination of the images formed on the retina.   The student must determine in each case whether the image is upright or inverted, distinct or blurred, and larger or smaller than the object.   Also to be noted is how the image moves as the object is moved up or down and from side to side.   The report should include a statement of the procedures followed and the results obtained. Drawings should be made to illustrate the principles demonstrated.   All quantitative measurements should be recorded.

Procedure: Fill the tank with clear water to within 2 cm of the top.

1. *Accommodation.*   Set up the model so that it is directed toward a window or other bright object 4 or 5 m away.   With the retina in the middle or normal eye position, insert the weaker (+7) of the two crystalline lenses, in the groove L, Fig. 41.9, in the water just above the low partition.   The image of the object at which the model is pointed should now be in focus on the retina.   Note the character of the image, whether erect or inverted, and its size as compared with the size of the object.

The blind spot in the eye is a small area in which the retina is not sensitive. Note the probable effect on one's perception of that part of the image which falls on the blind spot.   The physics text gives directions for demonstrating the existence of the blind spot in one's own eye.

Next use as the object the lamp box with the radially slotted pattern and place it 35 cm from the cornea. At this object distance, the image will be much blurred until the weaker crystalline lens is replaced by the stronger (+20). This illustrates the process of accommodation or focusing which in the eye is automatically accomplished by a set of muscles which change the curvature of the crystalline lens. This +20 lens is to be used hereafter in all of the experiments with a "normal" eye.

2. *Farsightedness and Nearsightedness.* Make the eye farsighted by moving the retina forward to the position R_h. Adjust the position of the object box until the image is sharp. Then move the object box until it is 35 cm from the eye. Note the appearance of the image. Examine the spherical lenses +2.00 and −1.75, and determine whether a converging or diverging lens should be used for correcting the defect and bringing the image to a sharp focus on the retina. Test the conclusion by placing the correcting lens which you select in front of the eye and note whether it sharpens the image.

Repeat the experiment by moving the retina to the rearmost position R_m which corresponds to the nearsighted eye. Find as before the best viewing distance. Correct the myopia by applying the proper spectacle lens.

3. *Effect of Pupil Size; Vision through a Small Aperture.* Select any case in which the image is not quite clear and insert the diaphragm with the 13-mm hole either just in front of, or just behind, the cornea, and note how the image is sharpened. This illustrates the fact that a very small aperture in front of the eye will sharpen the image. Persons who have minor eye defects may, by looking at objects through a small hole in a card, or by squinting, see them distinctly. Explain this fact. Compare the intensity of the image with that of the normal eye.

4. *Astigmatism.* Using the lamp box at 35 cm and the retina in normal eye position, insert immediately *behind* the cornea the cylindrical concave lens marked −5.50, thereby producing astigmatism. (In the human eye astigmatism is generally due to a slight cylindrical curvature of the cornea; so in the model a change of cornea would perhaps be the logical way of producing astigmatism. This being impracticable, the same purpose is accomplished by the insertion of an additional lens.) Turn the cylindrical lens a little to make sharp one line alone of the object pattern; the others are blurred. Now place in front of the cornea the correcting convex cylindrical lens marked +1.75 and turn it until the image is again sharp. Notice the relative directions of the cylindrical axes in the two lenses. Repeat with the rear lens at a different angle.

5. *Compound Defects.* Astigmatism is often accompanied by farsightedness or nearsightedness. Place the cylindrical lens No. 5 at G_1 immediately behind the cornea, with its axis vertical, and have the retina in the position to give myopia. Make the correction by choosing the proper combination of spectacle lenses. Since in actual practice the two correcting lenses are combined into a single compound eyeglass lens, what shape would this lens have?

6. *Removal of the Crystalline Lens.* The crystalline lens is only one part of the lens system of the eye. Sometimes (for instance, in the case of cataract) it has to be removed. Take out the crystalline lens and show, by using lens No. 1 as a spectacle lens in front of the eye, that vision is still possible. Note that the image is distinct only for very near objects.

7. *The Use of a Magnifier, or Reading Glass.* With the retina in the normal position, insert lens No. 2 as the crystalline lens. Determine approximately the

diameter of the image with the object box at the normal distance. Use lens No. 1 as a magnifier in front of the eye and adjust the viewing distance until the image is distinct. Compare the ratio of the image sizes with the ratio of the viewing distances. What does this experiment show about the use of the magnifying glass in enabling the eye to view distinctly a very close-up object?

Note that, for close work, the nearsighted eye has an advantage over the normal eye. With the stronger crystalline lens, lamp at 35 cm, and retina at normal eye position, note the approximate size of the image. Then make the eye nearsighted, and move the object until the image is again sharp; the image will be considerably larger.

8. *Eyeglass Lenses.* With one eye look in turn through the correcting spherical lenses +2.00 and −1.75 and move them from side to side. Note in which case objects appear to move *with* the lens and in which case *against*. How is the effect altered when cylindrical lenses are used? Using your own or borrowed glasses, test as above and decide the character of the eye defects they are intended to correct. Note particularly if the two lenses are the same, and if not, which is the stronger. Also look for astigmatism. Even if this is small it will be readily noted by distortion of the view through it, as the glass is rotated in its own plane.

Calculate the focal length of the +7 lens and measure it approximately, using a distant-source object. Is the focal length in water longer or shorter than in air? Test your conclusion by experiment.

In practice the strength of a lens is most quickly and accurately determined by "neutralizing" it with a lens of opposite sign. See whether you can, perhaps, neutralize a spectacle lens with one of the lenses in the set, and thus determine its power.

9. At the conclusion of the experiment *be sure that no lenses are left in the model* and then empty it. Clean the lenses with a little absorbent cotton before putting them away.

Experiment 41.3

OPTICAL INSTRUMENTS

Object: To study the physical principles of several optical instruments by arranging lenses on an optical bench to simulate a reading glass, a telescope, and a microscope.

Method: The action of a simple magnifier is studied by using a short-focus convex lens to view an object of known size. The apparent size of the image is measured and the magnification determined and compared with the theoretical value. An astronomical telescope is simulated by using an objective lens and an eyepiece lens to view an object of known dimensions. The apparent size of the image is measured and the magnification of the telescope is determined and compared with the theoretical value. This procedure is then repeated for a Galilean telescope. A microscope is roughly simulated by the use of a short-focus objective lens and an eyepiece lens. The magnification is determined experimentally and compared with the theoretical value.

Theory: *Magnifier.* A simple magnifier is a converging lens placed so that the object to be examined is a little nearer to the lens than its principal focus (Fig. 41.10). An enlarged, erect, virtual image of the object is then seen. The image should be at the distance of most distinct vision, which is about 25 cm from the eye, the magnifier being adjusted so that the image falls at this distance.

The *linear magnification M* is approximately given by

$$M = \frac{25 \text{ cm}}{f} + 1 \tag{5}$$

where f is in centimeters.

The magnifier in effect enables one to bring the object close to the eye and yet observe it comfortably. When the object is thus brought closer, it subtends a larger angle at the eye than it would at a greater distance.

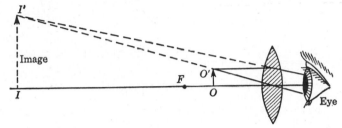

FIG. 41.10 Simple magnifier

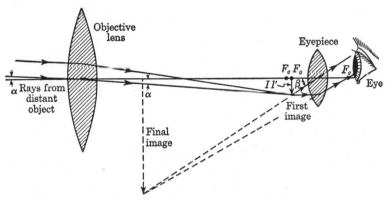

FIG. 41.11 Ray diagram for astronomical telescope

The Astronomical Telescope. The astronomical refracting telescope consists of an objective lens system and an eyepiece, Fig. 41.11. Light from a distant object enters the objective lens and forms a real image II'. The eyepiece, used as a simple magnifier, produces an inverted and enlarged final image.

Neglecting the length of the telescope, the first image subtends the same angle α at the center of the objective lens as the object does at the observer's naked eye, and similarly the first and second images subtend the same angle β at the optical center of the eyepiece. Hence

$$\text{Angular magnification} = \frac{\beta}{\alpha} \div \frac{II'/f_e}{II'/f_o} = \frac{f_o}{f_e} \tag{6}$$

Owing to the approximations made, Eq. (6) applies only for distant objects.

The Galilean Telescope.   A Galilean telescope (Fig. 41.12) consists of a converging objective lens, which alone would form, practically at the principal focus, a real inverted image QQ' of a distant object, and a *diverging* eyepiece lens.   In passing through the concave lens, rays that are converging as they enter are made

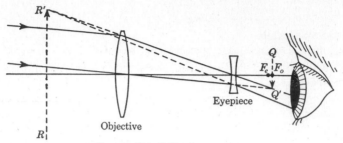

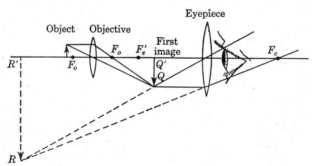

Fig. 41.12   Galilean telescope

to diverge as they leave.   To an observer the rays appear to come from RR', the enlarged virtual image.   With this design of telescope an erect image is obtained. For a distant object the magnification is

$$M = f_o/f_e \tag{7}$$

The Microscope.   This device for the production of high magnification consists of two converging lenses (in practice, lens systems), a so-called objective lens of very short focal length and an eyepiece of moderate focal length.   The objective forms within the tube of the instrument a somewhat enlarged, real image of the object.   This image is then magnified by the eyepiece.   Thus the final image seen by the eye is virtual and very much enlarged.

Fig. 41.13   Ray diagram for a microscope

Figure 41.13 shows the ray construction for determining the position and size of the image.   The object is placed just beyond the principal focus of the objective lens, and a real image is formed at QQ'.   This image is, of course, not caught on a screen but is merely formed in space.   It consists, as does any real image, of the points of intersection of rays coming from the object.   This image is examined by means of the eyepiece, which serves here as a simple magnifier.   The position of the eyepiece, then, should be such that the real image QQ' lies just within the principal focus F_e'.   Hence the final image RR' is virtual and enlarged, and it is inverted with respect to the object.

The magnification produced by a microscope is the product of the magnification M_e produced by the eyepiece and the magnification M_o produced by the objective lens. Hence, for a final image at the distance of most distinct vision (25 cm)

$$M = M_o M_e = \frac{q}{p}\left(\frac{25 \text{ cm}}{f_e} + 1\right) \tag{8}$$

where p and q are the distances of object and first image, respectively, from the objective, and f_e is the focal length of the eyepiece, all distances measured in centimeters.

Apparatus: Optical bench and accessories, Fig. 41.6, including: object box with illuminated screen; two lens holders; convex lenses, 4-cm anastigmatic, 5 cm, 10 cm, 20 cm; concave lens, −10 cm; iron washers to fit the 4- and 5-cm lenses; vernier caliper; half-meterstick; two caliper jaws; support rod and clamp for meterstick; extension cord and lamp.

Procedure: 1. *Simple Magnifier.* First determine the focal length of a short-focus (5-cm) convex lens by focusing rays parallel to the principal axis (from a distant object) on the screen and taking the lens-to-screen distance. Then mount the lens at the zero position on the optical bench. As the object to be magnified mount the hollow steel washer on another lens holder in the position OO' shown in Fig. 41.10. With the washer (and not the lens) well illuminated, adjust the position of the washer so that the object distance is about four-fifths of the focal length of the lens. Move the washer until it is brought into clear focus and clamp its holder in place. Then place a meterstick provided with caliper jaws horizontally at the 25-cm mark above the bench. While looking at the *image* through the lens with one eye, view the scale of the meterstick with the other eye. The image of the hole in the washer will be seen superimposed against the meterstick. Keep both eyes open and after a while both the magnified image of the washer and the actual scale can be viewed simultaneously. Measure the apparent size of the image on the meterstick by bringing the caliper jaws into coincidence with the circumference of the image of the hole. Try to keep the head fixed while the two caliper jaws are viewed at the same time. Let each member of the group take at least one view.

In this and the following parts of this experiment it is well to draw a diagram and to note on the figure all the pertinent data observed.

Measure the diameter of the hole in the washer with the vernier caliper and compare the measured magnification (image size/object size) with the theoretical value calculated from Eq. (5).

Repeat the observations, using the 10-cm lens and record the effect of using the lower-power lens for the magnifier.

2. *Astronomical Telescope.* Use the 5-cm convex lens in the zero position as the eyepiece L_e of Fig. 41.11. Use the 20-cm convex lens about 30 cm from L_e for the objective lens L_o. Place the illuminated object box on the opposite end of the bench from L_o. By a slight adjustment of L_o focus the image of the object so that when viewed through L_e it is clearly seen superimposed exactly upon the object. Arrange the meterstick with the caliper jaws to coincide with the image. Look through the lenses with one eye to see the magnified image, while the other eye looks directly at the object. Adjust the objective slightly so that there will be

no relative motion (parallax) between object and image when the eye is moved. Measure and record the length of the object and its image.

Compare the observed magnification I/O with the theoretical value calculated from Eq. (6).

Swing the optical bench so that a distant object may be viewed through the window, refocus the telescope, and observe its performance. Note the distance between L_o and L_e and compare it with the sum of the focal lengths of the objective and eyepiece.

3. *Galilean Telescope.* Replace the eyepiece L_e with a *concave* lens (focal length about -10 cm) and adjust the lenses to view again a distant object through a window. Compare the image with that observed with the astronomical telescope.

Compare the separation of the objective and eyepiece lenses with the difference of their focal lengths.

Comment on the differences between the astronomical and Galilean telescopes and point out where these are useful for certain applications. How could the final image of the Galilean telescope be photographed? What change would be necessary in the astronomical telescope so that its final image could be photographed on a film directly (without a camera)?

4. *Microscope.* Set up a simulated microscope, using approximately the dimensions shown in Fig. 41.14. For the objective lens L_o use a 4-cm anastigmatic

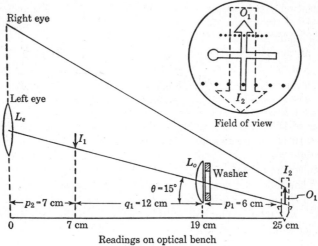

FIG. 41.14   Simulated microscope

convex lens. For the eyepiece lens L_e use the 10-cm convex lens. Place a washer in contact with L_o and note whether the distortion is reduced. Explain the effect observed. Use the illuminated source for the object and set it at such a distance from the objective lens that I_1, the image formed by the objective lens on a screen placed about 7 cm from L_e, is about two times the size of the object.

Remove the screen and view the image I_1 through the eyepiece. If it is not in clear focus, bring it into focus by *slowly* moving the *objective* lens, but not the eyepiece. Note that only a part of the object can be seen through the eyepiece.

This limitation of area of field is inherent in the microscope. Observe the chromatic and spherical aberrations that the uncorrected instrument produces.

To measure the magnification of the microscope, the object O_1 is viewed directly with the right eye. Turn the head sidewise (eyes vertical) so that the right eye looks straight over the top of L_e. (If the observer's left eye is dominant the eye directions given here may be reversed.) View the second image I_2 *through* both lenses with the left eye. The eyepiece lens should be higher than the objective, which is in turn higher than the object, as shown in the figure. When the apparatus is properly adjusted, the object and image I_2 appear to be superimposed, as indicated in the circular figure. The magnification may be measured directly by using some ink dots placed on the object screen a distance apart equal to the width of the arrow. View the arrow through the lenses by one eye and the dots by the other unaided eye. The image of the arrow is seen superimposed on the dots. Estimate the size of the image by counting the spaces between the dots. One edge of the arrow image should fall directly on a dot.

Compare the observed magnification with that calculated from Eq. (8).

Use a lower-powered (5-cm) lens for the objective. Without making any quantitative observations note the change in the magnification.

Review Questions: 1. What is a lens? Explain the difference between converging and diverging lenses. 2. Define principal focus; focal length. Explain the difference between real and virtual focal lengths. 3. Show by the aid of a sketch the ray-diagram method for locating the image of an object as formed by a positive lens; by a negative lens. 4. Define magnification. Show how it is related to image and object distances. 5. State the thin-lens equation. Give the conventions for the algebraic sign of each factor in the equation. 6. Describe several different methods used to measure the focal length of a convex lens. Justify the working equations used in each of these methods. 7. State the equation for the equivalent focal length of two thin lenses in contact. Give the convention for the algebraic sign of each factor in this equation. 8. Describe two methods for the measurement of the focal length of a concave lens. 9. Make a sketch to show the essential features of the human eye. Explain the working of each of these parts. 10. Describe the usual optical defects to which eyes are subject. Show how each of these may be corrected by spectacles. 11. What is meant by the optical power of a lens? 12. Describe the simple magnifier by the aid of a ray diagram. What is its magnification? 13. Draw the ray diagram for an astronomical telescope. What is the equation for its magnification? 14. Sketch the essential features of a Galilean telescope. What magnification does it produce? What advantage does it have over the astronomical telescope? 15. Draw a ray diagram for a microscope. State the equation for its magnification.

Questions and Problems: 1. In Exp. 41.1 the focal length of a convex lens was obtained by sighting the lens upon a distant object and locating the position of the image. If a 16.5-cm focal length lens is used and the "distant" object is a lamp 4.50 m away from the lens, what percentage error is made by assuming that the object is infinitely distant?

2. Draw a graph showing the variation of $1/p$ with $1/q$ for a converging lens. What is the significance of the Y intercept of this curve? of the X intercept?

3. An optical bench is 200 cm long. An object is placed at the zero mark and a screen at the other end. A simple lens in a certain position forms an image on the screen. Another clear image is formed when the lens is moved 30 cm nearer the screen. What is the focal length of the lens?

4. Two thin lenses are placed in contact. One has a focal length of 20.0 cm; the other is a diverging lens of 60.0 cm focal length. What is the equivalent focal length of the combination?

5. An optical bench is 125 cm long. An object is placed at the zero mark. With a simple converging lens at the 8.0-cm mark an image is formed at the 35.0-cm position. When a diverging lens is placed at the 20-cm mark, the image is found at the 85.0-cm position. What is the focal length of the diverging lens?

6. Would a person whose spectacles are corrected for both astigmatism and nearsightedness note the difference if his lenses should become rotated in the frame? for astigmatism only? for nearsightedness only?

7. A certain lens has a power of −1.33 diopters. What is the type and focal length of the lens? What kind and power of lens would be required to neutralize it?

8. What would be the equivalent focal length of the combination of lenses Nos. 3 and 4 in Exp. 41.2?

9. What power and type of spectacle lens would be required to change the distance for comfortable reading from 25 to 30 cm?

10. A certain farsighted person can read fairly clearly without spectacles at a distance of 80 cm. His glasses have a power of 2.5 diopters. What is his reading distance when wearing glasses?

11. Why do astronomical telescopes have objective lenses of such large diameters? Why is this not necessary for the Galilean telescopes used in opera glasses?

12. Cross hairs are sometimes put in astronomical telescopes. Where should they be placed? Explain.

13. An astronomical telescope has an objective of 50-cm focal length. The eyepiece has a focal length of 3.5 cm. How far must these lenses be separated when viewing an object 250 cm away from the objective of the telescope?

14. Describe some of the imperfections that one gets in a simple microscope made up of two uncorrected lenses.

15. A simple microscope has an eyepiece of focal length 12.00 cm and an objective with a focal length of 2.00 cm. An object is placed 25.0 mm from the objective. What magnification does this microscope produce? How far apart are the lenses for distinct vision?

CHAPTER 42

THE SPECTROMETER; INDEX OF REFRACTION

When light travels from one medium to another in which the speed is different, the direction of travel of the light is, in general, changed. When light travels from a medium of greater speed to one of lesser speed, the light is bent toward the normal. If light travels from the medium of lesser speed to the medium of greater speed, the light is bent away from the normal. The relation between the speed v_1 in the first medium, the speed v_2 in the second medium, the angle of incidence i, and the angle of refraction r is given by

$$n = \frac{\sin i}{\sin r} = \frac{v_1}{v_2} \tag{1}$$

where the constant n is the *index of refraction*.

If a ray of light, Fig. 42.1, is incident on one surface of a prism, the ray is bent at both the first and second surfaces. The emergent ray is not parallel to the incident ray but is deviated by an amount that depends upon the index of refraction n, the refracting angle A of the prism, and the angle of incidence i at the first surface. The angle of deviation is smallest when the ray passes through the prism in a direction perpendicular to the bisector of the refracting angle A of the prism. For this *angle of minimum deviation D* there

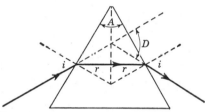

FIG. 42.1   Refraction at the angle of minimum deviation

is a simple relation between the refracting angle A, the angle of minimum deviation D, and the index of refraction n. In symbols

$$n = \frac{\sin \dfrac{A + D}{2}}{\sin \dfrac{A}{2}} \tag{2}$$

Experiment 42.1

INDEX OF REFRACTION BY PRISM SPECTROMETER

Object: To study the prism spectrometer and to use it to determine the index of refraction of a glass prism by the method of minimum deviation.

Method: The refracting angle of a prism is measured with the spectrometer, Fig. 42.2, by reflecting part of a beam of light from each of two refracting faces of the prism and measuring the angle between the two corresponding reflected

305

beams. Using a monochromatic source of light, the spectrometer is adjusted for minimum deviation of the beam and the angle of minimum deviation is measured. From the experimental values of the angle of the prism and the angle of minimum deviation, the index of refraction of the prism for the light used is calculated. In a similar manner the index of refraction is determined for several components of

FIG. 42.2   Prism spectrometer

a line spectrum and the dispersion curve is obtained by plotting the measured index of refraction against the known wavelengths.

Apparatus: Spectrometer; prism; sodium lamp; mercury, helium, neon, or hydrogen discharge tube; high-voltage transformer or induction coil, with terminals for discharge tubes; incandescent lamp.

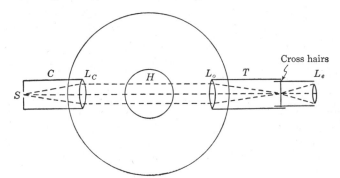

FIG. 42.3   Principal parts of a spectrometer

A spectrometer is an instrument for producing and measuring spectra. It consists of three essential parts: a collimator C, Fig. 42.3, whose function is to produce parallel rays; a prism, or grating, mounted on the prism table H; and a telescope T, to examine the spectrum. The instrument is shown diagrammatically in Fig. 42.4. Positions of the parts are indicated by a graduated platform R. In the type of spectrometer illustrated in Fig. 42.4, the collimator C is mounted in a bracket that is rigidly attached to the base of the instrument. The telescope T and the graduated platform R are capable of independent rotations about a common vertical axis through the center of the graduated circle. The relative angular positions of the telescope and the graduated platform are indicated by a circular scale and two vernier scales V_1 and V_2 on opposite sides of the platform.

The telescope can be locked in any position about the circumference of the circle by the lock nut L. Slight angular movement of the telescope is accomplished by the slow motion screw M which operates only when the lock nut L is set. In a similar manner, the rotating platform R can be locked in position by the nut K, and slow motion is provided by the screw N. The telescope and the collimator are supported on their respective brackets by the yokes Y. Slight

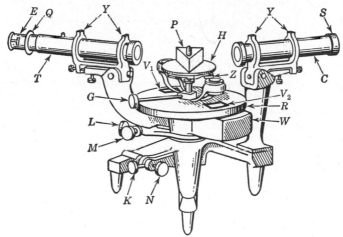

FIG. 42.4   Diagram of a prism spectrometer

adjustments of the axes of the telescope and collimator can be made by means of small supporting screws in the yokes. The rotating bracket which carries the telescope is provided with a counterweight W for relieving the strain on the bearing and ensuring proper alignment.

The prism P is mounted on a prism table H, which consists of a circular platform supported on a vertical shaft. The height of the prism table and its angular position can be adjusted and fixed by means of the setscrew G. Three leveling screws Z provide for leveling the prism table.

The collimator consists essentially of a tube fitted with an achromatic lens at one end and an adjustable slit at the other. The width of the slit S can be adjusted by rotating a knurled ring. (In some instruments this adjustment is made by a thumbscrew.) With the slit at the principal focus of the collimator lens, rays of light diverging from the slit are rendered parallel by the lens as shown in Fig. 42.3.

The telescope is equipped with an achromatic objective lens, an eyepiece, and a pair of cross hairs. The eyepiece is focused upon the cross hairs by sliding it in the barrel, and the telescope is focused by means of a knurled collar Q on the barrel.

The prism should have one face frosted. If all three faces are polished, great care must be taken to avoid light coming to the telescope through the wrong face.

If the prism is placed on the prism table H so that it is facing the collimator as in Fig. 42.5, a part of the light from the collimator is reflected by each face of the prism.   If the telescope is set successively, first in the position to receive the light from the right-hand face, and then that from the left-hand face, the telescope

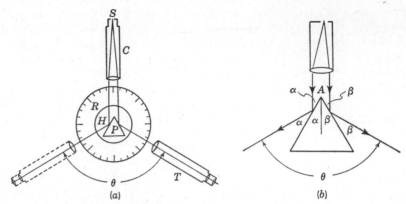

FIG. 42.5   Arrangement of spectrometer for measuring the angle of a prism

must be turned through an angle θ.   From the geometry of Fig. 42.5b and the law of reflection

$$\alpha + \beta = A$$
$$\theta = 2\alpha + 2\beta = 2A \tag{3}$$

and

Procedure: The spectrometer is reasonably rugged but it has carefully machined parts that can easily be injured if not properly handled.   The spectrometer has numerous adjustable parts that should be untouched by the student. Make only those adjustments called for in the procedure.   All parts that are expected to be movable will turn easily.   *Never force any movement.*

1. Turn on the sodium lamp.   It will require some time to come to full brightness.

2. Examine the spectrometer, noting the parts described above.   Study the scale on the platform R.   Note the value of the smallest scale division and the number of vernier divisions.   From these readings determine the least count of the vernier.   Ask the instructor to check your least count and method of reading the scale.   The two verniers provided on opposite sides of the scale can be used to correct for instrumental errors in the scale engravings.   Only one vernier need be used in this experiment; be careful to take all readings at the same vernier. Loosen the lock screw K (Fig. 42.4) and turn the platform until one window is near, but not under, the collimator.   Make all readings at this window.   *Lock the screw K and do not unlock it throughout the experiment.*

3. Focus the eyepiece (not the telescope) upon the cross hairs.   Pull the eyepiece almost out of the telescope by the ring at the end of the eyepiece, then insert it slowly until the cross hairs are in good focus with the eyes relaxed.

Focus the telescope (not the eyepiece) upon a distant object, viewed through an open window, by rotating the knurled ring Q until the image is in clear focus. In the final adjustment eliminate parallax between the image and the cross hairs. Adjust the knurled ring until the image does not move across the cross hairs as the eye is moved from side to side.   The telescope is then focused for parallel light.

Do not change the position of the eyepiece or the knurled ring after these adjustments are completed.

4. Illuminate the slit (open fairly wide) by means of the sodium lamp. If necessary adjust the position of the slit by sliding its assembly in the collimator tube until a sharp image of the slit is seen at the cross hairs. The best adjustment is obtained when parallax is eliminated. (Do not change the focus of the telescope in this process.) The collimator is then adjusted to give parallel light.

5. Measure the angle of the prism. Set the prism on the table with the ground-glass base near the screw clamp that holds the prism on the table. Loosen the lock screw G, rotate and raise or lower the prism table H until the prism is at the level of the collimator and the refracting angle of the prism is toward the collimator, Fig. 42.5a. Lock the screw G. In this position the light from the collimator falls on both reflecting faces of the prism.

View one of the reflecting faces of the prism with the unaided eye and note the reflected image of the slit. If you have difficulty locating the image, try moving the prism toward or away from the collimator. With the reflected image in sight, rotate the telescope until it is along your line of sight. Then look through the telescope and rotate it slightly until the image of the slit is seen near the intersection of the cross hairs. Lock the screw L (Fig. 42.4) and make the final adjustment by turning the slow-motion screw M, until the intersection of the cross hairs is at *one edge* of the slit. (In this and subsequent settings always use the same edge of the slit.) Record the scale reading, with fractional parts of the smallest scale division as indicated by the vernier scale.

Repeat this procedure for the image reflected by the second face. (*Caution: If the vernier scale passes the zero of the scale as the telescope is rotated, this fact should be noted and considered in determining the angle rotated.*)

6. Measure the angle of minimum deviation for sodium light. Loosen the lock screw G and rotate the prism table until the prism is in the position of Fig. 42.6. Locate with the unaided eye the *refracted* image of the slit. Slowly turn the prism table in such a direction as to diminish the angle of deviation, following the refracted beam with the unaided eye. A position is finally reached where further rotation of the prism in the same direction reverses the motion of the image of the slit. This is the position of minimum deviation. Lock the prism table. Bring the telescope to a position to view the refracted image of the slit and turn the telescope until the inter-

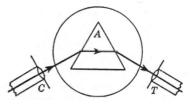

FIG. 42.6 Prism set for minimum deviation

section of the cross hairs is at the chosen edge of the slit. Lock the telescope and make the final adjustment by means of the slow-motion screw. Release the prism table and turn it through a small angle on either side of the position for minimum deviation. In this process the edge of the slit may not be exactly at the intersection of the cross hairs at reversal. If not, make further adjustment of the slow-motion screw until the desired condition is attained. Record the reading of the scale for the final setting.

7. Replace the sodium lamp by a gas-discharge tube containing mercury. (Or a helium or hydrogen tube may be used.) Observe the several lines that appear in the spectrum. This spectrum is characteristic of the gas in the tube.

Repeat Step 6 for each of the prominent lines of the spectrum. It is necessary to adjust the prism for minimum deviation for *each and every* line. Record the reading of the scale for each setting of the telescope.

8. Look up, in the Appendix or in a handbook, the wavelengths of the lines of the spectrum you used and record the values.

9. Replace the discharge tube by a neon tube and describe the spectrum obtained. Examine and describe the spectrum obtained when the source is an incandescent electric lamp.

10. Remove the prism and turn the telescope to recieve the light directly from the collimator. Set the intersection of the cross hairs on the chosen edge of the slit. Record the scale reading.

Computations and Analysis: 1. From the data of Step 5 calculate the angle θ through which the telescope moved. Use Eq. (3) to compute the angle A of the prism.

2. From the data of Steps 6 and 10 calculate the angle of minimum deviation D for sodium light. Use in Eq. (2) this value of D and the value of A determined in Part 1 to compute the value of n for sodium light.

3. Repeat Part 2 of the Analysis for each of the lines observed in Step 7.

4. Use the calculated values of n and the wavelengths recorded in Step 8 to plot a curve of index of refraction against wavelength. This is a dispersion curve for the glass of which the prism is composed. Comment on the usefulness of this curve.

Review Questions: 1. Define index of refraction in two ways. 2. How is a light ray affected as it goes from one medium to another? 3. What is meant by the deviation of a ray by a prism? by minimum deviation? 4. For what position of a prism is the deviation minimum? 5. State the relation between index of refraction, angle of deviation, and refracting angle of the prism. 6. What is a spectrometer? Make a sketch to show its principal parts and trace the paths of the rays through it. 7. Describe the technique for measuring the refracting angle of a prism and derive the equation used. 8. Describe the technique for measuring the angle of minimum deviation.

Questions and Problems: 1. From the geometry of Fig. 42.1 derive Eq. (2).

2. Why is it desirable that the rays coming from the collimator to the prism be parallel?

3. How could a prism spectrometer be used to measure wavelengths?

4. The speed of light in a vacuum is 1.000292 times as great as that in air. What percentage error is made in assuming that the index of refraction of air is unity?

5. What change would occur in the deviation if the angle i, Fig. 42.1, were (*a*) increased? (*b*) decreased? Give reasons.

6. If a prism is set for minimum deviation for yellow light, what change, if any, would be necessary in order to adjust it for minimum deviation for blue light?

7. Describe an experimental method of determining the angle of minimum deviation without observing the direct image, by making two observations of the refracted image.

8. A prism, $n = 1.65$, has a refracting angle of 60°. What must be the angle of incidence to produce minimum deviation? What is the angle of minimum deviation?

9. Find the index of refraction of a 60° prism that produces minimum deviation of 50°.

CHAPTER 43

INTERFERENCE AND DIFFRACTION

When two or more waves travel through the same medium, each wave proceeds as though the others were not present. In this sense there is no interference. However, at each point in the medium the disturbance produced is the resultant of all the disturbances that come to the point. This phenomenon is called *interference*. With light waves interference phenomena are observed only when light comes to the observer from a single source by different paths.

According to Huygens' principle, every point on an advancing wave front may be regarded as a new source of disturbance, sending out wavelets. The envelope of these wavelets constitutes a new wave front. If a part of the wave front is cut off by an obstacle the wavelets near the edge of the obstacle advance into the region behind the obstacle. This bending of light around an obstacle is called *diffraction*. Diffraction phenomena result from interference between wavelets from different parts of a common wave front.

A *diffraction grating* is made by ruling with a diamond point on glass or metal a large number of fine equally spaced parallel lines a few wavelengths apart. The space between the rulings will reflect light in the metallic grating or transmit it in the glass grating. The ruled spaces diffusely reflect light so that there is no regularity to this portion of the beam. Replicas of gratings may be cheaply made by taking a cast of the grating in collodion.

Consider a transmission grating, Fig. 43.1, illuminated by plane waves of monochromatic light. Each point of the transparent part of the grating acts as a

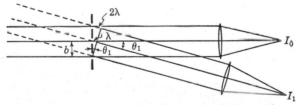

Fig. 43.1   Diffraction of light by a transmission grating

center of wavelets. In only certain directions will the wavelets from successive slits be in phase. In the direct forward directions the wavelets all combine to form a new wave front and a lens forms an image I_0. On either side of this central image there is a direction in which the wavelet from the second slit is one wave behind the wavelet from the first slit; the wavelet from the third is one wave behind that from the second and so on. At this angle θ_1, there will be reinforcement and the lens forms an image I_1. For this angle

$$\lambda = b \sin \theta_1 \tag{1}$$

where b is the distance between successive openings and is called the *grating space*.

311

At a second angle θ_2 for which the successive wavelets are two waves behind there will again be reinforcement and an image.   This is the second-order image. Again where the path difference is three or four or five wavelengths there will be reinforcement, each at a different angle.   In general

$$n\lambda = b \sin \theta_n \tag{2}$$

where n is the order of the spectrum.   When $n = 1$, Eq. (2) reduces to Eq. (1).

A diffraction grating may be used to determine the wavelength of light.   If the grating space b is known and the angle θ_n can be measured, the wavelength can be determined directly.

If the source, instead of being monochromatic, emits light of several different frequencies, the central image I_0 will be a single image, but the diffracted images I_1, I_2, I_3, etc., will consist of a group of lines, one for each frequency present. Each line is an image of the slit.

Experiment 43.1

WAVELENGTH BY SIMPLE DIFFRACTION GRATING

Object: To study the use of the diffraction grating in the formation of spectra and in measurement of wavelengths.

Method: An illuminated slit is viewed through a diffraction grating.   Measurements are made of the angles between the direct beam and the diffracted images.   From these angles and the known grating constant the wavelength of the light is determined.

Apparatus: Coarse diffraction grating, between 500 and 1000 lines/in.; replica diffraction grating; 2-meterstick; meterstick; sodium lamp (or sodium flame); slit; meterstick clamp; two tripod stands; two caliper jaws; electric lamp.

Procedure: 1. Mount the 2-meterstick in a horizontal position by means of a meterstick clamp and a tripod stand.   Mount the slit just above the middle of the 2-meterstick and place the sodium source behind the slit.   Set up the coarse grating with its plane parallel to the meterstick and on the line from the slit perpendicular to the meterstick (Fig. 43.2) about 50 cm from the slit.   Have the rulings on the grating in a vertical direction.

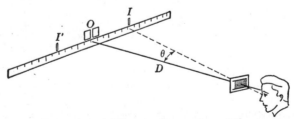

Fig. 43.2   Formation of diffraction images by a diffraction grating

2. Place the eye near the grating and view the central slit and the diffracted images of the slit on both sides.   Count the number of orders of images that can be seen.   What factors limit the number of orders that are visible?

3. Replace the coarse grating by the replica grating.   How many orders are now visible?   What other difference is observed between this case and that of

Step 2? Set the two caliper jaws at the positions of the two first-order diffracted images (*I* and *I'*, Fig. 43.2). Record the positions of the caliper jaws. Repeat these measurements for the second-order images and other orders if they are visible.

4. Measure the distance *D* (Fig. 43.2) from the scale to the grating.

5. Record the grating constant marked on the grating or given by your instructor.

6. Repeat Steps 3 and 4, using a distance *D* of 100 cm.

7. Replace the sodium source by a small electric lamp so shielded that light comes only through the slit. Observe the spectrum produced. Which color is diffracted the most and which the least? As in Step 3 measure the positions of the images for each end of the spectrum.

Computations and Analysis: 1. From the data of Step 3 compute the distance *II'* for each order. Calculate the average deviation of each image ($\frac{1}{2} II'$). Compute $\tan \theta = \frac{1}{2} II'/D$ for each order and use the tables in the Appendix to find $\sin \theta$. Compute the wavelength by use of Eq. (2).

2. Repeat Part 1 of the Analysis, using the data of Step 6. Determine the average of all the experimental values of the wavelength and compare this average value with the standard value, 5893 A.

3. From the data of Step 7 calculate the wavelength limits of the visible spectrum.

Experiment 43.2

DIFFRACTION GRATING SPECTROMETER

Object: To study the use of the diffraction grating for the formation and measurement of spectra.

Method: An illuminated slit is viewed through a diffraction grating. Measurements are made of the angles that the diffracted beams make with the direct beam. From these angles, using a known wavelength, the grating constant is determined. Several unknown wavelengths are determined by use of the grating and the measured diffraction angles.

Apparatus: Replica diffraction grating, Fig. 43.3; spectrometer; sodium lamp (or sodium flame); discharge tube of mercury, helium, or hydrogen; high-voltage transformer or induction coil, arranged with terminals for mounting discharge tubes; meterstick, with two caliper jaws; meterstick clamp; tripod support rod; electric lamp and cord.

Procedure: I. *Meterstick Grating Spectroscope.* 1. Arrange the apparatus as shown in Fig. 43.2. Mount a meterstick horizontally so that the 50-cm mark is immediately in front of and touching the center of the slit on the sodium-arc source. Place the replica grating, with its lines vertical, in the holder on the spec-

Fig. 43.3 Replica diffraction grating

trometer prism table. Have the grating about 50 cm distant from and directly on a line through the slit, perpendicular to the meterstick. Arrange the plane of the grating parallel to the meterstick.

2. Place the eye near the grating and view the central slit and the diffracted images of the slit on both sides. Both first- and second-order images should be seen. Here the lens and retina of the eye are taking the place of the lens and screen in Fig. 43.1. Move the caliper jaws until they coincide with the centers of the first-order diffracted images I and I'. If the diffracted images are not symmetrical with respect to the slit, the grating should be rotated slightly around a vertical axis. Record the positions of the caliper jaws.

3. Measure the distance D from the slit to the grating. The tangent of the angle θ is the ratio of $\frac{1}{2} II'$ to D. Calculate this ratio and find $\sin \theta$ from the tables in the Appendix.

4. Assume as correct the grating constant marked on the grating (or given by your instructor). Compute the wavelength of sodium light.

II. *Grating Spectrometer*. 5. Adjust the spectrometer as described in Steps 1 to 4 of Exp. 42.1.

6. Illuminate the slit by means of the sodium lamp. With the slit fairly wide turn the telescope until the intersection of the cross hairs is at one edge of the slit and read the position of the telescope. Mount the grating in the grating holder on the prism table with the rulings vertical. Turn the prism table until the plane of the grating is perpendicular to the rays from the collimator. To be sure that the grating is in this position note roughly the angular positons of the first-order diffracted images on both sides of the slit. Adjust grating until these angles are approximately equal; then lock prism table in position. After adjustments are complete do not change width of slit until all measurements are completed.

7. Set the telescope successively with the intersection of the cross hairs at the edge of the first- and second-order images on both sides of the central image and record the angular positions.

8. Replace the sodium lamp with an electric discharge tube containing mercury, helium, or hydrogen. Select one of the strong lines near the short-wave end of the spectrum and record the angular positions of the first-order images on both sides of the central image. Repeat observations for each of the other strong lines.

9. With the sodium lamp at the slit, set the telescope at the second-order image. Make the slit very narrow and observe that there are two lines. The wavelengths of the two lines are 5890 A and 5896 A.

Computations and Analysis: 1. From the data of Step 7 calculate the angular separation of the two first-order diffracted images and that of the two second-order images. These two angular separations are, respectively, $2\theta_1$, and $2\theta_2$. The standard wavelength of sodium light is 5893 A. Use this value and the measured angles θ_1 and θ_2 in Eq. (2) to determine the grating constant b. Compute also the number N of lines per *inch* for the grating.

2. For each of the lines measured in Step 8 determine the wavelength by use of Eq. (2) and the average value of b found in Part 1 of the Analysis. Compare the observed wavelengths with the standard values given in the tables in the Appendix.

Review Questions: 1. What is meant by interference of light waves? 2. What is Huygens' principle? 3. What is diffraction? How is it related to the Huygens' wavelets? 4. Describe the diffraction grating and explain how it produces a diffraction pattern. 5. Derive the working equation of a diffraction grating. What is the grating constant? 6. Describe the procedures needed for the measurement of wavelengths by means of a grating.

Questions and Problems: 1. When the eye is placed directly behind the grating, what takes the place of the lens of Fig. 43.1?

2. Will the angular separation of orders be greater for a grating with many lines per inch or for a grating with fewer lines?

3. Will the angular separation between red and blue rays be greater in the first-order or in the second-order spectrum?

4. Will the image of the slit be more sharply defined in the first-order spectrum or in the third-order spectrum?

5. Contrast the formation of a spectrum by a diffraction grating with the production of the spectrum of the same light source by a glass prism.

6. Imagine a light made up of wavelengths spaced at 100 A intervals. Make a rough sketch of the appearance of the spectrum viewed in a prism spectroscope and in a grating spectroscope.

7. What factors govern the choice between a prism and a diffraction grating for use in a spectrograph?

8. A student places a small sodium lamp just in front of a blackboard. Standing 20.0 ft away he views the light at right angles to the blackboard while holding in front of his eye a transmission grating ruled with 14,500 lines/in. He has his assistant mark on the board the positions of the first-order diffracted images on each side of the lamp. The distance between these marks is found to be 14 ft 2 in. Compute the wavelength of the light.

9. A monochromatic beam of light of wavelength 6000 A falling on a grating at normal incidence gives a first-order image at an angle of 30°0′. Find the grating constant.

10. The deviation of the second-order diffracted image formed by an optical grating of 6000 lines/cm is 30°. Calculate the wavelength of the light used.

11. A yellow line and a blue line of the mercury-arc spectrum have wavelengths of 5791 and 4358 A, respectively. In the spectrum formed by a grating that has 5000 lines/in., compute the separation of the two lines in the first-order spectrum and in the third-order spectrum. Compare the angle of diffraction for the yellow line in the third order with that for the blue line in the fourth order.

12. A grating with 15,000 lines to the inch is mounted on a spectrometer and used with a monochromatic source of wavelength 3900 A. What is the angle through which the telescope must be turned in moving from the first- to the second-order image?

13. What number of lines per centimeter would produce a distance of 50 cm between the two third-order images (Fig. 43.2) if the grating were 3.0 m from the sodium-flame-illuminated slit?

14. With the grating used in Exp. 43.2, what would be the angle through which the telescope would have to be turned when it is moved from the first to the second of the two sodium lines? (Not the first and second *orders*.)

15. Describe a concave reflection grating and its use. What is the chief advantage over a transmission grating?

CHAPTER 44

SPECTROSCOPY

Sir Isaac Newton discovered in 1666 the significant fact that a beam of sunlight passed through a prism is spread out into a number of constituent portions, now designated a *spectrum*, Fig. 44.1. This phenomenon of dispersion is a result of

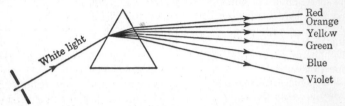

FIG. 44.1 Dispersion of white light by a prism

the fact that the index of refraction for the various wavelengths is different, so that they are variously refracted by the prism. These effects are widely used to analyze light from various sources by the art known as *spectroscopy*. Spectrum analysis provides an important means of identifying materials and of studying their atomic structure.

In order to produce a pure spectrum in which there is no large overlapping of wavelengths, a very narrow slit must be placed in front of the source and an image of the slit formed by a lens system, as shown in Fig. 44.2.

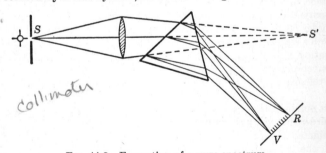

FIG. 44.2 Formation of a pure spectrum

In the prismatic spectrum, the angular separation of the colors is not uniform throughout the spectrum. Owing to the fact that the dispersive power is greatest in the short wavelength region, the angular separation of colors is greater at the violet end than at the red end of the visible spectrum. The prismatic spectrum departs, therefore, from the *normal* spectrum produced by a diffraction grating, in which the sine of the angular deviation is directly proportional to the wavelength. Also, the color sequence in the prismatic spectrum is the reverse of that in the grating spectrum.

An instrument employing a prism for the qualitative comparison of spectra is called a *prism spectroscope;* one which makes a photographic record of the spectrum is called a *spectrograph;* and an instrument provided with means for making quantitative measurements is a *spectrometer.*

Types of Spectra. There are three types of spectra which may be emitted by bodies, depending upon the state of the matter exciting the light.

1. *Continuous Spectrum.* Light from an *incandescent* solid or liquid is found to contain continuous gradations of color, merging with imperceptible shadings from red, on one extreme, to violet, on the other. No regions of darkness are observed. Such a spectrum is emitted from the filament of an electric lamp, for example. Different substances emit similar continuous spectra, and this type is, therefore, not so useful for spectroscopic analysis as the two other types.

2. *Bright-line Spectrum.* This type of spectrum, sometimes called the *characteristic spectrum,* is truly characteristic of the emitting substance and the means used for its excitation. It is obtained when the source of light is in the gaseous or vapor phase and has been rendered incandescent by heating or by an electric discharge. Such spectra are characterized by bright "lines" corresponding to light of definite colors, with dark spaces, or absence of light, between the lines. For example, NaCl gives only the familiar yellow line, with an instrument of low resolution. Lithium gives a single red line. Hydrogen, in an electric discharge tube, furnishes a spectrum consisting of an intense red line, a blue and two violet lines, as well as a number of fainter lines in the visible and a larger number of invisible frequencies in the ultraviolet and infrared regions. Iron has a most abundant spectrum, with thousands of lines; this feature makes the iron spectrum particularly valuable as reference lines.

3. *Absorption Spectrum.* When light from a source furnishing a continuous spectrum is passed through some cooler absorbing medium, the resulting spectrum is found to be crossed by dark spaces (lines), known as the absorption spectrum of the absorbing material. For example, the *Fraunhofer lines* in the solar spectrum are produced by absorption of light in the continuous spectrum emitted by the incandescent material in the interior of the sun. Since the vapors and gases constituting the atmosphere of the sun are at a much lower temperature (5000°C) than the interior of the sun (20,000,000°C), they absorb much of the light which they themselves emit if incandescent. It is by examining such spectra that we are able to identify the substances present in stars, though they are thousands of light-years distant from us.

A representation of a few characteristic spectra is shown in Fig. 44.3.

Experiment 44.1

A STUDY OF SPECTRA

Object: To study simple prismatic spectra, to calibrate a prism spectrometer, and to use the calibrated spectrometer to measure several unknown wavelengths.

Method: A prism spectrometer is adjusted for minimum deviation of the D lines in the sodium spectrum. Various sources containing lines of known wavelength throughout the visible spectrum are used, and the angular position of the telescope is observed for each known wavelength. A calibration curve is plotted showing the relationship between the wavelength and the spectrometer setting. The line spectrum of an unknown source is produced, and the wavelengths of the prominent lines are determined from the calibration curve. By comparison of the experimentally determined wavelengths with values given in wavelength tables, the source is identified.

Apparatus: Prism spectrometer, Fig. 42.2 and 42.4; sodium arc; high-voltage transformer (or spark coil), with terminals for discharge tubes; mercury and

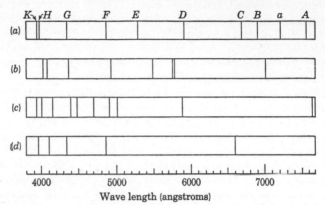

FIG. 44.3   Characteristic spectra: (a) Fraunhofer lines in solar spectrum; (b) mercury arc; (c) helium discharge tube; (d) hydrogen discharge tube

FIG. 44.4   Grating spectroscope

helium discharge tubes; discharge tubes or other sources of "unknown" materials; grating spectroscope.

Reference should be made to Exp. 42.1 for the description of the prism spectrometer.

The grating spectroscope shown in Fig. 44.4 consists of a replica diffraction grating through which light passes into the eye of the observer from a narrow slit. When the slit is illuminated by the light that is to be examined, the various lines appear superimposed on an empirically ruled wavelength scale. The scale is figured to read in centimeters $\times 10^{-5}$; the readings may be reduced to angstrom units by multiplying the main-scale numbers by 10^3.

Procedure: 1. Make the preliminary adjustments of the prism spectrometer, as described in the Procedure of Exp. 42.1. Adjust the spectrometer for minimum deviation for the sodium line, as indicated in Step 6 of Exp. 42.1. *In all of the subsequent observations the graduated platform and the prism table are to remain fixed, and only the telescope is to be moved.*

2. Calibrate the spectrometer, as follows. Use the mercury arc as a source. Set the cross hairs of the telescope on each prominent line and record the corresponding vernier readings. Look up the wavelengths of the mercury lines in the table in the Appendix. Using the known wavelengths in the spectra of sodium and mercury, plot a calibration curve showing the relationship between wavelength and spectrometer setting.

If there are any regions of uncertainty in drawing the curve, try to fill in the gap by the use of suitable spectra from other known sources.

It may be helpful to identify the lines in the known spectra by the aid of the calibrated diffraction spectroscope. View the source simultaneously with another observer using the prism spectrometer.

3. Replace the known source by an unknown source, as directed by the instructor. Locate a number of strong lines and measure and record the vernier settings for them. Determine the wavelengths corresponding to these settings, by using the calibration curve. Attempt to identify the source by comparing the measured wavelengths with those given in tables.

4. If sunlight is available, view the solar spectrum with the prism spectrometer. What type of spectrum is obtained? Try to narrow the slit until the Fraunhofer lines become visible; if this is possible, sketch the spectrum observed.

Review Questions: 1. What is dispersion? 2. What is a spectrum? How can a spectrum be produced? 3. What is meant by spectrum analysis? Why is it so important? 4. How does a prismatic spectrum differ from that produced by a diffraction grating? 5. Explain the difference between a spectroscope, a spectrometer, and a spectrograph. 6. Describe the three types of spectra, giving the types of source of each. Give an example of each type of spectrum. 7. State and describe the steps that are necessary in the calibration of a prism spectrometer. 8. Explain the procedure for the identification of an unknown material by the use of a calibrated spectrometer.

Questions and Problems: 1. What type of spectrum is produced by each of the following sources: neon sign; carbon-filament lamp; tungsten-filament lamp; spark between iron electrodes; molten iron; vapor from a sodium flame in front of a carbon-arc lamp?

2. Show how the prismatic spectrum of sunlight demonstrates the fact that the speed of red light in glass is greater than that of blue light.

3. State the correlation which exists between the emission spectrum of sodium and part *a* of Fig. 44.3.

4. A dispersion curve is frequently drawn for a prism. Such a curve shows the variation of the index of refraction with wavelength. What correlation exists between such a curve and the calibration curve obtained in this experiment?

5. What modifications of experimental procedures are necessary when a spectrometer must be used in the infrared and ultraviolet portions of the spectrum?

6. What is the frequency of the light from a sodium flame? How many such waves are there in one centimeter?

7. Light from a sodium source enters a prism of dense flint glass, $n = 1.65$. What is the frequency of the waves in the glass? the speed of the light? the wavelength?

8. The spectra of two stars are compared and it is observed that the spectrum of one star shows certain Fraunhofer lines but these lines are not present in the spectrum of the other star. Can these lines be caused by absorption in the earth's atmosphere? Suggest a method for distinguishing between absorption in the earth's atmosphere and that in the atmosphere of a star.

9. How would the spectrum of sunlight as observed on the moon differ from that ordinarily observed on the earth?

10. A spectroscopist observes the D absorption line of sodium from two stars. For one star the wavelength is found to be 5875 A, and for the other star, 5925 A. (These differences are not caused by experimental error.) What conclusions concerning the motions of the stars can be drawn from these data?

11. Explain how one might study by spectroscopic methods the rotation of two stars around a common center.

12. Mention some advantages which a spectrograph has over a spectrometer. What are some of its disadvantages?

CHAPTER 45

POLARIZED LIGHT

Light waves comprise a small part of the whole spectrum of electromagnetic waves, which includes radio waves, heat waves, ultraviolet waves, x rays, and γ rays. Electromagnetic waves consist of transverse vibrations of electric and magnetic fields. Since it is the electric field that produces the effects with which we are concerned here, we shall confine our attention to that factor.

In a transverse wave the vibrations are at right angles to the direction of propagation but the vibrations may be in any plane about that direction, Fig. 45.1.

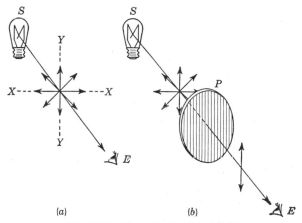

(a)   (b)

Fɪɢ. 45.1   The polarization of light

When the vibrations are uniformly distributed in all planes, the wave is *unpolarized*, Fig. 45.1a. If the vibrations are confined to a single path, the wave is *polarized*. If this path is confined to a single plane, the wave is said to be *plane-polarized*, Fig. 45.1b. For purposes of analysis it is often convenient to consider the transverse vibrations as resolved into two mutually perpendicular components along significant directions, such as XX and YY of Fig. 45.1a. In ordinary, unpolarized light the light energy is uniformly distributed among the transverse components in all possible directions; hence the two mutually perpendicular components are equal. If one component is greater than the other, the light is partly polarized; if one component is missing, the light is completely plane-polarized.

The eye is unable to distinguish between polarized light and unpolarized light. In order to detect polarized light it is necessary to use an *analyzer*, that is, some device that will transmit only those vibrations that are along a certain direction, the transmission axis, of the instrument.

An ordinary source of light emits unpolarized light. The light may be polarized by any one of a number of processes. Some of them are: reflection from a

321

nonmetallic surface; transmission through many transparent, nonmetallic plates; double refraction and absorption; and scattering by small particles.

When light is reflected from any nonmetallic surface, we may think of the components of the electric field parallel to the surface (represented by small circles in Fig. 45.2) and those perpendicular to this direction (represented by arrows). At any angle of incidence more of the parallel component is reflected and the reflected

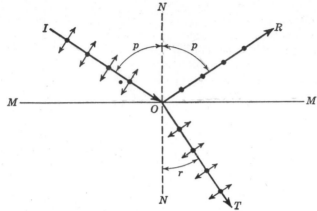

Fig. 45.2   Polarization by reflection. The double-headed arrows represent vibrations parallel to the plane of the paper; the dots represent vibrations perpendicular to the plane of the paper and hence parallel to the reflecting surface

beam is partly polarized. For each substance there is an angle of incidence p, called the *polarizing angle*, for which only the parallel component is reflected. The polarizing angle is related to the index of refraction n of the reflector by the equation

$$\tan p = n \tag{1}$$

At this angle of incidence the reflected beam is completely plane-polarized. The transmitted beam is partly polarized in a plane at right angles to the plane of the reflected beam. If a pile of transparent plates is used in place of one plate, more and more of the parallel component is removed at each successive reflection and finally the transmitted component is almost completely plane-polarized.

Many crystals have the property of breaking up a beam of light into two components, each of which is plane-polarized; the two planes of polarization are at right angles to each other. These two beams travel at different speeds through the crystal and hence are differently refracted. One beam obeys the regular laws of refraction and is called the *ordinary* ray; the other does not obey these laws and is called the *extraordinary* ray. When one looks through a doubly refracting crystal at a small object, he sees two images. One is formed by the ordinary ray, the other by the extraordinary ray.

Some crystals absorb one of the two polarized beams and transmit the other. One such material is quinine iodosulphate (herapathite). Many small crystals of this material embedded in a transparent cellulose film make a Polaroid sheet that may be used to produce plane-polarized light.

Another method of eliminating one of the rays is used in the Nicol prism, Fig. 45.3. The index of refraction for the ordinary ray is greater in calcite than in Canada balsam; the index of refraction for the extraordinary ray is greater in Canada balsam than in calcite. Hence the crystal is cut in a special manner, the surfaces are polished, and then cemented together with Canada balsam. The cut is made at such an angle that the ordinary ray is incident on the Canada

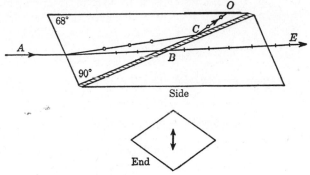

FIG. 45.3   A Nicol prism

balsam at greater than the critical angle and thus is totally reflected. The transmitted light is completely polarized.

Any of these polarizers may be used as an analyzer in studying polarized light.

Certain substances have the property of rotating the plane of polarization of plane-polarized light. These materials are said to be *optically active*. The rotation of the plane of polarization by sugar solutions is used commercially to determine the nature and degree of purity of various sugars.

When light passes through a doubly refracting material, one beam travels more slowly through the material than does the other. When the two recombine on emergence, they are out of phase. If an analyzer is inserted into the beam, the waves show interference effects. Some materials that are not normally doubly refracting become so when under stress. The interference effects may be used to study stresses in such transparent materials.

Experiment 45.1

PLANE-POLARIZED LIGHT

Object: To study phenomena of polarized light; in particular, methods of producing plane-polarized light, rotation of the plane of polarization by an optically active material, and interference effects.

Method: Plane-polarized light is produced by reflection, transmission by a pile of glass plates, a Polaroid disk, a Nicol prism, and by scattering. A Polaroid disk is used as an analyzer. Rotation of the plane of polarization by a sugar solution is observed. Interference effects in thin sheets and in stressed materials are examined.

Apparatus: Glass mirror; pile of glass plates; pair of Polaroid disks, Fig. 45.4; Nicol prism; calcite crystal, at least $\frac{1}{2}$ in. thick; tube of sugar solution; Cellophane sheets; sodium lamp, or sodium flame; electric lamp; protractor; screen with slit; sheet of mica; U-shaped specimen of transparent celluloid; optical bench or stands to hold optical parts.

Procedure: 1. View a source of light through one Polaroid disk; note the effect on the transmitted light caused by rotating the disk through 360°.   Set up the

slit illuminated by the sodium lamp or sodium flame.   Mount the glass mirror with its axis of rotation parallel to the slit.   Observe the reflected image of the slit through one of the Polaroid disks. The plane of transmission of the disk is marked by two scratches at opposite ends of a diameter.   Rotate the Polaroid disk. Is there a change of brightness of the image?   Is the brightness a maximum when the line of the scratches is parallel to the surface of the glass?   Where is it a minimum?   Hold the Polaroid disk in the

FIG. 45.4   Polaroid disks

minimum position and rotate the mirror.   Find the position of the mirror for which the reflected image can be extinguished by rotating the Polaroid disk. The angle of incidence is then the polarizing angle p.   Measure the polarizing angle roughly by means of a protractor.   By use of Eq. (1) calculate the index of refraction of the glass.

2. Replace the mirror by a pile of plates.   Adjust the position of the plates to the polarizing angle as in Step 1.   Observe the orientation of the Polaroid disk for which the brightness of the reflected image is a minimum.   Is the reflected beam completely plane-polarized?   Observe the transmitted beam through the Polaroid disk.   Rotate the disk to find the positions of maximum and minimum brightness.   Compare these orientations with the corresponding orientations for the reflected beam.   Is the transmitted beam completely plane-polarized?   Compare the planes of polarization of the reflected and transmitted beams.

3. Mount one of the Polaroid disks in front of the slit.   Rotate the disk.   Do you observe any change in intensity?   Set the disk with its axis of transmission vertical.   Observe the slit and Polaroid disk through the second Polaroid disk. Rotate the second Polaroid disk and note the change in intensity of the light. What do these observations indicate about the effect of polarized light on the eye? Note the position of the second disk for minimum and maximum transmission. What is the relation between the planes of transmission of the two disks for minimum and for maximum transmission?

4. Repeat Step 3 with the Nicol prism in place of the first Polaroid disk.

5. Make a small dot on a piece of paper and set the calcite crystal over it.   Do you observe two images?   Rotate the crystal about your line of sight and observe the motion of the images.   Do they both move?   Which image is formed by the ordinary ray and which by the extraordinary?   Observe these images through a Polaroid disk.   Rotate the disk until first one and then the other image disappears.   Explain the disappearances.   Observe and explain the relative orientation of the Polaroid disk for each disappearance.   at 90°

6. Mount the tube of sugar solution so that the light from the electric lamp

travels along its axis. Observe through a Polaroid disk the light scattered at right angles to the tube. Is there a change in intensity as the disk is rotated? Explain. What is the plane of vibration of the polarized component of the scattered light?

7. Set the tube of sugar solution between two Polaroid disks. Place the sodium lamp so that light passes through the disks and tube. Remove the tube and set the analyzing disk for extinction. Replace the sugar solution and observe the light. Rotate the analyzer until there is again extinction of the light. The angle through which the disk is rotated in this process is the angle by which the sugar solution rotated the plane of polarization.

8. Place the two Polaroid disks in line in front of the sodium lamp. Rotate the second disk and note the two positions for extinction. Insert a sheet of mica between the two disks and again rotate the analyzer. Is there any position for extinction? Explain. Rotate the mica sheet and repeat the observations. What change if any is produced? CIRCULAR POLARIZATION

9. With the disks in place as in Step 8 set the analyzer for extinction. Insert a Cellophane bag or several overlapping sheets of Cellophane. Describe and explain the appearance. Rotate the analyzer; describe and explain the result.

Replace the Cellophane by pieces of glass or transparent celluloid. Are these materials uniformly bright? Explain nonuniformities.

Place the U-shaped celluloid between the Polaroid disks. Gently press the ends of the U toward each other. Describe and explain the results.

10. Summarize the observations and conclusions reached in the various steps of the experiment.

Review Questions: 1. What is a light wave? 2. What is meant by a polarized wave? a plane-polarized wave? 3. Describe various ways in which plane-polarized light may be produced. 4. Why must an analyzer be used to study polarized light? 5. In the process of polarization by reflection what is the relationship between the angle of polarization and the index of refraction of the reflector? 6. Can a polished sheet of aluminum be used to produce polarization by reflection? 7. What is double refraction? Ordinary ray? Extraordinary ray? 8. What is meant by an optically active substance? 9. What equipment must one have to observe interference in polarized light?

Questions and Problems: 1. When light is polarized by reflection from the surface of a pool of water, what is the direction of vibration of the reflected light?
2. What is the polarizing angle for water?
3. What is the direction of the transmission axis of Polaroid sunglasses? Explain.
4. Is glass ever doubly refracting? Explain.
5. A certain glass has an index of refraction of 1.65. What is the polarizing angle?
6. In order that a beam of light from air, striking a certain surface, may be plane polarized it must have an angle of incidence of 53.18°. What is the index of refraction of the substance?
7. The phenomena of polarization afford experimental evidence that light waves are transverse. Give the reasoning that supports this statement.
8. Briefly describe some useful applications of polarized light.
9. Explain the use of Polaroid glasses in viewing 3-D movies.

APPENDIX

I. Adjustment of a Mercury Barometer

The pressure of the atmosphere may be most accurately determined by the mercurial barometer, illustrated in Fig. A.1. The glass tube G, closed at its upper

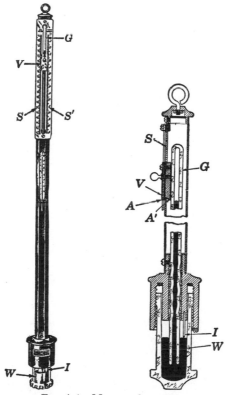

Fig. A.1 Mercury barometer

end, projects into the mercury cistern W. The metric scale S and the British scale S' have their zeros at the lower end of the ivory tip I. Consequently, it is necessary that the upper level of the mercury in the cistern always be brought to this point before a reading is taken. This is done by adjusting a screw at the bottom of the barometer. A vernier reading device V makes possible accurate settings and readings at the top of the mercury column.

The following procedure is used in making a measurement of atmospheric pressure by means of a mercurial barometer:

1. Adjust the mercury level to the tip of the ivory point by turning the well upward with great care until the tip of the ivory point just touches the mercury surface. When the mercury is clean, this contact may be judged by noting appar-

tact between the tip of the point and its image in the mercury. When the
y has become contaminated, the mercury level is adjusted upward until
the point has made the slightest depression discernible in the mercury surface.
Another way of making the setting is to hold a piece of white paper behind the
well as a background, holding the eye at the level of the ivory point and adjusting
the mercury surface until the light between the point and the mercury is just cut
off.

2. Move the vernier scale up until the top of the mercury column appears below
the sighting edges A and A'. Tap the barometer lightly to permit the mercury
to form a free meniscus. Then move the vernier scale downward until the sighting edges A and A' are in line with the uppermost point of the meniscus.

3. Read the scale in millimeters and determine tenths with the aid of the vernier
scale.

II. Method of Focusing a Reading Telescope

The reading telescope, Fig. A.2, has three essential parts: an objective lens O
that forms an image of the object being viewed, a set of cross hairs X, and an eyepiece E by which the image and cross hairs are viewed. The problem of adjusting

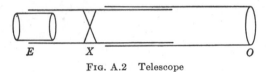

Fig. A.2   Telescope

the telescope is to adjust the image and cross hairs to the same place and to place
the eyepiece so that cross hairs and image are viewed simultaneously.

There are two independent adjustments of a telescope: the eyepiece can be
moved in the barrel with respect to the cross hairs, and the whole barrel can be
moved to focus on the image.

The steps in focusing the telescope are:

1. Move the eyepiece in or out until the cross hairs are most distinct. Maintain this position of the eyepiece throughout the remaining adjustment.

2. Move the barrel containing eyepiece and cross hairs until the cross hairs
coincide with the image formed by the objective. In this adjustment, point the
telescope at the object to be viewed and move the barrel until the image is distinct. Then make the final adjustment by eliminating parallax between the
image and the cross hairs.

Parallax is the apparent motion of two objects, with respect to each other, that
is seen when the *eye* is moved. Observe this effect by holding your two index
fingers at different distances from your eye, then move your eye from side to side.
Move one or both fingers until the apparent motion (parallax) disappears. This
condition is found only when the two fingers are at the same distance from the
eye. The elimination of parallax is a delicate test for the coincidence of two
objects.

When the image seen in the telescope is distinct, move the eye from side to side
(or up and down) slightly. If the cross hairs appear to move across the image,
adjust the barrel in or out and again test for parallax by moving the eye slightly.
Continue this process until parallax has been eliminated. The telescope is then
properly focused.

III. CONVENTIONAL SYMBOLS USED IN DIAGRAMS OF ELECTRIC CIRCUITS

Ammeter	
Galvanometer	
Voltmeter	
Cell	
Battery	
Resistor	
Variable resistor (rheostat)	
Inductor	
Capacitor	
Wires crossing but not connected	
Single pole single throw switch or tap key	
Double pole single throw switch	
Double pole double throw switch	
Reversing switch	

IV. TABLES

TABLE 1. ELASTIC CONSTANTS

Material	Young's modulus		Modulus of rigidity	
	lb/in²	dynes/cm²	lb/in²	dynes/cm²
Aluminum, rolled..........	10×10^6	7.0×10^{11}	3.4×10^6	2.4×10^{11}
Brass, yellow.............	13	9.0	5.1	3.5
Copper, rolled.............	18.6	12.8	6.1	4.2
Glass....................	10	7.0	3.5	2.4
Iron, wrought.............	27.5	19.3		
Lead....................	2.2	1.5	0.78	0.54
Steel, drawn.............	28	19	11.8	8.11

TABLE 2a. DENSITIES AND WEIGHT-DENSITIES OF SOLIDS AND LIQUIDS

Substance	Density gm/cm³	Weight-density lb/ft³
Alcohol (ethyl) at 20°C............	0.79	49.4
Water at 4°C....................	1.000	62.4
Water at 20°C...................	0.998	62.3
Gasoline........................	0.68	42
Mercury........................	13.6	850
Oak............................	0.8	50
Aluminum......................	2.7	169
Copper.........................	8.89	555
Ice............................	0.92	57
Iron, wrought...................	7.85	490

TABLE 2b. DENSITY OF WATER AT VARIOUS TEMPERATURES

Temperature °C	Density gm/cm³	Temperature °C	Density gm/cm³
0	0.999841	16	0.998943
1	900	17	774
2	941	18	595
3	965	19	405
4	973	20	203
5	965	21	0.997992
6	941	22	770
7	902	23	538
8	849	24	296
9	781	25	044
10	700	26	0.996783
11	605	27	512
12	498	28	232
13	377	29	0.995944
14	244	30	646
15	099		

TABLE 3. THERMAL CONSTANTS

Substance	Specific heat cal/gm°C or Btu/lb°F	Coefficient of linear expansion per °C
Alcohol (ethyl)..........	0.65	
Aluminum..............	0.22	2.4×10^{-5}
Brass.................	0.09	1.9
Copper................	0.093	1.7
Glass, crown...........	0.16	0.8
Iron..................	0.11	1.2
Lead..................	0.031	2.9
Mercury..............	0.033	
Steel.................	0.118	1.1
Zinc.................	0.092	2.6

TABLE 4. PRESSURE AND DENSITY OF SATURATED WATER VAPOR

t	P	d	t	P	d
°C	mm Hg	gm/liter	°C	mm Hg	gm/liter
−10	2.0	2.2×10^{-3}	80.0	355.1	293.8×10^{-3}
− 9	2.1	2.4	85.0	433.5	354.1
− 8	2.3	2.6	90.0	525.8	424.1
− 7	2.6	2.8	91.0	546.1	439.5
− 6	2.8	3.0	92.0	567.1	455.2
− 5	3.0	3.3	93.0	588.7	471.3
− 4	3.3	3.5	94.0	611.0	487.8
− 3	3.6	3.8	95.0	634.0	505
− 2	3.9	4.1	96.0	657.7	523
− 1	4.2	4.5	96.5	669.8	
0	4.6	4.9	97.0	682.1	541
1	4.9	5.2	97.5	694.5	
2	5.3	5.6	98.0	707.3	560
3	5.7	5.9	98.2	712.5	
4	6.1	6.4	98.4	717.6	
5	6.5	6.8	98.6	722.8	
6	7.0	7.3	98.8	728.0	
7	7.5	7.8	99.0	733.3	579
8	8.0	8.3	99.2	738.6	
9	8.6	8.8	99.4	743.9	
10	9.2	9.4	99.6	749.3	
11	9.8	10.0	99.8	754.7	
12	10.5	10.7	100.0	760.0	598
13	11.2	11.4	100.2	765.5	
14	12.0	12.1	100.4	770.9	
15	12.8	12.8	100.6	776.4	
16	13.6	13.6	100.8	781.9	
17	14.5	14.5	101	787.5	618
18	15.5	15.4	102	815.9	639
19	16.5	16.3	103	845.1	661
20	17.6	17.3	104	875.1	683
21	18.7	18.3	105	906.1	705
22	19.8	19.4	106	937.9	728
23	21.1	20.6	107	970.6	751
24	22.4	21.8	108	1004.3	776
25	23.8	23.0	109	1038.8	801
26	25.2	24.4	110	1074.5	827
27	26.8	25.8	112	1148.7	880
28	28.4	27.2	114	1227.1	936
29	30.1	28.8	116	1309.8	995
30	31.8	30.4	118	1397.0	1057
35	42.0	39.6	120	1489	1122
40	55.1	51.1	125	1740	1299
45	71.7	65.6	130	2026	1498
50	92.3	83.2	135	2348	1721
55	117.8	104.6	140	2710	1968
60	149.2	130.5	150	3569	2550
65	187.4	161.5	160	4633	3265
70	233.5	198.4	175	6689	4621
75	289.0	242.1	200	11650	7840

TABLE 5. SPEED OF SOUND AT 0°C (32°F) IN VARIOUS MEDIUMS

Medium	ft/sec	m/sec
Air	1,087	331.5
Hydrogen	4,167	1,270
Carbon dioxide	846	258.0
Water	4,757	1,450
Brass	11,500	3,500
Iron	16,730	5,100
Glass	18,050	5,500

TABLE 6. RESISTIVITY AND TEMPERATURE COEFFICIENT OF RESISTANCE

Material	Resistivity at 20°C		Temperature coefficient of resistance, based upon resistance at 0°C per °C
	μohm-cm	ohm-CM/ft	
Copper, commercial	1.72	10.5	0.00393
Silver	1.63	9.85	0.00377
Aluminum	2.83	17.1	0.00393
Iron, annealed	9.5	57.4	0.0052
Tungsten	5.5	33.2	0.0045
German (nickel) silver	20 to 33	121 to 200	0.0004
Manganin	44	266	0.00000
Carbon, arc lamp	6000	......	−0.0003

TABLE 7. ELECTROCHEMICAL DATA

Element	Atomic mass	Valence	Electrochemical equivalent, grams per coulomb
Aluminum	27.1	3	0.0000936
Copper	63.6	2	0.0003294
Copper	63.6	1	0.0006588
Gold	197.2	3	0.0006812
Hydrogen	1.008	1	0.0000105
Iron	55.8	3	0.0001929
Iron	55.8	2	0.0002894
Lead	207.2	2	0.0010736
Silver	107.9	1	0.00111800

TABLE 8. WAVELENGTHS OF PRINCIPAL LINES OF EMISSION SPECTRA

Helium	Hydrogen	Mercury
7065A red	6563A red	5791A yellow
6678 red	4861 blue-green	5770 yellow
5876 yellow	4340 blue	5461 green
5016 green	4102 violet	4916 blue-green
4922 blue-green	3970 violet	4358 blue
4713 blue		4078 violet
4471 blue		4046 violet
4387 blue		
4121 violet		
4026 violet		

Natural Sines

Angle	.0	.1	.2	.3	.4	.5	.6	.7	.8	.9	Complement difference
0°	0.0000	0017	0035	0052	0070	0087	0105	0122	0140	0157	0175	89°
1	0175	0192	0209	0227	0244	0262	0279	0297	0314	0332	0349	88
2	0349	0366	0384	0401	0419	0436	0454	0471	0488	0506	0523	87
3	0523	0541	0558	0576	0593	0610	0628	0645	0663	0680	0698	86
4	0698	0715	0732	0750	0767	0785	0802	0819	0837	0854	0872	85
5	0.0872	0889	0906	0924	0941	0958	0976	0993	1011	1028	1045	84
6	1045	1063	1080	1097	1115	1132	1149	1167	1184	1201	1219	83
7	1219	1236	1253	1271	1288	1305	1323	1340	1357	1374	1392	82
8	1392	1409	1426	1444	1461	1478	1495	1513	1530	1547	1564	81
9	1564	1582	1599	1616	1633	1650	1668	1685	1702	1719	1736	80
10	0.1736	1754	1771	1788	1805	1822	1840	1857	1874	1891	1908	79
11	1908	1925	1942	1959	1977	1994	2011	2028	2045	2062	2079	78
12	2079	2096	2113	2130	2147	2164	2181	2198	2215	2233	2250	77 17
13	2250	2267	2284	2300	2317	2334	2351	2368	2385	2402	2419	76
14	2419	2436	2453	2470	2487	2504	2521	2538	2554	2571	2588	75
15	0.2588	2605	2622	2639	2656	2672	2689	2706	2723	2740	2756	74
16	2756	2773	2790	2807	2823	2840	2857	2874	2890	2907	2924	73
17	2924	2940	2957	2974	2990	3007	3024	3040	3057	3074	3090	72
18	3090	3107	3123	3140	3156	3173	3190	3206	3223	3239	3256	71
19	3256	3273	3289	3305	3322	3338	3355	3371	3387	3404	3420	70
20	0.3420	3437	3453	3469	3486	3502	3518	3535	3551	3567	3584	69
21	3584	3600	3616	3633	3649	3665	3681	3697	3714	3730	3746	68
22	3746	3762	3778	3795	3811	3827	3843	3859	3875	3891	3907	67
23	3907	3923	3939	3955	3971	3987	4003	4019	4035	4051	4067	66 16
24	4067	4083	4099	4115	4131	4147	4163	4179	4195	4210	4226	65
25	0.4226	4242	4258	4274	4289	4305	4321	4337	4352	4368	4384	64
26	4384	4399	4415	4431	4446	4462	4478	4493	4509	4524	4540	63
27	4540	4555	4571	4586	4602	4617	4633	4648	4664	4679	4695	62
28	4695	4710	4726	4741	4756	4772	4787	4802	4818	4833	4848	61
29	4848	4863	4879	4894	4909	4924	4939	4955	4970	4985	5000	60
30	0.5000	5015	5030	5045	5060	5075	5090	5105	5120	5135	5150	59 15
31	5150	5165	5180	5195	5210	5225	5240	5255	5270	5284	5299	58
32	5299	5314	5329	5344	5358	5373	5388	5402	5417	5432	5446	57
33	5446	5461	5476	5490	5505	5519	5534	5548	5563	5577	5592	56
34	5592	5606	5621	5635	5650	5664	5678	5693	5707	5721	5736	55
35	0.5736	5750	5764	5779	5793	5807	5821	5835	5850	5864	5878	54
36	5878	5892	5906	5920	5934	5948	5962	5976	5990	6004	6018	53 14
37	6018	6032	6046	6060	6074	6088	6101	6115	6129	6143	6157	52
38	6157	6170	6184	6198	6211	6225	6239	6252	6266	6280	6293	51
39	6293	6307	6320	6334	6347	6361	6374	6388	6401	6414	6428	50
40	0.6428	6441	6455	6468	6481	6494	6508	6521	6534	6547	6561	49
41	6561	6574	6587	6600	6613	6626	6639	6652	6665	6678	6691	48 13
42	6691	6704	6717	6730	6743	6756	6769	6782	6794	6807	6820	47
43	6820	6833	6845	6858	6871	6884	6896	6909	6921	6934	6947	46
44°	6947	6959	6972	6984	6997	7009	7022	7034	7046	7059	7071	45°
Complement	.9	.8	.7	.6	.5	.4	.3	.2	.1	.0		Angle

Natural Cosines

Natural Sines

	.0	.1	.2	.3	.4	.5	.6	.7	.8	.9	Complement difference
45°	0.7071	7083	7096	7108	7120	7133	7145	7157	7169	7181	7193	44°
46	7193	7206	7218	7230	7242	7254	7266	7278	7290	7302	7314	43 12
47	7314	7325	7337	7349	7361	7373	7385	7396	7408	7420	7431	42
48	7431	7443	7455	7466	7478	7490	7501	7513	7524	7536	7547	41
49	7547	7559	7570	7581	7593	7604	7615	7627	7638	7649	7660	40
50	0.7660	7672	7683	7694	7705	7716	7727	7738	7749	7760	7771	39
51	7771	7782	7793	7804	7815	7826	7837	7848	7859	7869	7880	38 1
52	7880	7891	7902	7912	7923	7934	7944	7955	7965	7976	7986	37
53	7986	7997	8007	8018	8028	8039	8049	8059	8070	8080	8090	36
54	8090	8100	8111	8121	8131	8141	8151	8161	8171	8181	8192	35
55	0.8192	8202	8211	8221	8231	8241	8251	8261	8271	8281	8290	34 1
56	8290	8300	8310	8320	8329	8339	8348	8358	8368	8377	8387	33
57	8387	8396	8406	8415	8425	8434	8443	8453	8462	8471	8480	32
58	8480	8490	8499	8508	8517	8526	8536	8545	8554	8563	8572	31
59	8572	8581	8590	8599	8607	8616	8625	8634	8643	8652	8660	30
60	0.8660	8669	8678	8686	8695	8704	8712	8721	8729	8738	8746	29
61	8746	8755	8763	8771	8780	8788	8796	8805	8813	8821	8829	28
62	8829	8838	8846	8854	8862	8870	8878	8886	8894	8902	8910	27
63	8910	8918	8926	8934	8942	8949	8957	8965	8973	8980	8988	26
64	8988	8996	9003	9011	9018	9026	9033	9041	9048	9056	9063	25
65	0.9063	9070	9078	9085	9092	9100	9107	9114	9121	9128	9135	24
66	9135	9143	9150	9157	9164	9171	9178	9184	9191	9198	9205	23 7
67	9205	9212	9219	9225	9232	9239	9245	9252	9259	9265	9272	22
68	9272	9278	9285	9291	9298	9304	9311	9317	9323	9330	9336	21
69	9336	9342	9348	9354	9361	9367	9373	9379	9385	9391	9397	20 6
70	0.9397	9403	9409	9415	9421	9426	9432	9438	9444	9449	9455	19
71	9455	9461	9466	9472	9478	9483	9489	9494	9500	9505	9511	18
72	9511	9516	9521	9527	9532	9537	9542	9548	9553	9558	9563	17
73	9563	9568	9573	9578	9583	9588	9593	9598	9603	9608	9613	16 5
74	9613	9617	9622	9627	9632	9636	9641	9646	9650	9655	9659	15
75	0.9659	9664	9668	9673	9677	9681	9686	9690	9694	9699	9703	14
76	9703	9707	9711	9715	9720	9724	9728	9732	9736	9740	9744	13 4
77	9744	9748	9751	9755	9759	9763	9767	9770	9774	9778	9781	12
78	9781	9785	9789	9792	9796	9799	9803	9806	9810	9813	9816	11
79	9816	9820	9823	9826	9829	9833	9836	9839	9842	9845	9848	10
80	0.9848	9851	9854	9857	9860	9863	9866	9869	9871	9874	9877	9 3
81	9877	9880	9882	9885	9888	9890	9893	9895	9898	9900	9903	8
82	9903	9905	9907	9910	9912	9914	9917	9919	9921	9923	9925	7
83	9925	9928	9930	9932	9934	9936	9938	9940	9940	9943	9945	6 2
84	9945	9947	9949	9951	9952	9954	9956	9957	9959	9960	9962	5
85	0.9962	9963	9965	9966	9968	9969	9971	9972	9973	9974	9976	4
86	9976	9977	9978	9979	9980	9981	9982	9983	9984	9985	9986	3 1
87	9986	9987	9988	9989	9990	9990	9991	9992	9993	9993	9994	2
88	9994	9995	9995	9996	9996	9997	9997	9997	9998	9998	9998	1
89°	9998	9999	9999	9999	9999	1.0000	1.0000	1.0000	1.0000	1.0000	1.0000	0° 0
Complement	.9	.8	.7	.6	.5	.4	.3	.2	.1	.0		Angle

Natural Cosines

NATURAL TANGENTS

Angle	.0	.1	.2	.3	.4	.5	.6	.7	.8	.9	Complement difference
0°	0.0000	0017	0035	0052	0070	0087	0105	0122	0140	0157	0175	89°
1	0175	0192	0209	0227	0244	0262	0279	0297	0314	0332	0349	88
2	0349	0367	0384	0402	0419	0437	0454	0472	0489	0507	0524	87
3	0524	0542	0559	0577	0594	0612	0629	0647	0664	0682	0699	86
4	0699	0717	0734	0752	0769	0787	0805	0822	0840	0857	0875	85
5	0.0875	0892	0910	0928	0945	0963	0981	0998	1016	1033	1051	84
6	1051	1069	1086	1104	1122	1139	1157	1175	1192	1210	1228	83
7	1228	1246	1263	1281	1299	1317	1334	1352	1370	1388	1405	82
8	1405	1423	1441	1459	1477	1495	1512	1530	1548	1566	1584	81
9	1584	1602	1620	1638	1655	1673	1691	1709	1727	1745	1763	80
10	0.1763	1781	1799	1817	1835	1853	1871	1890	1908	1926	1944	79 18
11	1944	1962	1980	1998	2016	2035	2053	2071	2089	2107	2126	78
12	2126	2144	2162	2180	2199	2217	2235	2254	2272	2290	2309	77
13	2309	2327	2345	2364	2382	2401	2419	2438	2456	2475	2493	76
14	2493	2512	2530	2549	2568	2586	2605	2623	2642	2661	2679	75
15	0.2679	2698	2717	2736	2754	2774	2792	2811	2830	2849	2867	74
16	2867	2886	2905	2924	2943	2962	2981	3000	3019	3038	3057	73 19
17	3057	3076	3096	3115	3134	3153	3172	3191	3211	3230	3249	72
18	3249	3269	3288	3307	3327	3346	3365	3385	3404	3424	3443	71
19	3443	3463	3482	3502	3522	3541	3561	3581	3600	3620	3640	70
20	0.3640	3659	3679	3699	3719	3739	3759	3779	3799	3819	3839	69
21	3839	3859	3879	3899	3919	3939	3959	3979	4000	4020	4040	68 20
22	4040	4061	4081	4101	4122	4142	4163	4183	4204	4224	4245	67
23	4245	4265	4286	4307	4327	4348	4369	4390	4411	4431	4452	66
24	4452	4473	4494	4515	4536	4557	4578	4599	4621	4642	4663	65 21
25	0.4663	4684	4706	4727	4748	4770	4791	4813	4834	4856	4877	64
26	4877	4899	4921	4942	4964	4986	5008	5029	5051	5073	5095	63
27	5095	5117	5139	5161	5184	5206	5228	5250	5272	5295	5317	62 22
28	5317	5340	5362	5384	5407	5430	5452	5475	5498	5520	5543	61
29	5543	5566	5589	5612	5635	5658	5681	5704	5727	5750	5774	60 23
30	0.5774	5797	5820	5844	5867	5890	5914	5938	5961	5985	6099	59
31	6009	6032	6056	6080	6104	6128	6152	6176	6200	6224	6249	58 24
32	6249	6273	6297	6322	6346	6371	6395	6420	6445	6469	6494	57
33	6494	6519	6544	6569	6594	6619	6644	6669	6694	6720	6745	56 25
34	6745	6771	6796	6822	6847	6873	6899	6924	6950	6976	7002	55
35	0.7002	7028	7054	7080	7107	7133	7159	7186	7212	7239	7265	54 26
36	7265	7292	7319	7346	7373	7400	7427	7454	7481	7508	7536	53 27
37	7536	7563	7590	7618	7646	7673	7701	7729	7757	7785	7813	52 28
38	7813	7841	7869	7898	7926	7954	7983	8012	8040	8069	8098	51 28
39	8098	8127	8156	8185	8214	8243	8273	8302	8332	8361	8391	50 29
40	0.8391	8421	8451	8481	8511	8541	8571	8601	8632	8662	8693	49 30
41	8693	8724	8754	8785	8816	8847	8878	8910	8941	8972	9004	48 31
42	9004	9036	9067	9099	9131	9163	9195	9228	9260	9293	9325	47 32
43	9325	9358	9391	9424	9557	9490	9523	9556	9590	9623	9567	46 33
44°	9657	9691	9725	9759	9793	9827	9861	9896	9930	9965	1.0000	45° 34
Complement	.9	.8	.7	.6	.5	.4	.3	.2	.1	.0		Angle

NATURAL COTANGENTS

EXPERIMENTAL COLLEGE PHYSICS

NATURAL TANGENTS

Angle	.0	.1	.2	.3	.4	.5	.6	.7	.8	.9	Dif.
45°	1.0000	1.0035	1.0070	1.0105	1.0141	1.0176	1.0212	1.0247	1.0283	1.0319	36
46	1.0355	1.0392	1.0428	1.0464	1.0501	1.0538	1.0575	1.0612	1.0649	1.0686	37
47	1.0724	1.0761	1.0799	1.0837	1.0875	1.0913	1.0951	1.0990	1.1028	1.1067	38
48	1.1106	1.1145	1.1184	1.1224	1.1263	1.1303	1.1343	1.1383	1.1423	1.1463	40
49	1.1504	1.1544	1.1585	1.1626	1.1667	1.1708	1.1750	1.1792	1.1833	1.1875	41
50	1.1918	1.1960	1.2002	1.2045	1.2088	1.2131	1.2174	1.2218	1.2261	1.2305	43
51	1.2349	1.2393	1.2437	1.2482	1.2527	1.2572	1.2617	1.2662	1.2708	1.2753	45
52	1.2799	1.2846	1.2892	1.2938	1.2985	1.3032	1.3079	1.3127	1.3175	1.3222	47
53	1.3270	1.3319	1.3367	1.3416	1.3465	1.3514	1.3564	1.3613	1.3663	1.3713	49
54	1.3764	1.3814	1.3865	1.3916	1.3968	1.4019	1.4071	1.4124	1.4176	1.4229	52
55	1.4281	1.4335	1.4388	1.4442	1.4496	1.4550	1.4605	1.4659	1.4715	1.4770	54
56	1.4826	1.4882	1.4938	1.4994	1.5051	1.5108	1.5166	1.5224	1.5282	1.5340	57
57	1.5399	1.5458	1.5517	1.5577	1.5637	1.5697	1.5757	1.5818	1.5880	1.5941	60
58	1.6003	1.6066	1.6128	1.6191	1.6255	1.6319	1.6383	1.6447	1.6512	1.6577	64
59	1.6643	1.6709	1.6775	1.6842	1.6909	1.6977	1.7045	1.7113	1.7182	1.7251	68
60	1.7321	1.7391	1.7461	1.7532	1.7603	1.7675	1.7747	1.7820	1.7893	1.7966	72
61	1.8040	1.8115	1.8190	1.8265	1.8341	1.8418	1.8495	1.8572	1.8650	1.8728	77
62	1.8807	1.8887	1.8967	1.9047	1.9128	1.9210	1.9292	1.9375	1.9458	1.9542	82
63	1.9626	1.9711	1.9797	1.9883	1.9970	2.0057	2.0145	2.0233	2.0323	2.0413	88
64	2.0503	2.0594	2.0686	2.0778	2.0872	2.0965	2.1060	2.1155	2.1251	2.1348	94
65	2.145	2.154	2.164	2.174	2.184	2.194	2.204	2.215	2.225	2.236	10
66	2.246	2.257	2.267	2.278	2.289	2.300	2.311	2.322	2.333	2.344	11
67	2.356	2.367	2.379	2.391	2.402	2.414	2.426	2.438	2.450	2.463	12
68	2.475	2.488	2.500	2.513	2.526	2.539	2.552	2.565	2.578	2.592	13
69	2.605	2.619	2.633	2.646	2.660	2.675	2.689	2.703	2.718	2.733	14
70	2.747	2.762	2.778	2.793	2.808	2.824	2.840	2.856	2.872	2.888	16
71	2.904	2.921	2.937	2.954	2.971	2.989	3.006	3.024	3.042	3.060	17
72	3.078	3.096	3.115	3.133	3.152	3.172	3.191	3.211	3.230	3.250	19
73	3.271	3.291	3.312	3.333	3.354	3.376	3.398	3.420	3.442	3.465	22
74	3.487	3.511	3.534	3.558	3.582	3.606	3.630	3.655	3.681	3.700	25
75	3.732	3.758	3.785	3.812	3.839	3.867	3.895	3.923	3.952	3.981	28
76	4.011	4.041	4.071	4.102	4.134	4.165	4.198	4.230	4.264	4.297	32
77	4.331	4.366	4.402	4.437	4.474	4.511	4.548	4.586	4.625	4.665	37
78	4.705	4.745	4.787	4.829	4.872	4.915	4.959	5.005	5.050	5.097	44
79	5.145	4.193	5.242	5.292	5.343	5.396	5.449	5.503	5.558	5.614	52
80	5.67	5.73	5.79	5.85	5.91	5.98	6.04	6.11	6.17	6.24	7
81	6.31	6.39	6.46	6.54	6.61	6.69	6.77	6.85	6.94	7.03	8
82	7.12	7.21	7.30	7.40	7.49	7.60	7.70	7.81	7.92	8.03	10
83	8.14	8.26	8.39	8.51	8.64	8.78	8.92	9.06	9.21	9.36	14
84	9.51	9.68	9.84	10.0	10.2	10.4	10.6	10.8	11.0	11.2	
85	11.4	11.7	11.9	12.2	12.4	12.7	13.0	13.3	13.6	14.0	3
86	14.3	14.7	15.1	15.5	15.9	16.3	16.8	17.3	17.9	18.5	6
87	19.1	19.7	20.4	21.2	22.0	22.9	23.9	24.9	26.0	27.3	
88	28.6	30.1	31.8	33.7	35.8	38.2	40.9	44.1	47.7	52.1	
89°	57.	64.	72.	82.	95.	115.	143.	191.	286.	573.	
Angle	.0	.1	.2	.3	.4	.5	.6	.7	.8	.9	

NATURAL TANGENTS

INDEX

TABLE OF LOGARITHMS

Natural numbers	0	1	2	3	4	5	6	7	8	9	Proportional parts								
											1	2	3	4	5	6	7	8	9
10	0000	0043	0086	0128	0170	0212	0253	0294	0334	0374	4	8	12	17	21	25	29	33	37
11	0414	0453	0492	0531	0569	0607	0645	0682	0719	0755	4	8	11	15	19	23	26	30	34
12	0792	0828	0864	0899	0934	0969	1004	1038	1072	1106	3	7	10	14	17	21	24	28	31
13	1139	1173	1206	1239	1271	1303	1335	1367	1399	1430	3	6	10	13	16	19	23	26	29
14	1461	1492	1523	1553	1584	1614	1644	1673	1703	1732	3	6	9	12	15	18	21	24	27
15	1761	1790	1818	1847	1875	1903	1931	1959	1987	2014	3	6	8	11	14	17	20	22	25
16	2041	2068	2095	2122	2148	2175	2201	2227	2253	2279	3	5	8	11	13	16	18	21	24
17	2304	2330	2355	2380	2405	2430	2455	2480	2504	2529	2	5	7	10	12	15	17	20	22
18	2553	2577	2601	2625	2648	2672	2695	2718	2742	2765	2	5	7	9	12	14	16	19	21
19	2788	2810	2833	2856	2878	2900	2923	2945	2967	2989	2	4	7	9	11	13	16	18	20
20	3010	3032	3054	3075	3096	3118	3139	3160	3181	3201	2	4	6	8	11	13	15	17	19
21	3222	3243	3263	3284	3304	3324	3345	3365	3385	3404	2	4	6	8	10	12	14	16	18
22	3424	3444	3464	3483	3502	3522	3541	3560	3579	3598	2	4	6	8	10	12	14	15	17
23	3617	3636	3655	3674	3692	3711	3729	3747	3766	3784	2	4	6	7	9	11	13	15	17
24	3802	3820	3838	3856	3874	3892	3909	3927	3945	3962	2	4	5	7	9	11	12	14	16
25	3979	3997	4014	4031	4048	4065	4082	4099	4116	4133	2	3	5	7	9	10	12	14	15
26	4150	4166	4183	4200	4216	4232	4249	4265	4281	4298	2	3	5	7	8	10	11	13	15
27	4314	4330	4346	4362	4378	4393	4409	4425	4440	4456	2	3	5	6	8	9	11	13	14
28	4472	4487	4502	4518	4533	4548	4564	4579	4594	4609	2	3	5	6	8	9	11	12	14
29	4624	4639	4654	4669	4683	4698	4713	4728	4742	4757	1	3	4	6	7	9	10	12	13
30	4771	4786	4800	4814	4829	4843	4857	4871	4886	4900	1	3	4	6	7	9	10	11	13
31	4914	4928	4942	4955	4969	4983	4997	5011	5024	5038	1	3	4	6	7	8	10	11	12
32	5051	5065	5079	5092	5105	5119	5132	5145	5159	5172	1	3	4	5	7	8	9	11	12
33	5185	5198	5211	5224	5237	5250	5263	5276	5289	5302	1	3	4	5	6	8	9	10	12
34	5315	5328	5340	5353	5366	5378	5391	5403	5416	5428	1	3	4	5	6	8	9	10	11
35	5441	5453	5465	5478	5490	5502	5514	5527	5539	5551	1	2	4	5	6	7	9	10	11
36	5563	5575	5587	5599	5611	5623	5635	5647	5658	5670	1	2	4	5	6	7	8	10	11
37	5682	5694	5705	5717	5729	5740	5752	5763	5775	5786	1	2	3	5	6	7	8	9	10
38	5798	5809	5821	5832	5843	5855	5866	5877	5888	5899	1	2	3	5	6	7	8	9	10
39	5911	5922	5933	5944	5955	5966	5977	5988	5999	6010	1	2	3	4	5	7	8	9	10
40	6021	6031	6042	6053	6064	6075	6085	6096	6107	6117	1	2	3	4	5	6	8	9	10
41	6128	6138	6149	6160	6170	6180	6191	6201	6212	6222	1	2	3	4	5	6	7	8	9
42	6232	6243	6253	6263	6274	6284	6294	6304	6314	6325	1	2	3	4	5	6	7	8	9
43	6335	6345	6355	6365	6375	6385	6395	6405	6415	6425	1	2	3	4	5	6	7	8	9
44	6435	6444	6454	6464	6474	6484	6493	6503	6513	6522	1	2	3	4	5	6	7	8	9
45	6532	6542	6551	6561	6571	6580	6590	6599	6609	6618	1	2	3	4	5	6	7	8	9
46	6628	6637	6646	6656	6665	6675	6684	6693	6702	6712	1	2	3	4	5	6	7	7	8
47	6721	6730	6739	6749	6758	6767	6776	6785	6794	6803	1	2	3	4	5	5	6	7	8
48	6812	6821	6830	6839	6848	6857	6866	6875	6884	6893	1	2	3	4	4	5	6	7	8
49	6902	6911	6920	6928	6937	6946	6955	6964	6972	6981	1	2	3	4	4	5	6	7	8
50	6990	6998	7007	7016	7024	7033	7042	7050	7059	7067	1	2	3	3	4	5	6	7	8
51	7076	7084	7093	7101	7110	7118	7126	7135	7143	7152	1	2	3	3	4	5	6	7	8
52	7160	7168	7177	7185	7193	7202	7210	7218	7226	7235	1	2	2	3	4	5	6	7	7
53	7243	7251	7259	7267	7275	7284	7292	7300	7308	7316	1	2	2	3	4	5	6	6	7
54	7324	7332	7340	7348	7356	7364	7372	7380	7388	7396	1	2	2	3	4	5	6	6	7